Classical
Complex Analysis

Jones and Bartlett Books in Mathematics and Computer Science

Advanced Calculus,
Revised Edition
Lynn H. Loomis and Shlomo Sternberg

Algorithms and Data Structures in
Computer Engineering
Christopher H. Nevison, et al., Editors

Calculus with Analytic Geometry,
Fourth Edition
Murray H. Protter and Philip E. Protter

Classical Complex Analysis
Liang-shin Hahn and Bernard Epstein

College Geometry
Howard Eves

Differential Equations: Theory and
Applications
Ray Redheffer

Discrete Mathematics
James L. Hein

Discrete Structures, Logic, and
Computability
James L. Hein

Fundamentals of Modern Elementary
Geometry
Howard Eves

An Introduction to Computational
Science and Mathematics
Charles F. Van Loan

Introduction to Differential Equations
Ray Redheffer

Introduction to Fractals and Chaos
Richard M. Crownover

Introduction to Numerical Analysis
John Gregory and Don Redmond

Lebesgue Integration on Euclidean
Space
Frank Jones

The Limits of Computing
Henry M. Walker

Logic, Sets, and Recursion
Robert L. Causey

Mathematical Models in the Social and
Biological Sciences
Edward Beltrami

The Poincaré Half-Plane: A Gateway
to Modern Geometry
Saul Stahl

Selected Works in Applied
Mathematics and Mechanics
Eric Reissner

The Theory of Numbers: A Text and
Source Book of Problems
Andrew Adler and John E. Coury

Theory of Computation
James L. Hein

Wavelets and Their Applications
Mary Beth Ruskai, et al.

The Way of Analysis
Robert S. Strichartz

Classical Complex Analysis

Liang-shin Hahn

Mathematics Department
University of New Mexico

Bernard Epstein

Emeritus
Mathematics Department
University of New Mexico

Jones and Bartlett Publishers
Sudbury, Massachusetts

Boston   London   Singapore

Editorial, Sales, and Customer Service Offices
Jones and Bartlett Publishers
40 Tall Pine Drive
Sudbury, MA 01776
508-443-5000
1-800-832-0034

Jones and Bartlett Publishers International
Barb House, Barb Mews
London W6 7PA
United Kingdom

Library of Congress Cataloging-in-Publication Data

Hahn, Liang-shin.
    Classical complex analysis / Liang-shin Hahn and Bernard Epstein.
       p.   cm.
    Includes index.
    ISBN 0-86720-494-X
    1. Functions of complex variables. 2. Mathematical analysis.
    I. Epstein, Bernard. II. Title.
    QA331.7.H35 1996
    515'.9--dc20
                                                          96-3964
                                                           CIP

Acquisitions Editor: Carl J. Hesler
Production Editor: Anne S. Noonan
Manufacturing Buyer: Dana L. Cerrito
Editorial Production Service: Superscript Editorial Production Services
Typesetting: Scanway Graphics International
Printing and Binding: Book Press
Cover Design: Hannus Design Associates
Cover Printing: John P. Pow Company

Printed in the United States of America
00 99 98 97 96   10 9 8 7 6 5 4 3 2 1

To
Shin-Yi, Shin-Jen, Shin-Hong
and
David, Deborah, Ruth, Jeremy, Jonathan, Rebecca

Contents

Preface

We have both presented, over a period of many years and at a number of universities, a one-year graduate-level course in complex analysis. We feel that this course is an essential part of the education of any serious student of mathematics. We have sought to present the most basic material during the first semester, leaving the second semester (or at least most of it) for "selected topics" which would vary from year to year. We distributed handwritten notes to our students, and in this way we have polished our manuscript over the years and accumulated a large collection of chapters that were never presented in their entirety in any one year.

The book starts with power series so that the student gets concrete examples of analytic functions beyond the elementary ones. However, more importantly, the power series approach almost immediately lets the student appreciate the advantage of doing mathematics with complex numbers instead of being confined to real numbers. It also lets the student realize that complex analysis is a natural continuation of elementary calculus (using real numbers). In addition, the student has a chance to review elementary calculus and hence has a smooth introduction to complex analysis. Our experience shows that this approach has truly significant instructional merit. We then define analytic functions and introduce Möbius transformations as beautiful and useful examples, with emphasis on the geometric aspect. The Cauchy theorem is presented in the classical manner in order to preserve, insofar as possible, the introductory nature of the course. Chapter 5, "Singularities and Residues," and the first two sections of Chapter 6, "The Maximum Modulus Principle" and "The Schwarz Lemma," contain immediate consequences of the Cauchy theorem. For a one-semester course we suggest that these, plus Chapter 12, "The Picard Theorems," be covered. This combination should give the student a sense of achievement by mastering one of the highlights that were the stimulus for modern complex analysis and, at the same time, show how the basic theorems can be applied to prove deep results. Other than the notation for the supremum norm, the last four sections of Chapter

6 will not be needed for the rest of the book, and so may be skipped during the first reading.

The rest of the book consists of selected topics. For a one-year course, our experience shows that it is possible to cover most of the material in the book, except perhaps Section 7.3, "Extensions of Theorems of Mittag-Leffler and Weierstrass," Section 7.9, "The Runge Approximation Theorem," Section 9.2, "Univalent Functions," and Section 10.4 "Conformal Mapping of Multiply-Connected Regions." Probably at least one of these topics, but not all, can be covered.

As for Section 7.9, our original intention was to present an alternate proof of the Cauchy theorem via the Runge approximation theorem, but we were not quite able to do so without disturbing the balance of the book. Similar reasons caused us to omit both the proof of the Carathéodory theorem concerning the behavior of the Riemann mapping function near the boundary of a region and the expression of the Riemann mapping function in terms of the Bergman reproducing kernel in Chapter 10, "Conformal Mapping." In our handwritten versions there was also a chapter on elliptic functions, but in order to include its applications to number theory in a one-year course, many more elementary and basic topics would have had to be sacrificed. Thus, with regret, we decided to delete this beautiful topic.

The main merits we claim are clarity of presentation, careful selection of topics, and a generous supply of interesting exercises. We do not claim to have written a *Complex Analysis Made Easy*. The student who wants to achieve a thorough grasp of this subject must devote serious effort to it, and a major duty of the instructor and the author is to make that effort pleasurable. This book aims to "inoculate" readers with our enthusiasm for this fascinating subject.

A few words about prerequisites. This is an introductory course in complex analysis and no previous knowledge of the subject is assumed. The reader should, however, thoroughly understand "advanced calculus"—including the structure of the real number system, continuity of functions of one and two real variables, uniform continuity, sequences and series of real numbers and real functions, and the existence of the definite integral of a continuous function of one real variable on a bounded closed interval.

In publishing our manuscript in book form, the first author fondly recalls that in early handwritten versions his sons took turns drawing pictures, such as Charlie Brown in a baseball outfit with the caption: "Riemann ball is fun," or Snoopy dancing in the air chanting "Happiness is analyticity." So to his three sons, Shin-Yi, Shin-Jen, and Shin-Hong, the first author dedicates this book.

The second author also fondly recalls many incidents involving his children's interaction with mathematics. Two in particular come to mind. First, when Jeremy was about three years old, he expressed weariness with my extremely limited repertoire of bedtime stories (*Goldilocks* and *Little Red*

Riding Hood), and so I recited the Riemann Mapping Theorem (proof included!) to him one evening. The next night, to my amazement and amusement, he requested the Napping (!) Theorem. Since then, whenever I have presented Riemann's masterpiece in class I have been unable to refrain from telling this anecdote. Second, when I was trying to help a very unenthusiastic Rebecca with her high-school calculus, she would interrupt with: "Do you really like this stuff?" So, to the student who is about to embark on the study of complex analysis, I express the hope that you will "really like this stuff."

It is our pleasure to express appreciation to Ms. Susan Pinter for her work on the illustrations. Also we want to thank Mr. Peter Espen, Ms. Liz Frank, and Mr. John Vance, who bailed the first author out of computer trouble on numerous occasions while he was typing the manuscript. Last but not least, the authors express their heartfelt gratitude to their colleague and friend, Professor Jeff Davis, for his meticulous reading of the entire manuscript and for innumerable suggestions. His contributions have improved both the mathematics and the presentation decisively.

L.-s. H. and B. E.

Chapter 1

Complex Numbers

1.1 The Complex Field

Complex analysis is calculus with complex numbers instead of real numbers. But why make the move to complex numbers? What deficiency do the real numbers have? And just what *are* complex numbers? The name sounds unappealing. Historically, complex numbers were introduced so that all quadratic equations could be solved. After their introduction, it was found that using complex numbers makes it possible to show that *every* nonconstant polynomial has a root. The first letter i of the word *imaginary* was introduced and "defined" to have the property $i^2 = -1$. Complex numbers were numbers of the form $a + ib$, where a and b are real numbers, and the operations of addition, subtraction, multiplication, and division were carried out just as for the real numbers, with i^2 replaced by -1. For example,

$$(2 + 3i) \cdot (1 - i) = 2 + 3i - 2i - 3i^2 = 5 + i.$$

However, this "definition" appears to be much too casual. For one thing, is the root of the equation $x^2 + 1 = 0$ unique? If not, which one is i? For another, is it meaningful to add a real number a and an imaginary number ib?

To answer this criticism, we give a formal definition for complex numbers: A *complex number* is an ordered pair (a, b) of real numbers; in particular,

$$(a, b) = (c, d) \iff a = c \text{ and } b = d.$$

Addition and multiplication of two complex numbers are defined by

$$(a, b) + (c, d) = (a + c, b + d),$$
$$(a, b) \cdot (c, d) = (ac - bd, bc + ad).$$

By using these definitions and the field properties of the real numbers, it is straightforward to verify that the set

$$\mathbb{C} = \{(a, b)\,;\, a, b \in \mathbb{R}\}$$

of all complex numbers forms a *field*, with $(0, 0)$ as the zero element and $(1, 0)$ as the unit element of the field.

For example, one can show the existence of the multiplicative inverse of a nonzero complex number (a, b): Because, as just stated, the multiplicative identity is $(1, 0)$, we wish to solve

$$(a, b) \cdot (x, y) = (1, 0).$$

Then

$$(ax - by,\, bx + ay) = (1, 0).$$

$$\therefore\ ax - by = 1, \quad bx + ay = 0.$$

Because $\begin{vmatrix} a & -b \\ b & a \end{vmatrix} = a^2 + b^2 \neq 0$, this system of simultaneous equations has the *unique* solution

$$x = \frac{a}{a^2 + b^2}, \quad y = \frac{-b}{a^2 + b^2};$$

that is,

$$(a, b)^{-1} = \left(\frac{a}{a^2 + b^2},\, \frac{-b}{a^2 + b^2} \right).$$

Now, consider complex numbers of the form $(a, 0)$, $a \in \mathbb{R}$. Their operations are

$$(a, 0) \pm (b, 0) = (a \pm b, 0),$$
$$(a, 0) \cdot (b, 0) = (ab, 0),$$
$$\frac{(a, 0)}{(b, 0)} = \left(\frac{a}{b}, 0 \right) \quad \text{(provided } b \neq 0\text{)}.$$

These complex numbers behave exactly like real numbers. Therefore, in the future, we shall identify a complex number $(a, 0)$ with the real number a. It follows that $\mathbb{C}$ contains the real field $\mathbb{R}$ as a subfield.

Next, observe that

$$(0, 1)^2 = (0, 1) \cdot (0, 1) = (-1, 0).$$

Because $(-1, 0)$ is identified with -1 in the real field $\mathbb{R}$, we see that $(0, 1)$ satisfies the equation $x^2 + 1 = 0$. We denote the complex number $(0, 1)$ by the symbol i.

We have

$$
\begin{aligned}
(a, b) &= (a, 0) + (0, b) \\
&= (a, 0) + (0, 1)(b, 0) = (a, 0) + (b, 0)(0, 1).
\end{aligned}
$$

This equality justifies our denoting the complex number (a, b) by either $a + ib$ or $a + bi$.

Summing up, we have shown that

(a) $\mathbb{C}$ contains the real field $\mathbb{R}$ as a subfield (or, more precisely, $\mathbb{C}$ contains a subfield isomorphic to the real field $\mathbb{R}$);

(b) there exists an element $i \in \mathbb{C}$ such that $i^2 + 1 = 0$;

(c) $\mathbb{C}$ is generated by $\mathbb{R}$ and i.

Conversely, if K is a field that satisfies the three conditions above, then we can easily show that K is isomorphic to the field $\mathbb{C}$. Thus it is legitimate to name $\mathbb{C}$ *the* complex field.

We now introduce some terminology. The *real part* of a complex number $a + ib$ $(a, b \in \mathbb{R})$, denoted $\Re(a + ib)$, is the real number a. Thus, $\Re(a + ib) = a$. Similarly, the *imaginary part* of a complex number $a + ib$, denoted $\Im(a + ib)$, is the real (!) number b. Thus, $\Im(a + ib) = b$. For example, $\Im(2 + 3i)$ is 3, *not* $3i$.

A complex number is real if and only if its imaginary part is zero; a complex number is said to be *purely imaginary* if and only if its real part is zero. The number 0 is the only complex number that is both real and purely imaginary. The *(complex) conjugate* of a complex number $z = a + ib$ $(a, b \in \mathbb{R})$, denoted $\bar{z}$, is defined as $\bar{z} = a - ib$. Real numbers are characterized among complex numbers by $z = \bar{z}$; purely imaginary numbers, on the other hand, are characterized by $z = -\bar{z}$. The real and imaginary parts can be expressed in terms of the complex number z and its conjugate:

$$
\Re z = \frac{z + \bar{z}}{2}, \quad \Im z = \frac{z - \bar{z}}{2i}.
$$

We have the following elementary properties:

(a) If $z = a + ib$, $a, b \in \mathbb{R}$, then

$$
z\bar{z} = (a + ib)(a - ib) = a^2 + b^2;
$$

therefore, $z\bar{z}$ is always real and nonnegative. Moreover, $z\bar{z} = 0$ if and only if $z = 0$.

Remark. Here we have used the property that the sum of the squares of real numbers is zero if and only if each of these numbers is zero. This statement is no longer true for complex numbers: $1^2 + i^2 = 0$.

(b) $\overline{z + w} = \bar{z} + \bar{w}$;

(c) $\overline{z \cdot w} = \bar{z} \cdot \bar{w}$;

(d) $\overline{\left(\dfrac{z}{w}\right)} = \dfrac{\bar{z}}{\bar{w}}$ (provided $w \neq 0$);

(e) $\bar{\bar{z}} = z$.

Using complex conjugation, we can compute the multiplicative inverse of a nonzero complex number $a + ib$ more efficiently than before:

$$
\begin{aligned}
(a, b)^{-1} &= \frac{1}{a + ib} = \frac{a - ib}{(a + ib)(a - ib)} \\
&= \frac{a - ib}{a^2 + b^2} = \frac{a}{a^2 + b^2} - \frac{ib}{a^2 + b^2}.
\end{aligned}
$$

The real number system has no *zero divisor*; that is,

$$\text{if } a \cdot b = 0 \ (a, b \in \mathbb{R}), \text{ then } a = 0 \text{ or } b = 0.$$

This statement is also true of the complex numbers (in fact, it is one of the field properties), as we now show: Suppose $z \cdot w = 0$. Then

$$(z\bar{z}) \cdot (w\bar{w}) = (zw) \cdot \overline{(zw)} = 0.$$

But because (as we have seen) $z\bar{z}$ and $w\bar{w}$ are real numbers, either $z\bar{z} = 0$ or $w\bar{w} = 0$. By Property (a), this means that either $z = 0$ or $w = 0$.

1.2   Geometric Representation

Because a complex number is an ordered pair of real numbers, and the same is true for a point in a plane, a *bijection* (one-to-one, onto correspondence) exists between all complex numbers and all the points in a plane. Indeed, complex numbers may be represented as two-dimensional vectors with the real part corresponding to the horizontal component and the imaginary part to the vertical component (Figure 1.1). Because the points on the x-axis correspond to real numbers and those on the y-axis to purely imaginary

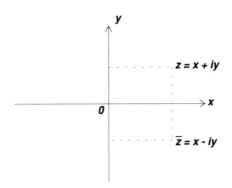

Figure 1.1

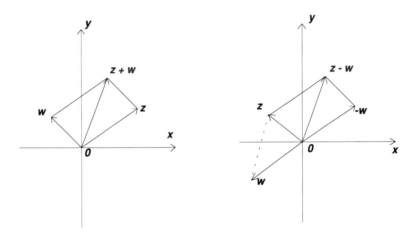

Figure 1.2

numbers, they are called the *real* and *imaginary axes*, respectively. A plane equipped with the real and imaginary axes is called the *complex plane* or the *Gaussian plane*.

The addition of two complex numbers z and w can be carried out geometrically by observing that the four points, 0, z, $z + w$, w are the vertices of a parallelogram (Figure 1.2). Similarly, $z - w$ is the vector with w as its initial point and z as its terminal point.

Very often, using a vector diagram makes the properties of expressions involving complex numbers easier to grasp. For example, the absolute value of a real number is the distance from the corresponding point to the origin. Thus we can extend the *absolute-value* concept to a complex number as the distance from the point representing the complex number to the origin, or equivalently, the length of the vector $z = a + ib\,(a, b \in \mathbb{R})$:

$$| z |= \sqrt{a^2 + b^2} = (z\bar{z})^{\frac{1}{2}}.$$

We have, for any two complex numbers z and w:

(a) $| \bar{z} |=| z |$;

(b) $| \Re z |\leq| z |, \quad | \Im z |\leq| z |$;

(c) $| zw |=| z | \cdot | w |$;

(d) $\left| \dfrac{z}{w} \right| = \dfrac{| z |}{| w |}$ (provided $w \neq 0$) ;

(e) $| z + w |\leq| z | + | w |.$

    The geometric interpretation of this last relation leads us to call it the *triangle inequality*. The *algebraic* proof of this inequality is as follows:

$$
\begin{aligned}
| z + w |^2 &= (z + w) \cdot \overline{(z + w)} \\
&= (z + w) \cdot (\bar{z} + \bar{w}) \\
&= z\bar{z} + z\bar{w} + \bar{z}w + w\bar{w} \\
&= | z |^2 + 2\Re(z\bar{w})+ | w |^2 \\
&\leq | z |^2 +2 | z\bar{w} | + | w |^2 \quad \text{(by (b) above)} \\
&= | z |^2 +2 | z | \cdot | w | + | w |^2 \quad \text{(by (c) and (a))} \\
&= (| z | + | w |)^2.
\end{aligned}
$$

    Taking square roots and recalling that absolute values are nonnegative, we obtain the desired result.

    As a consequence of (e), we have, by an obvious induction,

(f) $| z_1 + z_2 + \cdots + z_n |\leq| z_1 | + | z_2 | + \cdots + | z_n |$;

(g) $\left|| z | - | w |\right| \leq| z \pm w |\leq| z | + | w |.$

    The reader may find the left inequality of (g) not quite as simple as the remaining items in the above list.

    It is useful to visualize the product of two complex numbers by introducing polar coordinates (Figure 1.3). If the polar coordinates of point (a, b) are (r, θ), we know that

$$a = r \cos \theta, \quad b = r \sin \theta.$$

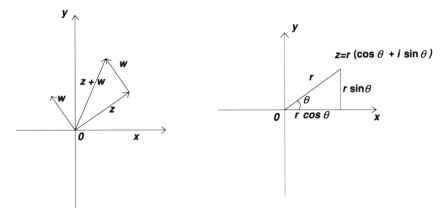

Figure 1.3

Thus we can write

$$z = a + ib = r(\cos\theta + i\sin\theta).$$

Of course, r is the absolute value, $|z|$, of z; the polar angle θ is called the *argument* of z, and is denoted arg z. Note that arg z is defined only up to integer multiples of 2π, and 0 is the only complex number whose absolute value is 0 and whose argument is not defined.

Now, if z and w have the polar representations

$$z = r(\cos\theta + i\sin\theta) \quad \text{and} \quad w = \rho(\cos\varphi + i\sin\varphi),$$

then, by the addition formulas of trigonometric functions,

$$\begin{aligned} zw &= r\rho\{(\cos\theta \cdot \cos\varphi - \sin\theta \cdot \sin\varphi) \\ &\quad + i(\sin\theta \cdot \cos\varphi + \cos\theta \cdot \sin\varphi)\} \\ &= r\rho\{\cos(\theta + \varphi) + i\sin(\theta + \varphi)\}. \end{aligned}$$

It follows that *the absolute value of a product is the product of the absolute values* (in agreement with an earlier result), and *the argument of a product is the sum of the arguments of the factors*. Recalling that arguments of complex numbers are defined only mod 2π, we have

$$\arg(zw) \equiv \arg z + \arg w \pmod{2\pi}.$$

In particular, if $z = w$, then

$$\left|z^2\right| = |z|^2, \quad \arg(z^2) \equiv 2\arg z \pmod{2\pi}.$$

By induction, we have, for any positive integer n,

$$|z^n| = |z|^n, \quad \arg(z^n) \equiv n \arg z \qquad (\mathrm{mod}\ 2\pi).$$

By combining these last two statements, we obtain

$$z^n = r^n(\cos n\theta + i \sin n\theta).$$

The case $r = 1$ is called the *deMoivre formula*:

$$(\cos\theta + i\sin\theta)^n = \cos n\theta + i\sin n\theta.$$

Because $\bar{z} = \dfrac{r^2}{z}$, we have

$$\arg(\bar{z}) \equiv \arg(r^2) + \arg\left(\frac{1}{z}\right) \equiv \arg\left(\frac{1}{z}\right) \qquad (\mathrm{mod}\ 2\pi).$$

On the other hand,

$$\bar{z} = r(\cos\theta - i\sin\theta) = r\{\cos(-\theta) + i\sin(-\theta)\}.$$

$$\therefore\ \arg(\bar{z}) \equiv -\arg z \qquad (\mathrm{mod}\ 2\pi).$$

Hence

$$\arg\left(\frac{1}{z}\right) \equiv -\arg z \quad (\mathrm{mod}\ 2\pi).$$

More generally, we have

$$\arg\left(\frac{z}{w}\right) \equiv \arg z - \arg w \quad (\mathrm{mod}\ 2\pi).$$

It follows immediately that the deMoivre formula is valid for all integers n (positive, negative, or zero).

We are now ready for a geometric construction of the product. Indeed, the triangle with vertices at 0, 1, z is similar to the one whose vertices are at 0, w, zw, with the same orientation, and because the points 0, 1, z and w are given, this similarity determines point zw uniquely (Figure 1.4).

Division can be handled similarly. Simply observe that the triangles with vertices at 0, 1, w and at 0, $\frac{z}{w}$, z are similar.

Remark. The polar representation of complex numbers is introduced here merely for pedagogical purposes. Later we shall return to this topic,

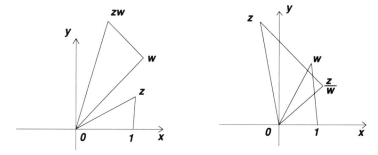

Figure 1.4

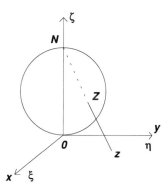

Figure 1.5

justifying rigorously the use of trigonometric functions without depending on geometric intuition.

1.3   The Riemann Sphere

We know that continuous functions on compact sets have many pleasant properties, and because $\mathbb{C}$ turns out to be a locally compact space, we can make it compact with a *one-point compactification*. But can we devise a nice model for the compactification?

Following Riemann, we construct a sphere of diameter 1 tangent to the complex plane $\mathbb{C}$ at the origin. Choose any point z in the plane and draw a line segment joining the "north pole" N of the sphere to z (Figure 1.5). This segment will intersect the surface of the sphere at one point other than the north pole N, say at Z. If we let point z correspond to point Z on the sphere, we obtain a bijection between $\mathbb{C}$ and the sphere with north pole deleted.

Now, if we let z move away from the origin (but, of course, remaining on the plane $\mathbb{C}$), then its image Z on the sphere approaches N. Conversely, all the points on the sphere that are close to N correspond to points on $\mathbb{C}$ with large absolute values. Riemann added an "ideal point" to $\mathbb{C}$, called it the *point at infinity*, denoted it by ∞, and let it correspond to the north pole N on the sphere. Thus we obtain a bijection between the *extended complex plane* $\hat{\mathbb{C}} = \mathbb{C} \cup \{\infty\}$ and the sphere. This correspondence is called the *stereographic projection* of the sphere onto the plane, and $\hat{\mathbb{C}}$ is often called the *Riemann sphere*.

Let the coordinates of points z, Z, and N be $(x,\ y,\ 0)$, $(\xi,\ \eta,\ \zeta)$ and $(0,\ 0,\ 1)$, respectively. Then these three points are collinear and Z is on the sphere with center at $(0,\ 0,\ \frac{1}{2})$ and radius $\frac{1}{2}$. Therefore, by considering similar triangles and the equation of the sphere, we obtain

$$\begin{cases} \dfrac{\xi}{x} = \dfrac{\eta}{y} = \dfrac{1-\zeta}{1}, \\[2mm] \xi^2 + \eta^2 + \left(\zeta - \dfrac{1}{2}\right)^2 = \left(\dfrac{1}{2}\right)^2. \end{cases}$$

Hence

$$\begin{cases} x = \dfrac{\xi}{1-\zeta}; \\[2mm] y = \dfrac{\eta}{1-\zeta}; \\[2mm] |z|^2 = \dfrac{\zeta}{1-\zeta}, \end{cases}$$

or

$$\begin{cases} \xi = \dfrac{x}{1+x^2+y^2} = \dfrac{z+\bar{z}}{2(1+|z|^2)}; \\[2mm] \eta = \dfrac{y}{1+x^2+y^2} = \dfrac{z-\bar{z}}{2i(1+|z|^2)}; \\[2mm] \zeta = \dfrac{x^2+y^2}{1+x^2+y^2} = \dfrac{|z|^2}{1+|z|^2}. \end{cases}$$

Next, we introduce a metric ρ on $\hat{\mathbb{C}}$ defined by the (Euclidean) distance on the Riemann sphere. More specifically, let z_1, $z_2 \in \hat{\mathbb{C}}$, and Z_1, Z_2 be the corresponding points on the Riemann sphere (Figure 1.6). We define

$$\rho(z_1,\ z_2) = \overline{Z_1 Z_2}.$$

We start with the particular case $Z_1 = N$, $Z_2 = Z\ (\neq N)$. Because

$$\frac{\overline{NZ}}{\overline{Nz}} = \frac{1-\zeta}{1} = \frac{1}{1+|z|^2},$$

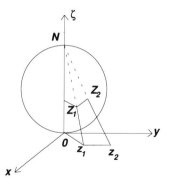

Figure 1.6

and $\overline{Nz} = \sqrt{1+ \mid z \mid^2}$, we obtain

$$\rho(z, \infty) = \overline{NZ} = \frac{1}{\sqrt{1+ \mid z \mid^2}}.$$

Now, suppose $z_1, z_2 \in \mathbb{C}$; because $\triangle 0NZ_1 \sim \triangle z_1 N0$, we obtain

$$\overline{NZ_1} \cdot \overline{Nz_1} = 1.$$

Similarly,

$$\overline{NZ_2} \cdot \overline{Nz_2} = 1. \quad \therefore \frac{\overline{Nz_1}}{\overline{Nz_2}} = \frac{\overline{NZ_2}}{\overline{NZ_1}}.$$

It follows that $\triangle NZ_1Z_2 \sim \triangle Nz_2z_1$, and so

$$\frac{\overline{Z_1Z_2}}{\overline{z_1z_2}} = \frac{\overline{NZ_1}}{\overline{Nz_2}}.$$

$$\therefore \rho(z_1, z_2) = \overline{Z_1Z_2} = \frac{\overline{NZ_1}}{\overline{Nz_2}} \cdot \overline{z_1z_2} = \frac{\mid z_1 - z_2 \mid}{\sqrt{(1+ \mid z_1 \mid^2)(1+ \mid z_2 \mid^2)}}.$$

Clearly, ρ satisfies the metric axioms:

(a) $\rho(z_1, z_2) \geq 0$ for any $z_1, z_2 \in \hat{\mathbb{C}}$;
   equality holds if and only if $z_1 = z_2$.
(b) $\rho(z_1, z_2) = \rho(z_2, z_1)$ for any $z_1, z_2 \in \hat{\mathbb{C}}$;

(c) The triangle inequality holds:

$$\rho(z_1, z_2) + \rho(z_2, z_3) \geq \rho(z_1, z_3)$$

for any $z_1, z_2, z_3 \in \hat{\mathbb{C}}$.

Note that

$$\rho(z_1, z_2) \leq 1 \quad \text{for any } z_1, z_2 \in \hat{\mathbb{C}}.$$

Thus, $\hat{\mathbb{C}}$, with the metric we have introduced, is a compact metric space (hence is bounded, which is, of course obvious; the distance between any two points is at most equal to 1, the diameter of the sphere), in contrast to $\mathbb{C}$ (with the absolute value metric), which is an unbounded, *locally* compact space.

Exercises

1. Determine when equality holds in the triangle inequality:

$$\mid z + w \mid \leq \mid z \mid + \mid w \mid .$$

2. Show that any complex number z with $\mid z \mid = 1$, other than $z = 1$, can be expressed as

$$z = \frac{t - i}{t + i}$$

with an appropriate choice of the *real* parameter t.

3. (a) Prove the *parallelogram law*:

$$\mid \alpha + \beta \mid^2 + \mid \alpha - \beta \mid^2 = 2(\mid \alpha \mid^2 + \mid \beta \mid^2)$$

for two arbitrary complex numbers α and β.

(b) Interpret the equality in (a) geometrically.

4. (a) Prove the *Lagrange identity*:

$$\left| \sum_{j=1}^{n} a_j b_j \right|^2 = \left(\sum_{j=1}^{n} \mid a_j \mid^2 \right) \cdot \left(\sum_{j=1}^{n} \mid b_j \mid^2 \right) - \sum_{1 \leq k < j \leq n} \mid a_j \bar{b}_k - a_k \bar{b}_j \mid^2 .$$

(b) Deduce the *Cauchy inequality*:

$$\left| \sum_{j=1}^{n} a_j b_j \right| \leq \left(\sum_{j=1}^{n} |a_j|^2 \right)^{\frac{1}{2}} \cdot \left(\sum_{j=1}^{n} |b_j|^2 \right)^{\frac{1}{2}}.$$

5. If $|\alpha| < 1$ and $|z| \leq 1$, show that

$$\left| \frac{z - \alpha}{1 - \bar{\alpha} z} \right| \leq 1.$$

When does equality hold?

6. Let ω be a nonreal complex number. Show that every complex number z can be written in the form

$$z = a + b\omega \quad (a, b \in \mathbb{R}).$$

Furthermore, a and b are uniquely determined by ω and z.

7. Show that in an arbitrary quadrangle the midpoints of the four sides are the vertices of a parallelogram.

8. Given two points z_1 and z_2, describe the locus of the points

$$z(t) = (1 - t)z_1 + tz_2$$

as the real parameter t increases from 0 to 1. What if t is permitted to assume *all* real values?

9. Show that three distinct points $\alpha, \beta, \gamma \in \mathbb{C}$ are collinear if and only if

$$\frac{\alpha - \beta}{\alpha - \gamma} \in \mathbb{R}.$$

10. (a) Show that $\triangle z_1 z_2 z_3 \sim \triangle w_1 w_2 w_3$ with the same orientation (i.e., both counterclockwise or both clockwise) if and only if

$$\begin{vmatrix} z_1 & w_1 & 1 \\ z_2 & w_2 & 1 \\ z_3 & w_3 & 1 \end{vmatrix} = 0.$$

(b) Show that $\triangle z_1 z_2 z_3$ is equilateral if and only if

$$z_1^2 + z_2^2 + z_3^2 - z_2 z_3 - z_3 z_1 - z_1 z_2 = 0.$$

11. Suppose that $z + \frac{1}{z} = 2 \cos \theta$, where θ is a real number. (Later we shall define $\cos \theta$ when θ is a complex number.) First show that $|z| = 1$ and then show that, for any integer n,

$$z^n + \frac{1}{z^n} = 2 \cos n\theta.$$

12. If α and β are two distinct fixed points, show that the locus of points γ for which

$$\arg\left\{\frac{\gamma - \alpha}{\gamma - \beta}\right\} = \text{constant}$$

is a circular arc with the points α and β as its two ends.

13. (Ptolemy-Euler)

(a) Show that for any $\alpha, \beta, \gamma, \delta \in \mathbb{C}$,

$$|\,\alpha - \beta\,| \cdot |\,\gamma - \delta\,| + |\,\alpha - \delta\,| \cdot |\,\beta - \gamma\,| \geq |\,\alpha - \gamma\,| \cdot |\,\beta - \delta\,|.$$

(b) Interpret the inequality geometrically.

(c) When does the equality hold?

14. Derive the identities:

$$\sum_{k=0}^{n} \cos(\varphi + k\theta) = \frac{\cos(\varphi + \frac{n\theta}{2}) \cdot \sin \frac{(n+1)\theta}{2}}{\sin \frac{\theta}{2}},$$

$$\sum_{k=0}^{n} \sin(\varphi + k\theta) = \frac{\sin(\varphi + \frac{n\theta}{2}) \cdot \sin \frac{(n+1)\theta}{2}}{\sin \frac{\theta}{2}} \qquad (0 < \theta < 2\pi).$$

15. Suppose $x_n + iy_n = (\sqrt{3} + i)^n$ $(x_n, y_n \in \mathbb{R}, n \in \mathbb{N})$. Show that

$$x_n y_{n+1} - x_{n+1} y_n = 2^{2n},$$
$$x_{n+1} x_n + y_{n+1} y_n = 2^{2n}\sqrt{3}.$$

16. Find the smallest positive integers m and n satisfying

$$(\sqrt{3} - i)^m = (1 + i)^n.$$

17. (a) Let $\zeta = e^{\frac{2\pi i}{n}}$ $(n \in \mathbb{N})$; show that

$$\prod_{k=0}^{n-1}(1 - \zeta^k z) = 1 - z^n,$$

$$\prod_{k=1}^{n-1}(1 - \zeta^k z) = 1 + z + \cdots + z^{n-1}.$$

(b) Show that

$$\sin\frac{\pi}{n} \cdot \sin\frac{2\pi}{n} \cdots \sin\frac{(n-1)\pi}{n} = \frac{n}{2^{n-1}} \quad (n \geq 2).$$

18. Show that if in the expression for $\rho(z_1, z_2)$ that we found in Section 1.3 we divide both the numerator and the denominator on the right by $| z_2 |$ and then let $| z_2 | \to \infty$, we obtain

$$\rho(z_1, \infty) = \frac{1}{\sqrt{1 + | z_1 |^2}}.$$

19. Show that

(a) $\rho(z_1, z_2) = \rho(\bar{z}_1, \bar{z}_2) = \rho\left(\dfrac{1}{z_1}, \dfrac{1}{z_2}\right)$;

(b) $\rho\left(z, -\dfrac{1}{\bar{z}}\right) = 1.$

20. Show that stereographic projection preserves angles.

Chapter 2

Power Series

2.1 Sequences

The definitions of a complex sequence and its limit are the same as for a real sequence. If to each natural number n there corresponds a complex number c_n, then

$$c_1, c_2, \cdots, c_n, \cdots,$$

or simply $\{c_n\}_{n=1}^{\infty}$, is called a *sequence*. Suppose

$$n_1, n_2, \cdots, n_k, \cdots$$

is a sequence of natural numbers satisfying

$$n_1 < n_2 < \cdots < n_k < \cdots;$$

then the sequence

$$c_{n_1}, c_{n_2}, \cdots, c_{n_k}, \cdots$$

is called a *subsequence* of $\{c_n\}_{n=1}^{\infty}$.

We say a sequence $\{c_n\}_{n=1}^{\infty}$ *converges* (to the *limit* c) if there exists a complex number c such that to every positive number ϵ there corresponds a natural number N with the property that

$$|c_n - c| < \epsilon \quad \text{whenever} \quad n \geq N.$$

In this case, we write

$$\lim_{n \to \infty} c_n = c, \quad \text{or simply } c_n \longrightarrow c \quad (\text{as } n \to \infty).$$

Clearly, if a sequence converges, then its limit is unique (by the triangle inequality); also, all its subsequences converge and they have the same limit.

A sequence that does not converge is said to be *divergent*. Of the divergent sequences, the most important are the ones where $|c_n| \longrightarrow \infty$; this means that to every $M > 0$ there corresponds a natural number N such that $|c_n| > M$ whenever $n \geq N$. In this case, we also write $c_n \longrightarrow \infty$ because c_n approaches the north pole of the Riemann sphere as $n \to \infty$. For example, $i^n n \longrightarrow \infty$ (as $n \to \infty$).

In the sequel, we assume that readers are familiar with the properties of real sequences.

If we denote

$$c_n = a_n + ib_n, \quad c = a + ib, \quad (a_n, b_n, a, b \in \mathbb{R}),$$

then the inequalities

$$\max\{|a_n - a|, |b_n - b|\} \leq |c_n - c| \leq |a_n - a| + |b_n - b|$$

show immediately that

THEOREM 2.1.1 *A sequence* $\{c_n\}_{n=1}^{\infty}$ *converges if and only if the sequences* $\{\Re c_n\}_{n=1}^{\infty}$ *and* $\{\Im c_n\}_{n=1}^{\infty}$ *converge.*

Just as for real sequences, the Cauchy criterion is fundamental.

THEOREM 2.1.2 (Cauchy Criterion) *A sequence* $\{c_n\}_{n=1}^{\infty}$ *converges if and only if to every* $\epsilon > 0$ *there corresponds a natural number* N *with the property that*

$$|c_m - c_n| < \epsilon \quad \text{whenever} \quad m, n \geq N.$$

Proof. The necessity is immediate: Suppose $\{c_n\}_{n=1}^{\infty}$ converges to some limit c; then, given $\epsilon > 0$, there corresponds a natural number N such that

$$|c_m - c| < \frac{\epsilon}{2}, \quad |c_n - c| < \frac{\epsilon}{2} \quad (m, n \geq N).$$

It follows that

$$|c_m - c_n| \leq |c_m - c| + |c_n - c| < \epsilon \quad (m, n \geq N).$$

Conversely, suppose a sequence $\{c_n\}_{n=1}^{\infty}$ satisfies the Cauchy criterion; that is, given $\epsilon > 0$, there corresponds a natural number N with the property that

$$|c_m - c_n| < \epsilon \quad \text{whenever} \quad m, n \geq N.$$

Then the inequalities

$$| a_m - a_n | \leq | c_m - c_n | < \epsilon, \quad | b_m - b_n | \leq | c_m - c_n | < \epsilon,$$

(where $a_n = \Re c_n$, $b_n = \Im c_n$, etc.), say that $\{a_n\}_{n=1}^{\infty}$ and $\{b_n\}_{n=1}^{\infty}$ are (real) Cauchy sequences. Hence, by the real version of the Cauchy criterion,

$$a_n \longrightarrow a, \quad b_n \longrightarrow b \quad (\text{as } n \to \infty),$$

for some suitable real numbers a and b. Set $c = a + ib$. Then

$$\begin{aligned} | c_n - c | &= | (a_n - a) + i(b_n - b) | \\ &\leq | a_n - a | + | b_n - b | \longrightarrow 0 \quad (\text{as } n \to \infty). \end{aligned}$$

It follows that $\{c_n\}_{n=1}^{\infty}$ converges to c.   □

It is easy to see that a convergent sequence $\{c_n\}_{n=1}^{\infty}$ is *bounded*; that is, there exists a positive number M such that

$$| c_n | < M \quad \text{for all } n \in \mathbb{N}.$$

The converse is not true, but a slightly weakened version is.

THEOREM 2.1.3 (Bolzano-Weierstrass) *A bounded sequence has a convergent subsequence.*
  Proof. This follows immediately from the reasoning used in the proof of the real version together with the inequalities that we have used repeatedly above.   □

Let E be a nonempty subset of $\mathbb{R}$. Suppose there exists a real number m satisfying

$$m \leq x \quad \text{for all } x \in E.$$

Then we say E is *bounded below*, and m is a *lower bound* of E. Similarly, if there exists $\ell \in \mathbb{R}$ satisfying

$$x \leq \ell \quad \text{for all } x \in E,$$

then we say E is *bounded above*, and ℓ is an *upper bound* of E. If E is bounded below *and* above; that is, there exist real numbers m and ℓ such that

$$m \leq x \leq \ell \quad \text{for all } x \in E,$$

then we say E is *bounded*.

The following theorem is an amplified version of the statement (as-sumed to be known to the reader) that $\mathbb{R}$ is complete, and so no proof is presented.

THEOREM 2.1.4 *If a subset E of $\mathbb{R}$ is bounded below, then there exists a largest among all the lower bounds. (It is called the* **greatest lower bound**, *or* **infimum** *of set E, and is denoted* inf E.) *Similarly, if E is bounded above, then there exists a smallest among all the upper bounds. (It is called the* **least upper bound**, *or* **supremum** *of set E, and is denoted* sup E.)

Given a sequence of real numbers $\{a_n\}_{n=1}^{\infty}$, for each $n \in \mathbb{N}$ let

$$m_n = \inf\{a_k; \; k \geq n\}$$

if the sequence is bounded below, and $m_n = -\infty$ if it is not bounded below. Similarly, let

$$\ell_n = \sup\{a_k; \; k \geq n\}$$

if the sequence is bounded above, and $\ell_n = \infty$ if it is not bounded above. Then

$$m_1 \leq m_2 \leq \cdots \leq m_n \leq \cdots \leq \ell_n \leq \cdots \leq \ell_2 \leq \ell_1.$$

Because a monotone sequence always has a limit (if we allow $\pm\infty$), $\lim_{n\to\infty} m_n$ and $\lim_{n\to\infty} \ell_n$ both exist. We call $\lim_{n\to\infty} m_n$ the *limes inferior* (denoted $\liminf_{n\to\infty} a_n$, or $\underline{\lim}_{n\to\infty} a_n$), and $\lim_{n\to\infty} \ell_n$ the *limes superior* (denoted $\limsup_{n\to\infty} a_n$, or $\overline{\lim}_{n\to\infty} a_n$) of the sequence $\{a_n\}_{n=1}^{\infty}$.

The following theorem characterizes the limes inferior and limes superior. Again, a formal proof is unnecessary.

THEOREM 2.1.5

$$\mu = \liminf_{n\to\infty} a_n$$

$\Longleftrightarrow$ *For an arbitrary $\epsilon > 0$, there corresponds a natural number*

N *such that*

$a_n > \mu - \epsilon$ *for all $n \geq N$, and*

$a_n < \mu + \epsilon$ *for infinitely many n.*

Similarly,

$$\lambda = \limsup_{n\to\infty} a_n$$

$\Longleftrightarrow$ *For an arbitrary* $\epsilon > 0$, *there corresponds a natural number*
    N *such that*
    $a_n < \lambda + \epsilon$ *for all* $n \geq N$, *and*
    $a_n > \lambda - \epsilon$ *for infinitely many* n.

Remark. In the terminology of Section 2.3, $\limsup_{n\to\infty} a_n$ and $\liminf_{n\to\infty} a_n$ are, respectively, the greatest and the least accumulation points of the set $\{a_n ; n \in \mathbb{N}\}$.

COROLLARY 2.1.6 *A real sequence* $\{a_n\}_{n=1}^{\infty}$ *is convergent if and only if*
$$\liminf_{n\to\infty} a_n = \limsup_{n\to\infty} a_n \; (= \lim_{n\to\infty} a_n).$$

The next result will be needed later.

LEMMA 2.1.7 *Let* $\{a_n\}_{n=1}^{\infty}$ *be a sequence of positive real numbers. Then*
$$\liminf_{n\to\infty} \frac{a_{n+1}}{a_n} \leq \liminf_{n\to\infty} \sqrt[n]{a_n} \leq \limsup_{n\to\infty} \sqrt[n]{a_n} \leq \limsup_{n\to\infty} \frac{a_{n+1}}{a_n}.$$

In particular, if $\lim_{n\to\infty} \frac{a_{n+1}}{a_n}$ *exists, then* $\lim_{n\to\infty} \sqrt[n]{a_n}$ *also exists, and they are equal.*

2.2 Series

Given a sequence $\{c_n\}_{n=0}^{\infty}$, the expression
$$\sum_{n=0}^{\infty} c_n = c_0 + c_1 + c_2 + c_3 + \cdots + c_n + \cdots$$

is called a *series*.

For each natural number n,
$$s_n = \sum_{k=0}^{n} c_k = c_0 + c_1 + c_2 + \cdots + c_n$$

is called the nth *partial sum* of the series.

If the sequence $\{s_n\}_{n=0}^{\infty}$ of partial sums converges, then we say the series is *convergent*, and the limit of $\{s_n\}_{n=0}^{\infty}$ is called the *sum* of the (infinite) series. A series that is not convergent is called *divergent*. Clearly, we have

THEOREM 2.2.1 *A series $\sum_{n=0}^{\infty} c_n$ is convergent if and only if both $\sum_{n=0}^{\infty} a_n$ and $\sum_{n=0}^{\infty} b_n$ are convergent (where $c_n = a_n + ib_n$, $a_n, b_n \in \mathbb{R}$), and in this case*

$$\sum_{n=0}^{\infty} c_n = \sum_{n=0}^{\infty} a_n + i \sum_{n=0}^{\infty} b_n.$$

The series version of the Cauchy Criterion is as follows:

THEOREM 2.2.2 (Cauchy Criterion)   *A series $\sum_{n=0}^{\infty} c_n$ converges if and only if to every $\epsilon > 0$ there corresponds a natural number N with the property that*

$$\mid c_{n+1} + c_{n+2} + \cdots + c_{n+p} \mid < \epsilon \quad \text{whenever} \quad n \geq N \text{ and } p > 0.$$

We emphasize that the integer p should remain free from any restriction other than its positivity.

By choosing $p = 1$ we obtain a necessary condition for convergence:

COROLLARY 2.2.3 *For a series $\sum_{n=0}^{\infty} c_n$ to be convergent it is necessary that $c_n \longrightarrow 0$ (as $n \to \infty$).*

Note carefully that this condition is not sufficient.

EXAMPLE. The general term of the harmonic series $\sum_{n=1}^{\infty} \frac{1}{n}$ tends to zero, yet the series diverges.

Given a series $\sum_{n=0}^{\infty} c_n$, we may form a new series

$$\sum_{n=0}^{\infty} \mid c_n \mid = \mid c_0 \mid + \mid c_1 \mid + \mid c_2 \mid + \cdots + \mid c_n \mid + \cdots$$

by replacing each term c_n by its absolute value $\mid c_n \mid$. If this new series is convergent, then we say the original series $\sum_{n=0}^{\infty} c_n$ is *absolutely convergent*. If a series is convergent but not absolutely convergent, then it is said to be *conditionally convergent*.

THEOREM 2.2.4 *An absolutely convergent series is convergent.*

Proof. This assertion follows immediately from the triangle inequality

$$|c_{n+1} + c_{n+2} + \cdots + c_{n+p}| \leq |c_{n+1}| + |c_{n+2}| + \cdots + |c_{n+p}|$$

and the Cauchy Criterion 2.2.2. □

The converse is not true. For example, the alternating series

$$1 - \frac{1}{2} + \frac{1}{3} - \frac{1}{4} + \frac{1}{5} - + \cdots$$

is convergent, but not absolutely.

The rearrangement of terms in an absolutely convergent series affects neither its convergence nor its sum. This seemingly obvious statement is *not* true for a conditionally convergent series, as Riemann showed:

If a series $\sum_{n=0}^{\infty} a_n$ of real numbers is conditionally convergent then the terms of the series can be rearranged so that the new series $\sum_{j=0}^{\infty} a_{n_j}$ converges to any preassigned real number, or even diverges.

Let $\sum_{n=0}^{\infty} c_n$ and $\sum_{n=0}^{\infty} c'_n$ be convergent with sums c and c', respectively. Then the series $\sum_{n=0}^{\infty} (c_n + c'_n)$ also converges and its sum is equal to $c + c'$. Similarly, $\sum_{n=0}^{\infty} k c_n$ converges to kc ($k \in \mathbb{C}$).

Let us now turn to the product of two convergent series $\sum_{n=0}^{\infty} c_n$ and $\sum_{n=0}^{\infty} c'_n$. Formally, the product leads to a double series $\sum_{m=0}^{\infty} \sum_{n=0}^{\infty} c_m c'_n$. One special way of defining the product of two convergent series is by grouping the terms of the formal double series as follows:

$$\begin{aligned} c_0 c'_0 &+ (c_0 c'_1 + c_1 c'_0) + (c_0 c'_2 + c_1 c'_1 + c_2 c'_0) + \cdots \\ &+ (c_0 c'_n + c_1 c'_{n-1} + \cdots + c_n c'_0) + \cdots. \end{aligned}$$

This series, whose nth term is $\sum_{k=0}^{n} c_k c_{n-k}$, is known as the *Cauchy product* of $\sum_{n=0}^{\infty} c_n$ and $\sum_{n=0}^{\infty} c'_n$. This is the natural definition of the product when we are working with power series, for the Cauchy product of two power series will again be a power series. However, for other types of series, for example *Dirichlet series*, the Cauchy product may not be the appropriate choice.

THEOREM 2.2.5 (Mertens) *Suppose $\sum_{n=0}^{\infty} c_n$ is absolutely convergent and $\sum_{n=0}^{\infty} c_n'$ is convergent; then their Cauchy product*

$$\sum_{n=0}^{\infty} (c_0 c_n' + c_1 c_{n-1}' + \cdots + c_n c_0')$$

is convergent, and its sum is equal to

$$\left(\sum_{n=0}^{\infty} c_n \right) \cdot \left(\sum_{n=0}^{\infty} c_n' \right).$$

Proof. Let s_n, s_n', and t_n be the partial sums of the c-series, c'-series, and the Cauchy product series, respectively. Set

$$c = \sum_{n=0}^{\infty} c_n, \quad c' = \sum_{n=0}^{\infty} c_n'.$$

Then

$$
\begin{aligned}
s_n s_n' - t_n &= c_1 c_n' + c_2 \left(c_{n-1}' + c_n' \right) + \cdots + c_n \left(c_1' + c_2' + \cdots + c_n' \right) \\
&= c_1 \left(s_n' - s_{n-1}' \right) + c_2 \left(s_n' - s_{n-2}' \right) + \cdots + c_n \left(s_n' - s_0' \right).
\end{aligned}
$$

Let k be defined by $k = \frac{n}{2}$ if n is even, and $k = \frac{n-1}{2}$ if n is odd. Then as $n \to \infty$ we have $k \to \infty$. Thus, by the Cauchy Criterion 2.2.2, for every positive number ϵ there corresponds an integer N such that

$$| s_n' - s_m' | < \epsilon \quad \text{if } m \geq k \geq N,$$

and

$$| c_{k+1} | + | c_{k+2} | + \cdots + | c_n | < \epsilon \quad \text{if } k \geq N.$$

Moreover, there exists a constant M satisfying the conditions

$$\sum_{j=1}^{\infty} | c_j | \leq M \quad \text{and} \quad | s_j' | \leq M \quad \text{(for all } j \in \mathbb{N}\text{).}$$

The existence of such a constant is assured by the absolute convergence of the c-series and the convergence (hence boundedness) of the c'-series.

$$\therefore \; |s_n s'_n - t_n| \leq \sum_{j=1}^{k} \left\{ |c_j| \cdot \left| s'_n - s'_{n-j} \right| \right\} + \sum_{j=k+1}^{n} \left\{ |c_j| \cdot \left| s'_n - s'_{n-j} \right| \right\}$$

$$\leq \epsilon \sum_{j=1}^{k} |c_j| + 2M \cdot \sum_{j=k+1}^{n} |c_j|$$

$$< \epsilon \cdot M + 2M \cdot \epsilon = 3M \cdot \epsilon \quad \text{if } k \geq N.$$

Because $\epsilon > 0$ is arbitrary and $s_n s'_n \longrightarrow cc'$ (as $n \to \infty$), we are done. $\square$

COROLLARY 2.2.6 *If both $\sum_{n=0}^{\infty} c_n$ and $\sum_{n=0}^{\infty} c'_n$ are absolutely convergent, then their Cauchy product is absolutely convergent.*

Proof. Apply the theorem to $\sum_{n=0}^{\infty} | c_n |$ and $\sum_{n=0}^{\infty} | c'_n |$. $\square$

Note that the Cauchy product of two conditionally convergent series may fail to be convergent.

EXAMPLE. Consider the Cauchy product of the alternating series

$$\sum_{n=0}^{\infty} \frac{(-1)^n}{\sqrt{n+1}}$$

with itself. The general term of the Cauchy product is of the form

$$c_{n+1} = (-1)^n \left(\frac{1}{\sqrt{1}\sqrt{n+1}} + \frac{1}{\sqrt{2}\sqrt{n}} + \cdots + \frac{1}{\sqrt{n+1}\sqrt{1}} \right).$$

$$\therefore \; |c_{n+1}| \geq \frac{1}{\sqrt{n+1}\sqrt{n+1}} + \frac{1}{\sqrt{n+1}\sqrt{n+1}} + \cdots$$

$$+ \frac{1}{\sqrt{n+1}\sqrt{n+1}}$$

$$= \frac{n+1}{n+1} = 1.$$

Therefore, the condition of Corollary 2.2.3 is violated.

2.3   Some Terminology about the Topology of the Complex Plane

Throughout this book we shall use the notation $D(z_0, r)$ to denote the *disc* with center at z_0 and radius r $(r > 0)$:

$$D(z_0, r) = \{z \in \mathbb{C}; \ |z - z_0| < r\}.$$

The corresponding closed disc, the punctured disc, and the circle will be denoted by $\bar{D}(z_0, r)$, $D'(z_0, r)$, and $C(z_0, r)$, respectively. Thus,

$$\begin{aligned}
\bar{D}(z_0, r) &= \{z \in \mathbb{C}; \ |z - z_0| \leq r\}, \\
D'(z_0, r) &= \{z \in \mathbb{C}; \ 0 < |z - z_0| < r\}, \\
C(z_0, r) &= \{z \in \mathbb{C}; \ |z - z_0| = r\}.
\end{aligned}$$

For simplicity, the *unit disc* will be denoted D:

$$D = D(0, 1) = \{z \in \mathbb{C}; \ |z| < 1\}.$$

A subset G of $\mathbb{C}$ is called *open* if for each point $z_0 \in G$ there is a disc $D(z_0, \epsilon)$ (called an *ϵ-neighborhood* of z_0) which is in G; that is,

$$D(z_0, \epsilon) \subset G \quad \text{for some } \epsilon > 0.$$

Note that discs *are* open (unless stated explicitly as closed disc). A subset F of $\mathbb{C}$ is called *closed* if its complement

$$F^c = \mathbb{C} \setminus F = \{z \in \mathbb{C}; \ z \notin F\}$$

is open. The smallest closed set, obtained by taking the intersection of all closed sets that contain set E, is called the *closure* of E, and is denoted $\bar{E}$. Thus E is closed if and only if $\bar{E} = E$.

Point $z_0 \in E$ is an *interior point* of E if there is an ϵ-neighborhood of z_0 that is in E:

$$D(z_0, \epsilon) \subset E \quad \text{for some } \epsilon > 0.$$

The set of all interior points of E is called the *interior* of E, and is denoted E^0. A point z_1 is an *exterior point* of set E if there is an ϵ-neighborhood of z_1 that does not meet E:

$$D(z_1, \epsilon) \cap E = \phi \quad \text{for some } \epsilon > 0,$$

where ϕ denotes, as is customary, the empty set.

The set of all exterior points of set E is called the *exterior* of E. A point $z_0 \in \bar{E}$ is a *boundary point* of E if every neighborhood of z_0 contains both a point in E and a point not in E: For every $\epsilon > 0$,

$$D(z_0, \epsilon) \cap E \neq \phi \quad \text{and} \quad D(z_0, \epsilon) \cap (\mathbb{C} \setminus E) \neq \phi.$$

The set of all boundary points of set E is called the *boundary* of E, and is denoted ∂E.

A point $z_0 \in E$ is an *isolated point* of set E if there is a neighborhood $D(z_0, \epsilon)$ of z_0 whose intersection with E is z_0 itself:

$$E \cap D(z_0, \epsilon) = \{z_0\} \quad \text{for some } \epsilon > 0;$$

that is,

$$E \cap D'(z_0, \epsilon) = \phi \quad \text{for some } \epsilon > 0.$$

An *accumulation point* of a set E is a point in $\bar{E}$ that is not an isolated point of $\bar{E}$.

A subset E of a set $G \subset \mathbb{C}$ is said to be *dense* in G if for every $z_0 \in G$ and $\epsilon > 0$, $E \cap D(z_0, \epsilon) \neq \phi$. A subset E of $\mathbb{C}$ is called *bounded* if it is contained in some disc.

An *open covering* of a set E is a collection of open sets whose union contains E. A *subcovering* is a subcollection with the same property, and a *finite covering* is a covering consisting of a finite number of sets. A subset E of $\mathbb{C}$ is said to have the *Heine-Borel property* if every open covering of E contains a finite subcovering.

The next theorem has far-reaching effects:

THEOREM 2.3.1 (Heine-Borel) *A subset K of $\mathbb{C}$ has the Heine-Borel property if and only if K is closed and bounded.*

Because of their importance, subsets of $\mathbb{C}$ with the Heine-Borel property are given a name: *compact*.[1]

---

[1] This is an example of Professor E. Hewitt's proposition: "Good theorems never die, they simply turn into definitions."

The next theorem, known as the *Cantor intersection theorem,* is needed in proving the all-important *Cauchy-Goursat Theorem 4.3.1* and the Monodromy Theorem 8.6.1.

THEOREM 2.3.2 (Cantor) *If* $\{K_n\}_{n=1}^{\infty}$ *is a nested sequence of nonempty compact subsets in* $\mathbb{C}$:

$$K_1 \supset K_2 \supset \cdots \supset K_n \supset \cdots,$$

then

$$K = \bigcap_{n=1}^{\infty} K_n \neq \phi.$$

Moreover, K consists of exactly one point if the diameter *of K_n tends to zero (as $n \to \infty$).*

Remark. The diameter of a nonempty set $E \subset \mathbb{C}$ is defined by

$$\text{diameter}\,(E) = \sup\{|\alpha - \beta|\;;\;\alpha,\,\beta \in E\}.$$

An open subset of $\mathbb{C}$ is *connected* if it cannot be expressed as a union of two nonempty disjoint open sets; and a closed subset of $\mathbb{C}$ is *connected* if it cannot be expressed as a union of two nonempty disjoint closed sets. (The concept of "connectedness" can be extended to subsets of $\mathbb{C}$ that are neither open nor closed; but this fact will not affect us.)

A subset E of $\mathbb{C}$ is *polygonally connected* if any two points of E can be joined by a polygon (a broken line) that is in E.

We remark that *a polygonally connected subset of $\mathbb{C}$ is connected*, but the converse is not necessarily true. However, we have the following

THEOREM 2.3.3 *An open subset of $\mathbb{C}$ is connected if and only if it is polygonally connected.*

A nonempty open connected subset of $\mathbb{C}$ is called a *region.* (Some authors use the term "domain" instead of "region"). A *component* of an open set G is a region in G which is not properly contained in a region in G.

THEOREM 2.3.4 *Every open set in $\mathbb{C}$ is uniquely expressible as a countable union of disjoint regions.*

The definition of continuity for a function $f(z)$ defined on a subset Ω of $\mathbb{C}$ is the same as in $\mathbb{R}$. A function $f(z)$ is *continuous at a point $z_0 \in \Omega$*

if, for every $\epsilon > 0$, there corresponds $\delta > 0$ with the property that

$$| f(z) - f(z_0) | < \epsilon \quad \text{whenever} \quad z \in \Omega \cap D(z_0, \delta).$$

If $f(z)$ is continuous at every point of Ω, we say that $f(z)$ is *continuous* on Ω. In general, this $\delta > 0$ depends both on $\epsilon > 0$ and on $z_0 \in \Omega$. In the special case that $\delta > 0$ can be chosen independently of $z_0 \in \Omega$ (for the same $\epsilon > 0$), we say $f(z)$ is *uniformly continuous* on Ω: Given $\epsilon > 0$, there corresponds $\delta > 0$, which depends only on ϵ, such that

$$| f(z_1) - f(z_2) | < \epsilon \quad \text{whenever} \quad | z_1 - z_2 | < \delta \quad \text{and} \quad z_1, z_2 \in \Omega.$$

EXAMPLE. $f(z) = z^2$ is continuous but not uniformly continuous on $\mathbb{C}$. It is, however, uniformly continuous on every compact subset of $\mathbb{C}$.

THEOREM 2.3.5 *A continuous function defined on a compact set is uniformly continuous.*

2.4   Uniform Convergence

Consider a sequence of functions

$$f_1(z), \ f_2(z), \ f_3(z), \cdots, f_n(z), \cdots,$$

or simply $\{f_n(z)\}_{n=1}^{\infty}$, all defined on a set $\Omega \subset \mathbb{C}$. For each point $z_0 \in \Omega$, $\{f_n(z_0)\}_{n=1}^{\infty}$ is a sequence of complex numbers. If for every point $z_0 \in \Omega$ the sequence $\{f_n(z_0)\}_{n=1}^{\infty}$ is convergent, then we say the sequence $\{f_n(z)\}_{n=1}^{\infty}$ of functions *converges pointwise* on Ω. In this case we define a function, $f(z)$, on Ω by

$$f(z) = \lim_{n \to \infty} f_n(z) \quad (z \in \Omega).$$

This function is called the *limit* (*function*) of the sequence $\{f_n(z)\}_{n=1}^{\infty}$. We restate this definition formally. A sequence $\{f_n(z)\}_{n=1}^{\infty}$ of functions is said to converge to the limit function $f(z)$ if, for each point $z \in \Omega$ and to every $\epsilon > 0$, there corresponds a natural number N such that

$$|f_n(z) - f(z)| < \epsilon \quad \text{whenever} \quad n \geq N.$$

In general, the natural number N depends not only on ϵ but also on $z \in \Omega$. However, if for every $\epsilon > 0$ a natural number N can be chosen *independently* of $z \in \Omega$; viz., for the same $\epsilon > 0$, there corresponds a natural number N that works for all $z \in \Omega$, then we say the sequence $\{f_n(z)\}_{n=1}^{\infty}$ converges *uniformly* to the limit function $f(z)$ on Ω.

EXAMPLE. Let $f_n(z) = \frac{z}{n}$ on $\mathbb{C}$. Then the sequence $\{f_n(z)\}_{n=1}^{\infty}$, converges to the limit $f(z) = 0$ for all $z \in \mathbb{C}$. However, in order that $\left|\frac{z}{n}\right| < \epsilon$, it is necessary that $n > \frac{|z|}{\epsilon}$. Thus, as $|z|$ increases, the corresponding N increases without bound. Thus, we say the sequence converges *pointwise*, but not uniformly, on $\mathbb{C}$. Note that in this example if we restrict the domain to a bounded set, say $\{z \in \mathbb{C}; \ |z| < 1996\}$, then the convergence is uniform.

The Cauchy Criterion has a counterpart for uniform convergence.

THEOREM 2.4.1 (Cauchy Criterion) *A sequence* $\{f_n(z)\}_{n=1}^{\infty}$ *of functions is uniformly convergent (to a limit function $f(z)$) on a set Ω if and only if for every $\epsilon > 0$ there corresponds a natural number N with the property that*

$$|f_m(z) - f_n(z)| < \epsilon \quad \text{whenever} \quad m, n \geq N \text{ and } z \in \Omega.$$

Proof. Necessity. Suppose $\{f_n(z)\}_{n=1}^{\infty}$ converges to $f(z)$ uniformly on Ω. Then, given any $\epsilon > 0$, there corresponds a natural number N such that

$$|f_n(z) - f(z)| < \frac{\epsilon}{2} \quad \text{whenever} \quad n \geq N \text{ and } z \in \Omega.$$

Now the triangle inequality

$$|f_m(z) - f_n(z)| \leq |f_m(z) - f(z)| + |f(z) - f_n(z)|$$

gives the desired result immediately.
Sufficiency. Suppose the condition is satisfied. Then, for an arbitrary $\epsilon > 0$, there corresponds N such that

$$|f_m(z) - f_n(z)| < \frac{\epsilon}{2} \quad \text{whenever} \quad m, n \geq N \text{ and } z \in \Omega.$$

Therefore, $\{f_n(z)\}_{n=1}^{\infty}$ has a limit for each $z \in \Omega$, by the sequence version of the Cauchy criterion. Set $f(z) = \lim_{n \to \infty} f_n(z)$ $(z \in \Omega)$. Fixing n and letting $m \to \infty$ in the inequality above, we obtain

$$|f(z) - f_n(z)| \leq \frac{\epsilon}{2} < \epsilon \quad \text{for all } z \in \Omega \text{ provided } n \geq N,$$

which establishes the uniform convergence on Ω. $\square$

If Ω is a region, the following concept turns out to be very useful in complex analysis.

DEFINITION 2.4.2 *A sequence* $\{f_n(z)\}_{n=1}^{\infty}$ *of functions defined in a region* Ω *is said to be* **locally uniformly convergent** *in* Ω *if the sequence* $\{f_n(z)\}_{n=1}^{\infty}$ *converges uniformly on every compact subset of* Ω.

EXAMPLE. We claim that the sequence $\{z^n\}_{n=1}^{\infty}$ is locally uniformly convergent in the unit disc D. Because every compact subset of D is a subset of some closed disc $\bar{D}(0,\, r)$ $(0 < r < 1)$, it is sufficient to prove uniform convergence on every closed disc $\bar{D}(0, r)$ $(0 < r < 1)$. Clearly, the limit function $f(z)$ is identically 0 in D. Given $\epsilon > 0$, the relation

$$|z^n| < \epsilon \quad \text{for all } z \in \bar{D}(0,\, r)$$

holds if and only if $n > \frac{\log \epsilon}{\log r}$. Thus, for each r $(0 < r < 1)$, choose the natural number N in the definition of the uniform convergence to satisfy $N > \frac{\log \epsilon}{\log r}$. This choice is possible for an arbitrary $\epsilon > 0$ and $0 < r < 1$. However, as $r \to 1-$ (with $\epsilon > 0$ fixed), N must be increased without bound, and so the convergence is *not* uniform on the unit disc D.

THEOREM 2.4.3 *Let* $\{f_n(z)\}_{n=1}^{\infty}$ *be a sequence of functions that are continuous on a region* Ω. *If the sequence converges locally uniformly to a function* $f(z)$ *on* Ω, *then* $f(z)$ *is continuous on* Ω. *Restated: The limit function of a locally uniformly convergent sequence of continuous functions on a region is continuous.*

Proof. Because Ω is open, for any $z_0 \in \Omega$ there is a positive number r satisfying $\bar{D}(z_0,\, r) \subset \Omega$. Because $\bar{D}(z_0,\, r)$ is compact, the convergence is uniform on $\bar{D}(z_0,\, r)$. Thus, for an arbitrary $\epsilon > 0$, there corresponds a natural number N with the property that

$$|f_n(z) - f(z)| < \frac{\epsilon}{3} \quad \text{whenever } n \geq N \text{ and } z \in \bar{D}(z_0,\, r).$$

Choose a natural number $m \geq N$. Because $f_m(z)$ is continuous at z_0, we can choose δ $(0 < \delta < r)$ such that

$$|f_m(z) - f_m(z_0)| < \frac{\epsilon}{3} \quad \text{whenever } z \in \bar{D}(z_0,\, \delta).$$

Therefore, for $z \in D(z_0,\, \delta)$, we have

$$|f(z) \; - \; f(z_0)| \leq |f(z) - f_m(z)| + |f_m(z) - f_m(z_0)|$$
$$+ \; |f_m(z_0) - f(z_0)| < \frac{\epsilon}{3} + \frac{\epsilon}{3} + \frac{\epsilon}{3} = \epsilon.$$

This result shows that $f(z)$ is continuous at z_0, but z_0 is an arbitrary point in Ω, and so we are done.   □

The following simple lemma turns out to be very convenient.

LEMMA 2.4.4 *A sequence of functions is locally uniformly convergent on a region Ω if and only if each point of Ω has a neighborhood where the sequence converges uniformly.*
Proof. This is an immediate consequence of the Heine-Borel Theorem 2.3.1.   □

Naturally, the concept of (local) uniform convergence has a counterpart for series whose terms are functions, and series versions of all the theorems and the lemma in this section are true.
One of the most commonly used tests for the uniform convergence of series of functions is the following:

THEOREM 2.4.5 (Weierstrass M-test) *Let all the functions in the sequence $\{f_n(z)\}_{n=0}^{\infty}$ be defined on $\Omega \subset \mathbb{C}$. If a series $\sum_{n=0}^{\infty} M_n$ of positive terms converges and*
$$|f_n(z)| \leq M_n$$
*for all $z \in \Omega$ and all sufficiently large n, then the series $\sum_{n=0}^{\infty} f_n(z)$ converges uniformly and absolutely on Ω. The series $\sum_{n=0}^{\infty} M_n$ is called a **majorant** for $\sum_{n=0}^{\infty} f_n(z)$.*
Proof. By the triangle inequality, we have

$$|f_{n+1}(z) + f_{n+2}(z) + \cdots + f_{n+p}(z)| \leq M_{n+1} + M_{n+2} + \cdots + M_{n+p}.$$

It follows that if the majorant converges, then the minorant converges uniformly and absolutely, by the Cauchy Criterion 2.2.2.   □

Unfortunately, this test has a slight drawback in that it applies only to series that are also absolutely convergent.

2.5   Geometric Series

Before we discuss general power series, let us consider the convergence of the *geometric series*:

$$1 + z + z^2 + \cdots + z^n + \cdots.$$

For this series it is easy to compute the partial sums:

$$S_n(z) = 1 + z + z^2 + \cdots + z^n = \begin{cases} \dfrac{1 - z^{n+1}}{1 - z}, & z \neq 1; \\ n + 1, & z = 1. \end{cases}$$

Therefore, if $z \in D$ we have

$$\left| S_n(z) - \frac{1}{1 - z} \right| = \frac{|z|^{n+1}}{|1 - z|},$$

and the right member can be made arbitrarily small by taking sufficiently large n.

$$\therefore\ 1 + z + z^2 + \cdots + z^n + \cdots = \frac{1}{1 - z} \quad \text{if } z \in D.$$

Note that by replacing z^n by $|z|^n$, exactly the same argument demonstrates the absolute convergence of the series inside the unit disc. Actually, a slight modification of the argument above also shows that the convergence is uniform on any closed disc inside the unit circle. For $z \in \bar{D}(0, \delta)$, where $0 < \delta < 1$,

$$\left| S_n(z) - \frac{1}{1 - z} \right| \leq \frac{\delta^{n+1}}{1 - \delta}.$$

Thus, given $\epsilon > 0$ and $0 < \delta < 1$, it is possible to choose N satisfying $\frac{\delta^N}{1-\delta} < \epsilon$. Then we have

$$\left| S_n(z) - \frac{1}{1 - z} \right| < \epsilon$$

for all $z \in \bar{D}(0, \delta)$ and $n > N$. On the other hand, if $|z| \geq 1$, the general term in the series is bounded away from zero, hence the series will not converge for such values of z (by Corollary 2.2.3).

We have shown:

THEOREM 2.5.1

(a) The geometric series

$$1 + z + z^2 + \cdots + z^n + \cdots$$

converges absolutely and locally uniformly in the unit disc to $\frac{1}{1-z}$.

(b) the series diverges on and outside the unit circle.

2.6   Circle of Convergence

Of all the series whose terms are functions, the simplest and most important are *power series*. They are of the form

$$\sum_{n=0}^{\infty} a_n(z - c)^n = a_0 + a_1(z - c) + a_2(z - c)^2 + \cdots + a_n(z - c)^n + \cdots,$$

where c is the *center* and a_n are the coefficients of the power series. For simplicity, we discuss the case in which the center c is at the origin. The corresponding results for general power series can easily be obtained by translation.

First we ask where this series converges. This question is answered by the *Abel power series theorem*:

THEOREM 2.6.1 (Abel) *To every power series*

$$\sum_{n=0}^{\infty} a_n z^n = a_0 + a_1 z + a_2 z^2 + \cdots + a_n z^n + \cdots$$

there corresponds a unique extended real number R $(0 \le R \le \infty)$, called the **radius of convergence** *of the power series, such that*
(a) the series converges absolutely and locally uniformly in $D(0, R)$, if $R > 0$;
(b) the series diverges for all z outside the closed disc $\bar{D}(0, R)$, if $R < \infty$.
Moreover, the radius of convergence R is given by the **Cauchy-Hadamard formula***:*

$$\frac{1}{R} = \limsup_{n \to \infty} \sqrt[n]{|a_n|}.$$

Proof. Let $\frac{1}{R} = \limsup_{n \to \infty} \sqrt[n]{|a_n|}$.
    (a) If $z \in D(0, R)$ $(R \ne 0)$, we may choose r and ρ $(0 < r < \rho < R)$ such that $z \in \bar{D}(0, r)$. Then, because $\frac{1}{\rho} > \frac{1}{R}$, there exists, by Theorem 2.1.5, a natural number N such that

$$|a_n|^{\frac{1}{n}} < \frac{1}{\rho}; \quad \text{that is,} \quad |a_n| < \frac{1}{\rho^n} \quad \text{for all } n \ge N.$$

Hence, for $z \in \bar{D}(0, r)$,

$$|a_n z^n| < \left(\frac{r}{\rho}\right)^n \quad \text{for all } n \ge N.$$

Because (the tail part of) the power series has a convergent geometric series as a majorant, we conclude, by the Weierstrass M-test 2.4.5, that the power series converges absolutely and uniformly on $\bar{D}(0, r)$.

(b) If $z \notin \bar{D}(0, R)$ ($R \neq \infty$), we choose ρ so that $R < \rho < |z|$. Because $\frac{1}{\rho} < \frac{1}{R}$, it follows from Theorem 2.1.5 that

$$|a_n|^{\frac{1}{n}} > \frac{1}{\rho}; \quad \text{that is,} \quad |a_n| > \frac{1}{\rho^n} \quad \text{for infinitely many } n,$$

and so

$$|a_n z^n| > \left(\frac{|z|}{\rho}\right)^n > 1 \quad \text{for infinitely many } n.$$

Hence, there is no possibility of convergence for $z \notin \bar{D}(0, R)$, by Corollary 2.2.3.

The uniqueness of the radius of convergence is clear. $\square$

COROLLARY 2.6.2 *Inside the* **circle of convergence** *$C(0, R)$, the sum of a power series is a continuous function.*

Remark. Nothing about convergence is claimed on the circle of convergence $C(0, R)$. Consider the power series:

$$z + z^2 + \cdots + z^n + \cdots,$$
$$z + \frac{z^2}{2} + \cdots + \frac{z^n}{n} + \cdots,$$
$$z + \frac{z^2}{2^2} + \cdots + \frac{z^n}{n^2} + \cdots.$$

All have the unit circle as their circle of convergence. Yet the first series diverges at every point of the unit circle; the second series converges (conditionally) at every point of the unit circle except at $z = 1$ (see Exercise 2.44); the third series converges absolutely and uniformly on the closed unit disc.

In general, it is rather difficult to evaluate the limes superior in the Cauchy-Hadamard formula. Fortunately, Lemma 2.1.7, combined with the Abel Theorem 2.6.1, gives the useful test:

THEOREM 2.6.3 (Ratio Test) *If* $\lim_{n\to\infty} \left| \frac{a_{n+1}}{a_n} \right| = \ell$ *exists* (∞ *allowed*), *then the radius of convergence* R *of the power series* $\sum_{n=0}^{\infty} a_n z^n$ *is given by* $R = \frac{1}{\ell}$.

Note that the existence of the limit of ratios implies, in particular, that there cannot be infinitely many zero coefficients. Note also that the Cauchy-Hadamard formula works for *every* power series, but the ratio test is valid only for series such that $\lim_{n\to\infty} \left| \frac{a_{n+1}}{a_n} \right|$ exists.

EXAMPLE. Find the disc of convergence of the power series

$$\sum_{n=0}^{\infty} \{3 + (-1)^n\}(z - 2)^n.$$

Solution. We have

$$a_n = \begin{cases} 4 & \text{if } n \text{ is even;} \\ 2 & \text{if } n \text{ is odd,} \end{cases}$$

$$\therefore \frac{a_{n+1}}{a_n} = \begin{cases} \frac{1}{2} & \text{if } n \text{ is even;} \\ 2 & \text{if } n \text{ is odd.} \end{cases}$$

Hence $\lim_{n\to\infty} \left| \frac{a_{n+1}}{a_n} \right|$ does not exist; thus the ratio test is not applicable. However, because $2 \leq a_n \leq 4$ for all $n \in \mathbb{N}$,

$$1 = \limsup_{n\to\infty} \sqrt[n]{2} \leq \limsup_{n\to\infty} \sqrt[n]{a_n} \leq \limsup_{n\to\infty} \sqrt[n]{4} = 1.$$

Thus, the radius of convergence is 1, and the disc of convergence is $D(2, 1)$. Note that (with $\zeta = z - 2$)

$$\sum_{n=0}^{\infty} \{3 + (-1)^n\}(z - 2)^n = 4 + 2\zeta + 4\zeta^2 + 2\zeta^3 + \cdots$$

$$= 4(1 + \zeta^2 + \zeta^4 + \cdots) + 2\zeta(1 + \zeta^2 + \zeta^4 + \cdots)$$
$$= (4 + 2\zeta)(1 + \zeta^2 + \zeta^4 + \cdots)$$
$$= \frac{2(2 + \zeta)}{1 - \zeta^2} = -\frac{2z}{3 - 4z + z^2}.$$

The series converges absolutely for $|\zeta| < 1$; that is, for $|z - 2| < 1$, and so rearranging the series is justified.

EXAMPLE. Find the radius of convergence of the power series

$$\sum_{n=0}^{\infty} \left(\cos \frac{n\pi}{3} \right) z^n.$$

Solution. Because $\left|\cos \frac{n\pi}{3}\right| \leq 1$ for all $n \in \mathbb{N}$, we have

$$\frac{1}{R} = \limsup_{n \to \infty} \sqrt[n]{\left|\cos \frac{n\pi}{3}\right|} \leq 1.$$

On the other hand, $\left|\cos \frac{n\pi}{3}\right| = 1$ for infinitely many n, therefore, $R = 1$.

Remark. As in the preceding example, the sum of this series can be evaluated and is, in fact, equal to

$$\frac{1 + \frac{\sqrt{3}}{2}z - \frac{\sqrt{3}}{2}z^2 - z^3 - \frac{\sqrt{3}}{2}z^4 + \frac{\sqrt{3}}{2}z^5}{1 - z^6} \quad (z \in D).$$

In general, the behavior of a power series on the circle of convergence is complicated. We do, however, have the following theorem, known as the *Abel continuity theorem*. Before we state the theorem we need some terminology.

Suppose z_0 is a point on the boundary circle of a disc. A *Stolz region* of point z_0 is the intersection of the disc and a region formed by an angular opening of size 2φ $(0 < \varphi < \frac{\pi}{2})$ symmetric with respect to the radius at z_0 (Figure 2.1). If point z approaches z_0 from a Stolz region of point z_0, then we say z approaches z_0 *nontangentially*.

THEOREM 2.6.4 (Abel) *Suppose a power series* $f(z) = \sum_{k=0}^{\infty} a_k z^k$ *converges at point z_0 on the circle of convergence $C(0, R)$. Then*

$$\lim f(z) = f(z_0)$$

as point z approaches z_0 nontangentially.

Proof. By a transformation $z = z_0 \zeta$, point $z = z_0$ is transformed to $\zeta = 1$, and the circle of convergence $C(0, R)$ is transformed to the unit circle. Therefore, without loss of generality we may assume that $z_0 = 1$

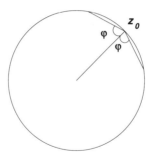

Figure 2.1

and $R = 1$ from the beginning. Moreover, replacing the constant term a_0 by $a_0 - f(1)$ if necessary, we may further assume that

$$f(1) = a_0 + a_1 + a_2 + \cdots + a_n + \cdots = 0.$$

Set $s_n = a_0 + a_1 + \cdots + a_n$. Then, by our assumption, we have $\lim_{n \to \infty} s_n = 0$; that is, given $\epsilon > 0$, there corresponds N such that

$$|s_n| < \epsilon \quad \text{for all} \quad n \geq N.$$

$$\begin{aligned}
\therefore f(z) &= \sum_{k=0}^{\infty} a_k z^k = \sum_{k=0}^{\infty} (s_k - s_{k-1}) z^k \qquad (s_{-1} = 0) \\
&= \lim_{n \to \infty} \left\{ s_0 + (s_1 - s_0) z + (s_2 - s_1) z^2 + \cdots \right. \\
&\qquad \left. + (s_n - s_{n-1}) z^n \right\} \\
&= \lim_{n \to \infty} \left\{ s_0(1 - z) + s_1 \left(z - z^2 \right) + s_2 \left(z^2 - z^3 \right) + \cdots \right. \\
&\qquad \left. + s_{n-1} \left(z^{n-1} - z^n \right) + s_n z^n \right\} \\
&= (1 - z) \cdot \lim_{n \to \infty} \left\{ s_0 + s_1 z + s_2 z^2 + \cdots + s_{n-1} z^{n-1} \right\} \\
&\qquad (\because \lim_{n \to \infty} s_n z^n = 0) \\
&= (1 - z) \cdot \sum_{k=0}^{N-1} s_k z^k + (1 - z) \cdot \sum_{k=N}^{\infty} s_k z^k.
\end{aligned}$$

Set $m = \max\{|s_k| \; ; \; k = 0, 1, \cdots, N - 1\}$. Then

$$\left| (1 - z) \cdot \sum_{k=0}^{N-1} s_k z^k \right| \leq |1 - z| \cdot mN < \epsilon \quad \text{for all} \quad z \in D\left(1, \frac{\epsilon}{mN} \right).$$

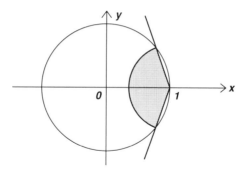

Figure 2.2

$$\therefore |f(z)| < \epsilon + |1 - z| \cdot \epsilon \cdot \sum_{k=N}^{\infty} |z|^k$$

$$= \epsilon \cdot \left\{ 1 + |1 - z| \cdot \frac{|z|^N}{1 - |z|} \right\}$$

$$< \epsilon \cdot \left\{ 1 + \frac{|1 - z|}{1 - |z|} \right\}.$$

If we can demonstrate that as z approaches 1 *nontangentially*, $\frac{|1-z|}{1-|z|}$ remains bounded, say by K, then we will have

$$|f(z)| < \epsilon \cdot (1 + K),$$

which means

$$\lim f(z) = 0 = f(1),$$

and this is what we want to show.

It remains to prove the boundedness of $\frac{|1-z|}{1-|z|}$ for z in the set

$$E_\varphi = \{ z \in D \,;\, |1 - z| \leq \cos \varphi, \ |\text{Arg}\,(1 - z)| \leq \varphi \} \quad \left(0 < \varphi < \frac{\pi}{2}\right).$$

For $z \in E_\varphi$, set $1 - z = r e^{i\theta}$ (Figure 2.2). Then

$$|\theta| \leq \varphi < \frac{\pi}{2}, \quad r = |1 - z| \leq \cos \varphi \leq \cos \theta.$$

$$\therefore \frac{|1 - z|}{1 - |z|} = \frac{r(1 + |z|)}{1 - |z|^2} \leq \frac{2r}{1 - (1 - re^{i\theta})(1 - re^{-i\theta})}$$

$$= \frac{2r}{2r \cos \theta - r^2} \leq \frac{2}{2 \cos \varphi - \cos \varphi}$$

$$\leq \frac{2}{\cos \varphi} \quad (\text{for all } z \in E_\varphi),$$

and we are done.   □

The converse of the Abel continuity theorem is not true. Consider

$$f(z) = \frac{1}{1+z} = 1 - z + z^2 - z^3 + - \cdots + (-1)^k z^k + \cdots \quad (z \in D).$$

We have

$$\lim_{z \to 1} f(z) = \frac{1}{2} \quad (\text{even tangentially}),$$

but the series $1 - 1 + 1 - 1 + - \cdots + (-1)^{k-1} + \cdots$ is obviously divergent.

2.7   Uniqueness

If

$$f(z) = a_0 + a_1 z + a_2 z^2 + \cdots + a_n z^n + \cdots$$

converges somewhere other than $z = 0$, then there is a disc (of positive radius) where the power series converges locally uniformly, hence $f(z)$ is a continuous function in this disc. We shall assume that this disc is of finite radius; however, the results will be all the more true if $R = \infty$.

Suppose there exists a sequence $\{z_k\}_{k=1}^\infty$ of distinct complex numbers converging to 0, and at each of these points

$$f(z_k) = \sum_{n=0}^\infty a_n z_k^n = 0 \quad (k \in \mathbb{N}).$$

Then, by continuity,

$$a_0 = f(0) = 0,$$

hence

$$f(z) = z \cdot \sum_{n=1}^\infty a_n z^{n-1}.$$

Because $f(z_k) = 0$ for all $k \in \mathbb{N}$, we have

$$\sum_{n=1}^\infty a_n z_k^{n-1} = 0 \quad \text{for all } k \in \mathbb{N}.$$

Because the sum of the power series $\sum_{n=1}^{\infty} a_n z^{n-1}$ is continuous at the origin, we obtain $a_1 = 0$.

Continuing this process, we obtain

LEMMA 2.7.1 *Suppose a power series* $f(z) = \sum_{n=0}^{\infty} a_n z^n$ *has a positive radius of convergence. If there exists a sequence* $\{z_k\}_{k=1}^{\infty}$ *of distinct complex numbers converging to zero such that*

$$f(z_k) = 0 \quad \text{for all } k \in \mathbb{N},$$

then

$$a_n = 0 \quad \text{for all } n = 0, 1, 2, \cdots,$$

and

$$f(z) = \sum_{n=0}^{\infty} a_n z^n \equiv 0.$$

COROLLARY 2.7.2 *If*

$$f(z) = \sum_{n=0}^{\infty} a_n z^n \equiv 0$$

in some neighborhood of the origin and $R > 0$, *then*

$$a_n = 0 \quad \text{for all } n = 0, 1, 2, \cdots.$$

Can we get a similar result if the sequence $\{z_k\}_{k=1}^{\infty}$ converges to some point other than the center? To solve this problem we first discuss the reexpansion of power series.

Let us assume that we have a power series $\sum_{n=0}^{\infty} a_n z^n$ and a point z_0 inside the circle of convergence; that is, $z_0 \in D(0, R)$. Can we set up a power series convergent in some disc centered at z_0 whose sum agrees with the original power series (about the origin) in the region common to both discs of convergence? We shall show that we can indeed find such a series and that it converges at least inside the circle that is internally tangent to the original circle. We remark that we could just as well have started with a power series about any point instead of a series about the origin.

We proceed in a purely formal manner to substitute in the original series. Let $z - z_0 = h$, where z is inside a small circle about z_0. (The ambiguous word "small" will be made precise in the development that follows.) Then

$$f(z) = \sum_{n=0}^{\infty} a_n z^n = \sum_{n=0}^{\infty} a_n (z_0 + h)^n$$
$$= \left\{ a_0 + a_1 z_0 + a_2 z_0^2 + \cdots \right\}$$
$$+ \left\{ a_1 + 2a_2 z_0 + 3a_3 z_0^2 + \cdots \right\} h$$
$$+ \left\{ a_2 + 3a_3 z_0 + 6a_4 z_0^2 + \cdots \right\} h^2 + \cdots.$$

Thus, we have arrived *formally* at a series about z_0. It is necessary to show that the formal reexpansion of the series is justified. Now, because $\sum_{n=0}^{\infty} a_n z^n$ converges absolutely for all $z \in D(0, R)$, the series

$$\sum_{n=0}^{\infty} |a_n| (|z_0| + |h|)^n$$

converges if $|z_0| + |h| < R$, hence we can rearrange the latter series at will under this condition. Now we note that this series dominates our new series for $f(z)$ about z_0, consequently that series is absolutely convergent and the formal rearrangement is justified provided z is inside the smaller circle; that is, $|z - z_0| = |h| < R - |z_0|$. Therefore, we have

$$f(z) = \left\{ a_0 + a_1 z_0 + a_2 z_0^2 + \cdots \right\}$$
$$+ \left\{ a_1 + 2a_2 z_0 + 3a_3 z_0^2 + \cdots \right\} (z - z_0)$$
$$+ \left\{ a_2 + 3a_3 z_0 + 6a_4 z_0^2 + \cdots \right\} (z - z_0)^2 + \cdots.$$

Now, suppose that the sequence $\{z_k\}_{k=1}^{\infty}$ of distinct points converges to a point z_0 inside the circle of convergence and

$$\sum_{n=0}^{\infty} a_n z_k^n = 0 \quad \text{for } k \in \mathbb{N}.$$

In the smaller circle about z_0, we can write

$$f(z) = b_0 + b_1(z - z_0) + b_2(z - z_0)^2 + \cdots$$

and, by Lemma 2.7.1, the sum $f(z)$ is everywhere zero in the smaller disc (Figure 2.3). Now choose a point ζ_1 inside the small disc but closer to the

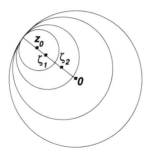

Figure 2.3

origin than z_0. There is a disc about this point in which the function $f(z)$ is identically zero.

Rewriting the power series about this point, we obtain

$$f(z) = c_0 + c_1(z - \zeta_1) + c_2(z - \zeta_1)^2 + \cdots.$$

But this series has a larger radius of convergence than the one about z_0. In fact, as drawn, its radius is greater by $| z_0 - \zeta_1 |$. If we continue to construct discs in this manner, then after a finite number of steps we will have a disc that contains the origin. Hence, by Corollary 2.7.2, we have shown:

THEOREM 2.7.3 *If a power series with positive radius of convergence vanishes on a sequence of distinct points converging to a point inside the circle of convergence, then all the coefficients of the series are zero and the sum of the series is identically zero.*

Remark. The conclusion is *not* true if the sequence converges to a point *on* the circle of convergence. (See the Blaschke Theorem 7.5.3.)

Combining this theorem with the Bolzano-Weierstrass Theorem 2.1.3, we obtain

COROLLARY 2.7.4 *A nonconstant power series having radius of convergence $R > 0$ can have at most a finite number of roots inside a circle of radius R_0, $R_0 < R$, concentric with the circle of convergence.*

COROLLARY 2.7.5 *If two power series with the same center agree at an infinite number of distinct points with an accumulation point in the interior of the smaller of the two circles of convergence, then the coefficients are all equal (therefore the series have the same radii of convergence).*

2.8   Differentiation of Power Series

The definition of differentiation of functions of a complex variable is formally the same as for functions of a real variable. We say that a function $f(z)$ is *differentiable* at a point $z_0 \in \Omega$ if

$$\lim_{h \to 0} \frac{f(z_0 + h) - f(z_0)}{h}$$

exists (as a finite number) and is independent of how the complex increment h tends to zero. The limit, when it exists, will be denoted $f'(z_0)$ and be called the *derivative* of $f(z)$ at $z = z_0$. The function $f(z)$ is *differentiable* on Ω if it is differentiable at every point of Ω.

Because, formally, the definition of the derivative is identical with that of functions of a real variable, it follows that the usual rules of differentiation, such as the formulas for the derivatives of sums, products, and quotients, are all valid. The derivative of a composite function is given by the chain rule.

LEMMA 2.8.1 *A power series*

$$a_0 + a_1 z + a_2 z^2 + \cdots + a_n z^n + \cdots$$

and its "derived series"

$$a_1 + 2a_2 z + 3a_3 z^2 + \cdots + na_n z^{n-1} + \cdots$$

have the same radii of convergence.

Proof. Clearly, multiplication by z does not affect the radius of convergence, and so the derived series and the series

$$a_1 z + 2a_2 z^2 + 3a_3 z^3 + \cdots + na_n z^n + \cdots$$

have the same radii of convergence. Now, because $\lim_{n \to \infty} \sqrt[n]{n} = 1$ (by Exercise 2.4(b)), we have

$$\limsup_{n \to \infty} | na_n |^{\frac{1}{n}} = \left(\lim_{n \to \infty} n^{\frac{1}{n}} \right) \cdot \left(\limsup_{n \to \infty} | a_n |^{\frac{1}{n}} \right) = \left(\limsup_{n \to \infty} | a_n |^{\frac{1}{n}} \right).$$

Thus, by the Cauchy-Hadamard formula in Theorem 2.6.1, we obtain the desired result. $\square$

Remark. These two series may not have the same convergence set. For example, the power series $\sum_{n=1}^{\infty} \frac{z^n}{n}$ converges for all z on the unit circle except at $z = 1$ (see Exercise 2.44), but its derived series $\sum_{n=1}^{\infty} z^{n-1}$ diverges there.

THEOREM 2.8.2 *If a function $f(z)$ is defined by a power series*

$$a_0 + a_1 z + a_2 z^2 + a_3 z^3 + \cdots + a_n z^n + \cdots$$

with a positive radius of convergence, then it is differentiable everywhere inside the circle of convergence and its derivative $f'(z)$ is given by the termwise differentiation of the series:

$$f'(z) = a_1 + 2a_2 z + 3a_3 z^2 + \cdots + na_n z^{n-1} + \cdots.$$

Furthermore, these two series have the same radius of convergence.

Proof. Let the radius of convergence of the power series $\sum_{n=0}^{\infty} a_n z^n$ be $R > 0$, and let $z_0, z_0 + h$ be two points in the disc of convergence; that is, $z_0, z_0 + h \in D(0, \rho)$ for some ρ $(0 < \rho < R)$. Then

$$\frac{f(z_0 + h) - f(z_0)}{h} = \sum_{n=0}^{\infty} a_n \frac{(z_0 + h)^n - z_0^n}{h}$$

$$= \sum_{n=1}^{\infty} a_n \left\{ (z_0 + h)^{n-1} + (z_0 + h)^{n-2} \cdot z_0 + \cdots + z_0^{n-1} \right\},$$

where the absolute value of the general term does not exceed $| na_n \rho^{n-1} |$. Because $\rho < R$, the series $\sum_{n=1}^{\infty} na_n \rho^{n-1}$ converges absolutely by Lemma 2.8.1. Thus, the series

$$\sum_{n=1}^{\infty} a_n \left\{ (z_0 + h)^{n-1} + (z_0 + h)^{n-2} \cdot z_0 + \cdots + z_0^{n-1} \right\}$$

converges uniformly for $| h | \le \rho - | z_0 |$, by the Weierstrass M-test 2.4.5. Hence it is a continuous function of h near $h = 0$. Taking the limit, we obtain

$$f'(z_0) = \lim_{h \to 0} \frac{f(z_0 + h) - f(z_0)}{h} = \sum_{n=1}^{\infty} na_n z_0^{n-1}. \qquad \square$$

Repeated application of the theorem gives us the following

COROLLARY 2.8.3 *A power series*

$$f(z) = \sum_{n=0}^{\infty} a_n z^n$$

with radius of convergence $R > 0$ *has derivatives of all orders, and for* $z \in D(0, R)$ *we have*

$$f^{(k)}(z) = \sum_{n=k}^{\infty} n(n-1)\cdots(n-k+1)a_n z^{n-k} \quad (k \in \mathbb{N}).$$

In particular,

$$f^{(k)}(0) = k!a_k,$$

hence

$$f(z) = \sum_{k=0}^{\infty} \frac{f^{(k)}(0)}{k!} z^k.$$

Of course, a similar result holds for power series about z_0. Suppose

$$f(z) = \sum_{n=0}^{\infty} a_n(z - z_0)^n.$$

Then

$$a_n = \frac{f^{(n)}(z_0)}{n!},$$

and so the given power series $\sum_{n=0}^{\infty} a_n(z - z_0)^n$ is the *Taylor series* of its sum $f(z)$; that is,

$$f(z) = \sum_{n=0}^{\infty} \frac{f^{(n)}(z_0)}{n!}(z - z_0)^n.$$

2.9 Some Elementary Functions

We introduce three important power series:

$$E(z) = 1 + \frac{z}{1!} + \frac{z^2}{2!} + \frac{z^3}{3!} + \cdots + \frac{z^n}{n!} + \cdots,$$

$$C(z) = 1 - \frac{z^2}{2!} + \frac{z^4}{4!} - \frac{z^6}{6!} + \cdots + (-1)^n \frac{z^{2n}}{(2n)!} + \cdots,$$

$$S(z) = \frac{z}{1!} - \frac{z^3}{3!} + \frac{z^5}{5!} - \frac{z^7}{7!} + \cdots + (-1)^n \frac{z^{2n+1}}{(2n+1)!} + \cdots.$$

By the ratio test we can easily verify that the radius of convergence of the first series is equal to ∞. It follows that the other two series also have this same radius of convergence, because the terms in the second and third series are majorized in absolute value by the corresponding terms of the first, which converges absolutely throughout the complex plane $\mathbb{C}$. Hence all these three series define *entire functions*; that is, functions defined and differentiable throughout the complex plane $\mathbb{C}$. Moreover, because all the coefficients are real, each of these functions is real-valued when z is real. Note also that $C(z)$ is even and $S(z)$ is odd; that is,

$$C(-z) = C(z), \quad S(-z) = -S(z) \quad \text{for all } z \in \mathbb{C}.$$

2.9.1. THE EXPONENTIAL FUNCTION

(a) The function $E(z) = \sum_{n=0}^{\infty} \frac{z^n}{n!}$ is characterized by the property

$$E(z + w) = E(z) \cdot E(w) \quad \text{for all } z, \, w \in \mathbb{C}$$

(and the initial condition $E'(0) = 1$). The following beautiful proof of this addition formula is due to Takagi. Because a power series can be differentiated termwise within the disc of convergence, we see immediately that

$$E^{(n)}(z) = E(z) \quad \text{for all } z \in \mathbb{C} \text{ and all } n \in \mathbb{N}.$$

If we consider $E(z + w)$ as a function of w, its Taylor-series expansion around $w = 0$ is given by

$$
\begin{aligned}
E(z + w) &= E(z) + \frac{E'(z)}{1!} w + \frac{E''(z)}{2!} w^2 + \cdots + \frac{E^{(n)}(z)}{n!} w^n + \cdots \\
&= E(z) \cdot \left\{ 1 + \frac{w}{1!} + \frac{w^2}{2!} + \cdots + \frac{w^n}{n!} + \cdots \right\} \\
&= E(z) \cdot E(w),
\end{aligned}
$$

which is what we want to show.

An immediate consequence of this addition formula is that

(b) $E(z) \neq 0$ for any $z \in \mathbb{C}$.

Indeed, because $E(z) \cdot E(-z) = E(0) = 1$, $E(z)$ can never vanish. If we now note that $E(z)$ is a continuous function of z, *a fortiori* a continuous function of the real variable x, we see that

(c) $E(x) > 0$ for all real x.

Because $E(0) = 1$ and $E(x)$ is real for $x \in \mathbb{R}$, if $E(x)$ were negative for some $x_0 \in \mathbb{R}$, then it would have to take on the value 0 for some $x' \in \mathbb{R}$ between x_0 and 0, by the intermediate value property.

We can now get a good idea of what the graph of $E(x)$ $(x \in \mathbb{R})$ looks like.

1^0 $E(0) = 1$.

2^0 $E(x) > 1$ if $x > 0$.

This is true because every term in the series is positive (when $x > 0$).

3^0 $E(x)$ is a strictly monotone increasing function for all $x \in \mathbb{R}$.

Indeed, $E'(x) = E(x) > 0$ for all $x \in \mathbb{R}$, by Property (c).

4^0 $E(x)$ is unbounded as $x \to \infty$.

This follows from

$$E(x) = 1 + \frac{x}{1!} + \frac{x^2}{2!} + \cdots > 1 + x \quad \text{for } x > 0.$$

5^0 $E(x) \longrightarrow 0$ as $x \to -\infty$.

This is a consequence of

$$E(-x) = \frac{1}{E(x)}.$$

6^0 The function $E(x)$ is a bijection of $\mathbb{R}$ to the set of all positive real numbers. That $E(x)$ is a surjection follows from the continuity of the function $E(x)$ together with (4^0) and (5^0) (using the intermediate value property); that it is an injection follows from (3^0).

These facts enable us to draw the graph of $E(x)$ $(x \in \mathbb{R})$ (Figure 2.4).

We now return to complex z. The coefficients of the power series defining the function $E(z)$ are all real. If $E_n(z)$ denotes the nth partial sum, then we have

$$\overline{E_n(z)} = E_n(\bar{z}).$$

$$\therefore \ \overline{E(z)} = \lim_{n \to \infty} \overline{E_n(z)} = \lim_{n \to \infty} E_n(\bar{z}) = E(\bar{z}).$$

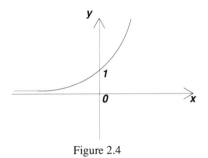

Figure 2.4

The first equality is justified because the mapping of a complex number to its complex conjugate is continuous ($\because |\,\bar\alpha\,| = |\,\alpha\,|$ for every $\alpha \in \mathbb{C}$). We obtain

(d) $\overline{E(z)} = E(\bar z)$ for all $z \in \mathbb{C}$.

Note that the above reasoning, hence the result, is valid for every power series with real coefficients (within the circle of convergence). In particular, this is true for $C(z)$ and $S(z)$ also.

(e) If $z = x + iy$ ($x,\ y \in \mathbb{R}$), then $|E(z)| = E(x)$.

Indeed,

$$\begin{aligned}
|E(z)|^2 &= E(z) \cdot \overline{E(z)} = E(z) \cdot E(\bar z) = E(z + \bar z) \\
&= E(2x) = E(x + x) = \{E(x)\}^2.
\end{aligned}$$

Because both $|E(z)| > 0$ and $E(x) > 0$, we obtain the desired result. Graphically, this means that the value of $|E(z)|$ remains constant as z traverses any vertical line in the complex plane. In particular,

(f) $|E(iy)| = |E(0 + iy)| = E(0) = 1$ for all $y \in \mathbb{R}$.

We now define the number e by

$$\begin{aligned}
e &= E(1) = 1 + \frac{1}{1!} + \frac{1}{2!} + \frac{1}{3!} + \cdots + \frac{1}{n!} + \cdots \\
&= 2.71828182845904523536\cdots.
\end{aligned}$$

Using the addition formula and induction, we obtain

$$\begin{aligned}
E(2) &= E(1 + 1) = \{E(1)\}^2 = e^2, \\
E(3) &= E(2 + 1) = E(2) \cdot E(1) = e^2 \cdot e = e^3, \\
&\quad \cdots \cdots \cdots
\end{aligned}$$

$$\therefore\ E(n) = e^n \quad \text{for every positive integer } n.$$

Moreover,
$$E(-n) \cdot E(n) = E(-n+n) = E(0) = 1$$

gives
$$E(-n) = \frac{1}{E(n)} = \frac{1}{e^n} = e^{-n}.$$

Furthermore,

$$\left\{ E\left(\frac{1}{2}\right) \right\}^2 = E\left(\frac{1}{2} + \frac{1}{2}\right) = E(1) = e$$

(and $E\left(\frac{1}{2}\right) > 0$) implies

$$E\left(\frac{1}{2}\right) = e^{\frac{1}{2}}.$$

For any rational number $\frac{p}{q}$ ($p, q \in \mathbb{Z}$, $q > 0$),

$$\left\{ E\left(\frac{p}{q}\right) \right\}^q = E(p) = e^p, \quad \text{and hence} \quad E\left(\frac{p}{q}\right) = e^{\frac{p}{q}}.$$

Similarly, for any *irrational* real number x we *define* e^x to be $E(x)$. In general, e^z (or $\exp z$) stands for the series

$$e^z = \exp z = E(z) = 1 + \frac{z}{1!} + \frac{z^2}{2!} + \cdots + \frac{z^n}{n!} + \cdots.$$

This function $E(z)$ is, for obvious reasons, called the *exponential function*.

2.9.2. THE TRIGONOMETRIC FUNCTIONS

Let us now turn to the functions $C(z)$ and $S(z)$. We already know that both are continuous throughout the complex plane $\mathbb{C}$ and that

$$C(-z) = C(z), \quad S(-z) = -S(z) \quad \text{for all } z \in \mathbb{C}.$$

Termwise differentiation of the defining series (which we justified earlier) gives

(a) $C'(z) = -S(z), \quad S'(z) = C(z) \quad$ for all $z \in \mathbb{C}$.

In particular, both of these functions satisfy the differential equation

$$f''(z) + f(z) = 0.$$

Now, by simply substituting iz into the series defining the exponential function (and, because the series converge absolutely, it is legitimate to rearrange the terms), we obtain

(b) $e^{iz} = C(z) + iS(z)$ for all $z \in \mathbb{C}$ (Euler).

Replacing z by $-z$ and using the evenness and oddness of $C(z)$ and $S(z)$, respectively, we obtain

$$e^{-iz} = C(z) - iS(z) \quad \text{for all } z \in \mathbb{C}.$$

Adding the last two equalities, we obtain

(c)

$$C(z) = \frac{e^{iz} + e^{-iz}}{2} \quad \text{for all } z \in \mathbb{C}.$$

Subtracting, we obtain

(d)

$$S(z) = \frac{e^{iz} - e^{-iz}}{2i} \quad \text{for all } z \in \mathbb{C}.$$

Multiplying the two equations in (b), we obtain

(e)

$$\{C(z)\}^2 + \{S(z)\}^2 = 1 \quad \text{for all } z \in \mathbb{C}.$$

If x is real both $C(x)$ and $S(x)$ are real, and then we obtain from (e)

(f)

$$-1 \leq C(x) \leq 1, \quad -1 \leq S(x) \leq 1 \quad \text{for all } x \in \mathbb{R}.$$

However, the boundedness of $C(z)$ and $S(z)$ does not hold true for nonreal z, as we shall soon see.

(g) From the addition formula for the exponential function and (b), (c) above, we have

$$
\begin{aligned}
C(z + w) &= \frac{1}{2}\left\{e^{i(z+w)} + e^{-i(z+w)}\right\} = \frac{1}{2}\left\{e^{iz} \cdot e^{iw} + e^{-iz} \cdot e^{-iw}\right\} \\
&= \frac{1}{2}[\{C(z) + iS(z)\} \cdot \{C(w) + iS(w)\} \\
&\quad + \{C(z) - iS(z)\} \cdot \{C(w) - iS(w)\}] \\
&= C(z) \cdot C(w) - S(z) \cdot S(w).
\end{aligned}
$$

This is just the familiar addition formula for the function $C(z)$ that we learn (for real z) in trigonometry. In a similar manner, we can work out the addition formula for the function $S(z)$:

(h)

$$S(z + w) = S(z) \cdot C(w) + C(z) \cdot S(w).$$

We could also obtain this identity by differentiating both sides of the addition formula for $C(z)$ with respect to z.

(i) There exists a real number x_0 such that $C(x_0) = 0$.

Proof. Because $C(x)$ is continuous and $C(0) = 1$, it is sufficient to show that $C(2) < 0$. Indeed, the intermediate-value property will then guarantee the existence of at least one such x_0 between 0 and 2.

$$\begin{aligned}
C(2) &= 1 - \frac{2^2}{2!} + \frac{2^4}{4!} - \frac{2^6}{6!} + \frac{2^8}{8!} - + \cdots \\
&= \left(1 - 2 + \frac{2}{3}\right) - \frac{2^6}{6!}\left(1 - \frac{2^2}{7 \cdot 8}\right) \\
&\quad - \frac{2^{10}}{10!}\left(1 - \frac{2^2}{11 \cdot 12}\right) - \cdots,
\end{aligned}$$

where we are allowed to group terms because the series is convergent. (Cf. Exercise 2.6.) Moreover, because each term is negative (including the sign at the front), the series must converge to a negative number; that is, $C(2) < 0$. □

(j) Because the roots of a power series cannot accumulate inside the disc of convergence, there can be only a finite number of points in the interval $(0, 2)$ for which $C(x) = 0$. It is easy to convince ourselves that there are none in the interval $(0, \sqrt{2})$, because $x = \sqrt{2}$ is the point for which the sum of the first two terms is zero. Indeed, for $0 < x < \sqrt{2}$, we have

$$\begin{aligned}
C(x) &= 1 - \frac{x^2}{2!} + \frac{x^4}{4!} - \frac{x^6}{6!} + \frac{x^8}{8!} - + \cdots \\
&= \left(1 - \frac{x^2}{2!}\right) + \frac{x^4}{4!}\left(1 - \frac{x^2}{5 \cdot 6}\right) \\
&\quad + \frac{x^8}{8!}\left(1 - \frac{x^2}{9 \cdot 10}\right) + \cdots > 0,
\end{aligned}$$

because each term is positive. Let the smallest of all the roots of $C(x)$ between 0 and 2 be named $\frac{\pi}{2}$; $C\left(\frac{\pi}{2}\right) = 0$. From the identity (e), we obtain $S\left(\frac{\pi}{2}\right) = \pm 1$. We determine the sign of $S\left(\frac{\pi}{2}\right)$ as follows:

$$
\begin{aligned}
S(x) &= \frac{x}{1!} - \frac{x^3}{3!} + \frac{x^5}{5!} - \frac{x^7}{7!} + - \cdots \\
&= \frac{x}{1!}\left(1 - \frac{x^2}{2 \cdot 3}\right) + \frac{x^5}{5!}\left(1 - \frac{x^2}{6 \cdot 7}\right) \\
&\quad + \frac{x^9}{9!}\left(1 - \frac{x^2}{10 \cdot 11}\right) + \cdots .
\end{aligned}
$$

Hence, $S(x) > 0$ for $0 < x \le \sqrt{6}$. Because $0 < \sqrt{2} < \frac{\pi}{2} < 2 < \sqrt{6}$, we have $S\left(\frac{\pi}{2}\right) > 0$, which implies that $S\left(\frac{\pi}{2}\right) = 1$.

Now, by the addition formulas obtained in (g) and (h),

(k)

$$
\begin{aligned}
S\left(\frac{\pi}{2} - z\right) &= S\left(\frac{\pi}{2}\right) \cdot C(-z) + C\left(\frac{\pi}{2}\right) \cdot S(-z) = C(z), \\
C\left(\frac{\pi}{2} - z\right) &= C\left(\frac{\pi}{2}\right) \cdot C(-z) - S\left(\frac{\pi}{2}\right) \cdot S(-z) = S(z).
\end{aligned}
$$

These identities also show that the graph of the function $C(z)$ can be obtained from that of the function $S(z)$, and vice versa. In a similar way,

$$
\begin{aligned}
S(\pi) &= S\left(\frac{\pi}{2}\right) \cdot C\left(\frac{\pi}{2}\right) + C\left(\frac{\pi}{2}\right) \cdot S\left(\frac{\pi}{2}\right) = 0, \\
C(\pi) &= C\left(\frac{\pi}{2}\right) \cdot C\left(\frac{\pi}{2}\right) - S\left(\frac{\pi}{2}\right) \cdot S\left(\frac{\pi}{2}\right) = -1,
\end{aligned}
$$

and hence

$$
\begin{aligned}
S(\pi - z) &= S(\pi) \cdot C(-z) + C(\pi) \cdot S(-z) = S(z), \\
C(\pi - z) &= C(\pi) \cdot C(-z) - S(\pi) \cdot S(-z) = -C(z).
\end{aligned}
$$

Applying this technique once more, we obtain

$$
S(2\pi) = 0, \qquad C(2\pi) = 1,
$$

$$
S(2\pi + z) = S(z), \quad C(2\pi + z) = C(z);
$$

and so both $S(z)$ and $C(z)$ are 2π-*periodic*.

From now on, the functions $C(z)$ and $S(z)$ will be called the *cosine* and *sine function*, respectively, and denoted $\cos z$ and $\sin z$.

(1) Because $(\sin x)' = \cos x$, and $\cos x > 0$ for $0 \leq x < \frac{\pi}{2}$, we see that $\sin x$ is strictly monotone increasing for $0 \leq x < \frac{\pi}{2}$. Using the formula $\sin(\pi - x) = \sin x$, we see that $\sin x$ is monotonically decreasing for $\frac{\pi}{2} < x \leq \pi$. This, together with periodicity, implies that the only real roots of the sine function are multiples of π.

Let us now examine the distribution of roots of the sine function in the complex plane $\mathbb{C}$. Are there nonreal values of z such that $\sin z = 0$? Before we answer this question, let us define the *hyperbolic functions*:

$$\cosh z = 1 + \frac{z^2}{2!} + \frac{z^4}{4!} + \cdots,$$

$$\sinh z = z + \frac{z^3}{3!} + \frac{z^5}{5!} + \cdots.$$

The following relations follow immediately from the definitions:

(a) $\cosh z = \dfrac{e^z + e^{-z}}{2}$;

(b) $\sinh z = \dfrac{e^z - e^{-z}}{2}$;

(c) $\cos(iz) = \cosh z$;

(d) $\sin(iz) = i \sinh z$;

(e) $\cosh^2 z - \sinh^2 z = 1$.

Returning to our question, we have, for $z = x + iy$ ($x,\ y \in \mathbb{R}$),

$$\sin z = \sin(x + iy) = \sin x \cdot \cos(iy) + \cos x \cdot \sin(iy)$$
$$= \sin x \cdot \cosh y + i \cos x \cdot \sinh y.$$

It follows directly from the definitions that the hyperbolic functions are real if z is real. Thus the terms $\sin x \cdot \cosh y$ and $\cos x \cdot \sinh y$ are both real.

Therefore, if $\sin z = 0$, then we have two equations involving the real quantities x and y

$$\sin x \cdot \cosh y = 0 \quad \text{and} \quad \cos x \cdot \sinh y = 0.$$

From the definition of $\cosh z$, it is immediately clear that for $z = y$ (real), $\cosh y \geq 1$. Thus, the first equation gives $\sin x = 0$. But this implies that $\cos x = \pm 1$. Hence, the second equation gives $\sinh y = 0$. But $\sinh y = 0$ if and only if $y = 0$, and so we have the result:

(m) $\sin z = 0$ if and only if z is real and a multiple of π.

Because $\cos z = \sin\left(\frac{\pi}{2} - z\right)$, this result implies:

(n) $\cos z = 0$ if and only if z is real and an odd multiple of $\frac{\pi}{2}$.

In particular, we have shown that neither the sine nor the cosine vanishes off the real axis.

Are there any nonreal periods? Suppose

$$\sin(z + p) = \sin z \quad \text{for all} \ z \in \mathbb{C}.$$

Then substituting $z = 0$, we obtain $\sin p = \sin 0 = 0$, and so p must be real by (m). The same is true for the cosine function.

Is the exponential function periodic? That is, is there a complex number p such that $e^{z+p} = e^z$ for all $z \in \mathbb{C}$? If such a period p exists, then, upon substituting z by 0, we obtain $e^p = e^0 = 1$, hence $|e^p| = \exp(\Re p) = 1$. But we have shown (Section 2.9.1(c)6°) that there is exactly one real solution to $e^x = a$ (for any $a > 0$), and, in this case, the unique real number x such that $e^x = 1$ is 0. Hence the real part of p is zero, and so any period must be on the imaginary axis; that is, p is of the form $p = i\alpha$ ($\alpha \in \mathbb{R}$), and our equation becomes $1 = e^p = e^{i\alpha} = \cos\alpha + i\sin\alpha$. Equating the real and imaginary parts, we obtain

$$\cos\alpha = 1 \quad \text{and} \quad \sin\alpha = 0.$$

Thus $\alpha = 0, \pm 2\pi, \pm 4\pi, \cdots$, and we obtain

$$p = 2n\pi i \quad (n = \pm 1, \pm 2, \cdots).$$

It is simple to verify that these are actually the periods of e^z.

Now we return to the trigonometric functions.

$$
\begin{aligned}
|\sin z|^2 &= |\sin(x + iy)|^2 = |\sin x \cdot \cosh y + i\cos x \cdot \sinh y|^2 \\
&= \sin^2 x \cdot \cosh^2 y + \cos^2 x \cdot \sinh^2 y \\
&= \sin^2 x \cdot (1 + \sinh^2 y) + (1 - \sin^2 x) \cdot \sinh^2 y \\
&= \sin^2 x + \sinh^2 y.
\end{aligned}
$$

Noting that $|\sinh y| \approx \frac{e^{|y|}}{2}$ (as $|y| \to \infty$), we see that $|\sin z|$ can become arbitrarily large. Thus, the sine, hence also the cosine, are unbounded in the complex plane. This is in sharp contrast to the behavior of these functions on the real axis. (Cf. the Liouville Theorem 4.4.9.)

2.9.3. THE LOGARITHMIC FUNCTION

We now turn to the logarithmic function. Let $w = \alpha + i\beta$ be a complex number with modulus 1. First, suppose $\alpha > 0$, $\beta > 0$. Then, because $|w| = 1$, we have $0 < \alpha < 1, 0 < \beta < 1$. Hence we can write $\alpha = \cos\theta$ for some unique θ ($0 < \theta < \frac{\pi}{2}$). Then $\sin^2\theta = 1 - \cos^2\theta = 1 - \alpha^2 = \beta^2$ and, because $0 < \theta < \frac{\pi}{2}$, we have $\sin\theta > 0$. Now, by assumption, $\beta > 0$, and so it follows that $\beta = \sin\theta$. Thus, for $|w| = 1$ and $\alpha > 0$, $\beta > 0$, we can write w uniquely as

$$w = \cos\theta + i\sin\theta = e^{i\theta} \quad \left(0 < \theta < \frac{\pi}{2}\right).$$

In a similar manner, if $\alpha > 0$, $\beta < 0$, we find a unique θ such that $w = e^{i\theta}$ for $-\frac{\pi}{2} < \theta < 0$. If $\alpha < 0$, $\beta > 0$, then $w = e^{i\theta}$ for $\frac{\pi}{2} < \theta < \pi$; and if $\alpha < 0$, $\beta < 0$, then $w = e^{i\theta}$ for $-\pi < \theta < -\frac{\pi}{2}$.

Checking the remaining four points 1, i, -1, $-i$, we see that we can write

$$\begin{aligned}
1 &= e^{i\theta}, & \theta &= 0; & i &= e^{i\theta}, & \theta &= \frac{\pi}{2}; \\
-1 &= e^{i\theta}, & \theta &= \pm\pi; & -i &= e^{i\theta}, & \theta &= -\frac{\pi}{2}.
\end{aligned}$$

By convention, we choose $\theta = \pi$ for $w = -1$.

Any complex number w whose modulus is 1 can be expressed in the form $w = e^{i\theta}$ for exactly one real value of θ satisfying $-\pi < \theta \leq \pi$. This number θ is called the **principal argument** *of w.*

Now, take any nonzero complex number w. Then

$$w = |w| \cdot \left(\frac{w}{|w|}\right) \quad \text{and} \quad \left|\frac{w}{|w|}\right| = 1.$$

Thus we can express *any* complex number $w \neq 0$ as

$$w = |w| \cdot e^{i\theta} \quad (-\pi < \theta \leq \pi),$$

where θ is again called the *principal argument* of w. Because $|w| > 0$, we can find a unique real number c such that $|w| = e^c$, and then $w = e^c \cdot e^{i\theta} = e^{c+i\theta}$. This justifies the polar representation

$$w = re^{i\theta} = r(\cos\theta + i\sin\theta)$$

of a complex number $w \neq 0$ ($r = |w|, -\pi < \theta \leq \pi$).

We have shown: *Every nonzero complex number w can be expressed in at least one way as the exponential of some complex number $c + i\theta$. This number is unique if the imaginary part, θ, is restricted to the real interval $-\pi < \theta \le \pi$.* This unique number is called the *principal logarithm* of w and will be denoted $\mathrm{Log}\, w$. If $w = e^z$, then z is called a *logarithm* of w ($w \ne 0$) and we write $z = \log w$.

Note that because the exponential function never vanishes, 0 has no logarithm.

If z and ζ are logarithms of w; that is, $w = e^z = e^\zeta$, then $e^{z-\zeta} = 1$, and hence $z - \zeta = 2\pi n i$ ($n \in \mathbb{Z}$). Conversely, if $e^z = w$, then $e^{z+2\pi n i} = w$ ($n \in \mathbb{Z}$); hence, given any logarithm of w, say z, the set of all logarithms of w is given by $z + 2\pi n i$ ($n \in \mathbb{Z}$). It follows that by restricting $\Im z$ to an interval $I = [a, a + 2\pi)$ (for some $a \in \mathbb{R}$) of length 2π, we get a *branch* of the logarithmic function; that is, for any $w \in \mathbb{C}$ ($w \ne 0$), there exists a unique z satisfying $e^z = w$ (hence $\log w = z$) with $\Im z \in I$.

2.9.4. The General Powers

Next we discuss the function $f(z) = z^\alpha$, where α is an arbitrary complex number and $z \ne 0$. We define

$$z^\alpha = e^{\alpha \log z},$$

where $\log z$ denotes an arbitrary branch of the logarithm of z. Thus

$$z^\alpha = e^{\alpha(\mathrm{Log}\, z + 2\pi n i)} = e^{\alpha \mathrm{Log}\, z} \cdot e^{2\pi n \alpha i} \quad (n \in \mathbb{Z}).$$

We see that z^α will, in general, have infinitely many distinct values. However, there are cases in which z^α has only finitely many values. This occurs if $\{e^{2\pi n \alpha i};\ n \in \mathbb{Z}\}$ has only finitely many distinct values. Then there must be two distinct integers m and n such that $e^{2\pi m \alpha i} = e^{2\pi n \alpha i}$; that is, $e^{2\pi(m-n)\alpha i} = 1$. Thus $(m - n)\alpha$ must be an integer, and this condition implies that α is a real rational number. Conversely, if α is a real rational number, p/q say (with $(p, q) = 1$ and $q \ge 1$), then $\{e^{2\pi n \frac{p}{q} i};\ n \in \mathbb{Z}\}$ clearly consists of q distinct values only, and we may take those corresponding to $n = 0, 1, \ldots, q - 1$. Thus, the function z^α is q-valued if and only if α is a rational number p/q, where $p, q \in \mathbb{Z}$, $(p, q) = 1$, $q \ge 1$. In particular, z^α is single-valued if α is an integer. If α is irrational then z^α assumes infinitely many distinct values.

Remark. If $\alpha \in \mathbb{R}$ and z is real and positive, then there is a value of z^α, called the *principal value*, which is real and positive.

2.10   The Maximum and Minimum Modulus Principles

THEOREM 2.10.1 *Let $f(z) = \sum_{n=0}^{\infty} a_n z^n$ be a nonconstant power series with radius of convergence $R > 0$. Then*

(a) the modulus $|f(z)|$ has no local maximum at any point inside the circle of convergence; that is, in any neighborhood of any point z_0 in $D(0, R)$ there exists a point z_1 such that

$$|f(z_1)| > |f(z_0)|;$$

(b) and if $f(z_0) \neq 0$, then $|f(z_0)|$ cannot be a local minimum, either; that is, in any neighborhood of z_0 in $D(0, R)$ there is also a point z_2 such that

$$|f(z_2)| < |f(z_0)|.$$

Proof. First, consider the special case that $z_0 = 0$ and

$$f(z) = 1 + a_k z^k + a_{k+1} z^{k+1} + \cdots,$$

where k is the smallest positive integer such that $a_k \neq 0$; such an integer exists because, by assumption, $f(z)$ is not constant. Now,

$$\left| \sum_{n=k+1}^{\infty} a_n z^n \right| = |z^k| \cdot \left| \sum_{n=k+1}^{\infty} a_n z^{n-k} \right| = |z|^k \cdot \left| \sum_{n=0}^{\infty} a_{n+k+1} z^{n+1} \right|,$$

and the series in the second factor is a continuous function of z that tends to 0 as z tends to 0. Because $a_k \neq 0$, we have

$$\left| \sum_{n=k+1}^{\infty} a_n z^n \right| < \frac{|a_k|}{2} |z|^k$$

for $|z|$ sufficiently small. Then

$$|f(z)| \geq \left| 1 + a_k z^k \right| - \left| \sum_{n=k+1}^{\infty} a_n z^n \right|$$

$$> \left| 1 + a_k z^k \right| - \frac{|a_k|}{2} |z|^k \quad \text{for sufficiently small } |z|.$$

For $z \neq 0$, we can write $z = r e^{i\theta}$, $r = |z|$, $\theta \in \mathbb{R}$, and because $a_k \neq 0$, $a_k = |a_k| \cdot e^{i\varphi}$ ($\varphi \in \mathbb{R}$),

$$\therefore |f(z)| > \left| 1 + |a_k| r^k e^{i(\varphi + k\theta)} \right| - \frac{|a_k|}{2} r^k \quad \text{for sufficiently small } r.$$

If we choose the direction θ of approach to the origin in such a way that $a_k z^k$ remains real and positive (by letting $\varphi + k\theta = 0$, which is possible because $k \neq 0$), then

$$| f(z) | > 1 + | a_k | r^k - \frac{| a_k |}{2} r^k = 1 + \frac{| a_k |}{2} r^k > 1 = f(0) = | f(0) |$$

for $| z |$ sufficiently small. Thus there are values of z, with $|z|$ arbitrarily small, where $|f(z)| > |f(0)|$.

Similarly, for small $| z |$, we have

$$| f(z) | \leq | 1 + a_k z^k | + \left| \sum_{n=k+1}^{\infty} a_n z^n \right|$$

$$< | 1 + a_k z^k | + \frac{| a_k z^k |}{2}$$

$$= \left| 1 + | a_k | r^k e^{i(\varphi + k\theta)} \right| + \frac{| a_k | r^k}{2}.$$

Thus, if we choose the approach to the origin so that $a_k z^k$ remains real and *negative* (by letting $\varphi + k\theta = \pi$), then

$$| f(z) | < 1 - | a_k | r^k + \frac{| a_k |}{2} r^k < 1 = | f(0) |$$

for $| z |$ sufficiently small. Thus $|f(0)| = 1$ is neither a local maximum nor a local minimum.

Now, suppose

$$f(z) = a_0 + a_1 z + \cdots + a_n z^n + \cdots.$$

Thanks to Corollary 2.7.4 of the Uniqueness Theorem 2.7.3, if $a_0 = f(0) = 0$ then there are values of z in every neighborhood of 0 where $|f(z)| > 0$ and we are done. Note that in this case Part (b) does not apply. Thus we may assume, without loss of generality, that $a_0 \neq 0$. Then

$$| f(z) | = | a_0 | \cdot \left| 1 + \frac{a_1}{a_0} z + \cdots + \frac{a_n}{a_0} z^n + \cdots \right|.$$

Applying the first part of the proof to the power series in the second factor and noting that $a_0 = f(0)$, we obtain the desired result for $z_0 = 0$.

Now, let z_0 be any point in $D(0, R)$. We make the substitution $z = z_0 + h$, and obtain a power series in h:

$$g(h) = f(z) = b_0 + b_1 h + b_2 h^2 + \cdots,$$

where $b_0 = a_0 + a_1 z_0 + a_2 z_0^2 + \cdots = f(z_0)$. Applying our previous result to this series, we see that in any neighborhood of $h = 0$ there is a point h_1 such that $|g(h_1)| > |g(0)|$; that is, $|f(z_1)| > |f(z_0)|$, where $z_1 = z_0 + h_1$. Thus $|f(z_0)|$ is not a local maximum. Similarly, $|f(z_0)|$ is not a local minimum provided that $f(z_0) \neq 0$.   □

COROLLARY 2.10.2 *Let $f(z) = \sum_{n=0}^{\infty} a_n (z - c)^n$ be a nonconstant power series with radius of convergence $R > 0$. Then there is no point z_1 inside the circle of convergence for which*

$$| f(z_1) |= \sup\{| f(z) |; z \in D(c, R)\}$$

and there is no point z_2 inside the circle of convergence for which

$$| f(z_2) |= \inf\{| f(z) |; z \in D(c, R)\},$$

provided $f(z)$ does not vanish in $D(c, R)$.

Proof. Because a (global) maximum point must be a local maximum point, nonexistence of the latter implies that of the former.   □

At the beginning of this book we mentioned that the complex numbers were introduced to enable us to solve all quadratic equations (with real coefficients). Later it was found that every nonconstant polynomial with complex coefficients (not just those with real coefficients) has a root in the complex field. We now prove this result, known as the *fundamental theorem of algebra*, as an application of the Minimum Modulus Principle 2.10.1:

THEOREM 2.10.3 (Gauss) *The complex field is algebraically closed; that is, every nonconstant polynomial equation (with complex coefficients) has at least one root in the complex field $\mathbb{C}$.*

Proof. Let

$$p(z) = a_n z^n + a_{n-1} z^{n-1} + \cdots + a_1 z + a_0, \quad a_n \neq 0, \ n \geq 1.$$

For sufficiently large $| z |$,

$$| p(z) |=| a_n z^n | \cdot \left| 1 + \frac{a_{n-1}}{a_n} \left(\frac{1}{z}\right) + \cdots + \frac{a_0}{a_n} \left(\frac{1}{z}\right)^n \right| > \frac{1}{2} |a_n z^n|.$$

Because $n \geq 1$ and $a_n \neq 0$, we can choose a radius R so large that

$$| p(z) | > | a_0 | \text{whenever } z \in C(0, R).$$

Because the closed disc $\bar{D}(0, R)$ is compact, the real continuous function $| p(z) |$ must attain its minimum somewhere on $\bar{D}(0, R)$. But this minimum cannot be attained on the circumference $C(0, R)$, for

$$| p(z) | > | a_0 | = | p(0) | \text{for } z \in C(0, R).$$

But then the minimum modulus principle (which is applicable because a polynomial is a particular case of a power series) implies that $p(z) = 0$ for some $z \in D(0, R)$.   □

If z_1 is a root of the polynomial $p(z)$, then we can use long division to get $p(z) = (z - z_1) \cdot q(z)$, where $q(z)$ is a polynomial and $\deg q = \deg p - 1$. Thus, if $\deg p > 1$, $q(z)$ also has a root, say $z = z_2$, where it is possible that $z_1 = z_2$. Proceeding in this manner, we obtain the factorization

$$p(z) = a_n(z - z_1) \cdot (z - z_2) \cdots (z - z_n).$$

The factorization is unique, except for the order of the factors.

Remark. Later we shall give additional proofs of the fundamental theorem of algebra, but every proof uses the property that

$$| p(z) | \longrightarrow \infty   \text{as } | z | \to \infty.$$

In fact, this property characterizes nonconstant polynomials among all entire functions. (See end of Section 5.2.)

Exercises

1. If $\lim_{n \to \infty} c_n = \ell$, show that

$$\lim_{n \to \infty} \frac{1}{n}(c_1 + c_2 + \cdots + c_n) = \ell.$$

Is the converse true?

2. Let $\{a_n\}_{n=1}^{\infty}$ and $\{b_n\}_{n=1}^{\infty}$ be sequences of complex numbers. Show that

(a)

$$\limsup_{n\to\infty} |a_n b_n| \le \left(\limsup_{n\to\infty} |a_n|\right) \cdot \left(\limsup_{n\to\infty} |b_n|\right),$$

$$\liminf_{n\to\infty} |a_n b_n| \ge \left(\liminf_{n\to\infty} |a_n|\right) \cdot \left(\liminf_{n\to\infty} |b_n|\right).$$

(b) Give examples for which the equalities are not true.

(c) Explain why the equality in the proof of Lemma 2.8.1 is justified.

3. (a) Prove Lemma 2.1.7.

(b) Give examples of sequences for which the equalities are not true.

(c) Is it true that the existence of $\lim_{n\to\infty} \sqrt[n]{a_n}$ implies that of $\lim_{n\to\infty} \frac{a_{n+1}}{a_n}$?

4. Find the limit of each of the following sequences :

(a) $a_n = \sqrt[n]{a}$ $(a > 0)$;

(b) $b_n = \sqrt[n]{n}$;

(c) $c_n = \dfrac{\sqrt[n]{n!}}{n}$;

(d) $d_n = \dfrac{\sqrt[n]{(n+1)(n+2)\cdots(2n)}}{n}$.

Hint: One way to do this problem is to appeal to Lemma 2.1.7, but there are other ways too. Actually, (a) is needed in the proof of Lemma 2.1.7. Can you supply an independent solution to (a)?

5. Suppose $c_n \ne 0$ for infinitely many $n \in \mathbb{N}$ and the series $\sum_{n=1}^{\infty} c_n$ converges absolutely. Show that the set

$$\left\{ z \in \mathbb{C} ; z = \sum_{n \in S} c_n, \ S \subset \mathbb{N} \right\}$$

is a compact perfect subset of the complex plane. (Convention: $\sum_{n \in \phi} c_n = 0$.) A subset E of $\mathbb{C}$ is said to be *perfect* if it is closed and every point in E is an accumulation point of E.

6. (a) Show that the convergence and the sum of an absolutely convergent series are not affected by rearranging its terms.

(b) Show that the convergence and the sum of a convergent series are not affected by inserting (possibly infinitely many) parentheses.

What if we remove parentheses (assuming the original series has infinitely many parentheses)?

7. Discuss the validity of the complex version of the Riemann theorem on conditionally convergent series mentioned after Theorem 2.2.4.

8. Find the set of all accumulation points for the following sets:

(a) $\left\{ \dfrac{1}{m} + \dfrac{i}{n}; \ m, \ n \in \mathbb{N} \right\}$;

(b) $\left\{ \dfrac{1}{m} + \dfrac{1}{n}; \ m, \ n \in \mathbb{N} \right\}$;

(c) $\left\{ \dfrac{1}{m} + \dfrac{(-1)^{\ell}}{2^n}; \ \ell, \ m, \ n \in \mathbb{N} \right\}$.

9. Let E be a nonempty subset of $\mathbb{C}$. Define

$$f(z) = \inf\{ |z - w| ; w \in E \} \quad (z \in \mathbb{C}).$$

Show that
(a) $f(z)$ is a continuous function in $\mathbb{C}$;
(b) $f(z) = 0 \iff z \in \bar{E}$.

10. (a) Suppose E and F are nonempty closed bounded subsets of $\mathbb{C}$. Show that there exist $z_0 \in E$ and $w_0 \in F$ such that

$$| z_0 - w_0 | = \inf\{ |z - w| ; z \in E, \ w \in F \}.$$

(b) Show that this is not true if the boundedness condition on E and F is dropped.

(c) What if only one of E or F is bounded (but both are still closed)?

11. Show that in the Cantor Intersection Theorem 2.3.2, "*compact*" cannot be replaced by "*closed*"; that is, find a nested sequence $\{F_n\}_{n=1}^{\infty}$ of nonempty closed sets in $\mathbb{C}$ such that

$$\bigcap_{n=1}^{\infty} F_n = \phi.$$

12. (a) Let a_n and b_n be nonzero complex numbers for $n = 1, 2, 3, \ldots$. Suppose $\lim_{n\to\infty} \left| \dfrac{a_n}{b_n} \right| = \ell$ exists, and $\ell \neq 0, \infty$. Show that if one of the series $\sum_{n=1}^{\infty} a_n$ and $\sum_{n=1}^{\infty} b_n$ converges absolutely, then so too does the other. In particular, the power series $\sum_{n=1}^{\infty} a_n z^n$ and $\sum_{n=1}^{\infty} b_n z^n$ have the same radii of convergence.

(b) What if $\ell = 0$ or $\ell = \infty$?

13. State the Cauchy Criterion for the uniform convergence of series of functions.

14. Give an example of a sequence of continuous functions that converges nonuniformly to a continuous function.

15. (a) Discuss the convergence and uniform convergence of the sequence $\{nz^n\}_{n=1}^{\infty}$.

(b) The same problem for the series $\sum_{n=1}^{\infty} \frac{z^n}{n^2(1-z^n)}$.

16. Show that the series

$$\frac{1}{1+|z|} - \frac{1}{2+|z|} + \frac{1}{3+|z|} - \frac{1}{4+|z|} + \cdots + \frac{(-1)^{n-1}}{n+|z|} + \cdots$$

is not absolutely convergent but is uniformly convergent in the whole complex plane (hence the Weierstrass M-test 2.4.5 is not applicable).

17. Show that $\sum_{n=1}^{\infty} \frac{z}{(1+|z|)^n}$ converges (absolutely) pointwise but not locally uniformly on $\mathbb{C}$.

18. Show that the series $\sum_{n=0}^{\infty} e^{-nz}$ converges locally uniformly in the open right half-plane $\{z \in \mathbb{C} ; \Re z > 0\}$. Find the sum of the series.

19. (Pólya-Szegö)
(a) Let $\theta, \varphi \in \mathbb{R}, 0 < \theta < \frac{\pi}{2}$. Suppose $\{z_k\}_{k=1}^{\infty}$ is a sequence of points in the sector

$$\Omega = \{z \in \mathbb{C} ; \varphi - \theta < \arg z < \varphi + \theta\}.$$

Show that the series

$$\sum_{k=1}^{\infty} z_k \quad \text{and} \quad \sum_{k=1}^{\infty} |z_k|$$

either both converge or both diverge.

(b) What if $\theta = \frac{\pi}{2}$?

20. (Pólya-Szegö) Let $\{z_k\}_{k=1}^{\infty}$ be a sequence of points in some closed half-plane. Show that the convergence of the series

$$\sum_{k=1}^{\infty} z_k \quad \text{and} \quad \sum_{k=1}^{\infty} z_k^2$$

implies that of

$$\sum_{k=1}^{\infty} |z_k|^2.$$

Hint: Where the half-plane is

$$\{z \in \mathbb{C}; \Re z \geq 0\} \quad \text{or} \quad \{z \in \mathbb{C}; \Im z \geq 0\},$$

the convergence of $\sum_{k=1}^{\infty} z_k$ implies that of $\sum_{k=1}^{\infty}(z_k \pm \bar{z}_k)$, which in turn implies the convergence of

$$\sum_{k=1}^{\infty}(z_k \pm \bar{z}_k)^2 = \left(\sum_{k=1}^{\infty} z_k^2\right) \pm 2\sum_{k=1}^{\infty} |z_k|^2 + \overline{\left(\sum_{k=1}^{\infty} z_k^2\right)}.$$

21. We have seen examples of power series whose radii of convergence R are 1 or ∞. Give an example of a power series for which $R = \pi$. How about $R = 0$?

22. Find the disc of convergence of the following power series:

(a) $\sum_{n=1}^{\infty} \dfrac{(z - i)^{2n}}{3^n n}$;

(b) $\sum_{n=1}^{\infty} \dfrac{n! z^n}{n^n}$;

(c) $\sum_{n=1}^{\infty} \dfrac{(z - 3)^n}{n(n + 1)}$;

(d) $\sum_{n=1}^{\infty} \dfrac{z^n}{n\sqrt{n}}$;

(e) $\sum_{n=2}^{\infty} \dfrac{z^n}{n(\operatorname{Log} n)^p}$ $\quad (p > 0)$;

(f) $\sum_{n=0}^{\infty} \dfrac{z^n}{\{\operatorname{Log}(n + 5)\}^n}$;

(g) $\sum_{n=0}^{\infty} z^{n^3}$;

(h) $\sum_{n=0}^{\infty} \{3^n + i^n\}(z - 2)^n$;

(i) $\sum_{n=3}^{\infty} \left(1 - \dfrac{1}{n^2}\right)^{-n^3} z^n$;

(j) $\sum_{n=1}^{\infty} \left(\sin \dfrac{ne}{17}\right) z^n$.

23. (a) Suppose

$$a_{3m} = 1, \quad a_{3m+1} = \frac{(-1)^m}{m}, \quad a_{3m+2} = \frac{1}{m^2}.$$

Find the radius of convergence of the power series $\sum_{n=7}^{\infty} a_n z^n$ and show that the power series diverges at every point of the circle of convergence.

(b) Same problem for

$$a_{2m} = 2^m, \quad a_{2m+1} = \frac{(-1)^m}{m}.$$

24. Discuss the convergence (pointwise, absolute, uniform) of the series:

$$\sum_{n=0}^{\infty} \left(\frac{z+i}{z-1}\right)^n.$$

25. (Fibonacci) Suppose $a_0 = 0$, $a_1 = 1$, $a_n = a_{n-1} + a_{n-2}$ $(n \geq 2)$.
  (a) Show that

$$\sum_{n=0}^{\infty} a_n z^n = \frac{z}{1 - z - z^2}.$$

  (b) What is the radius of convergence?
  (c) Show that

$$a_n = \frac{1}{\sqrt{5}} \left\{ \left(\frac{1+\sqrt{5}}{2}\right)^n - \left(\frac{1-\sqrt{5}}{2}\right)^n \right\}.$$

26. (Pólya-Szegö) What is the relation between the following two questions?
  (a) What are the coefficients (say, of z^{500}) in the power-series expansion of

$$\frac{1}{(1-z)(1-z^5)(1-z^{10})(1-z^{25})}?$$

  (b) In how many ways can a specified amount of money (say, 5 dollars) be changed into pennies, nickels, dimes, and quarters?

27. (a) Suppose $f(z) = \sum_{n=0}^{\infty} a_n(z - c)^n$ has the property that the series $\sum_{k=0}^{\infty} f^{(k)}(c)$ converges. Show that $f(z)$ is an entire function. (A function that can be expressed as a power series with the radius of convergence $R = \infty$ is called an *entire function*.)

(b) What if the convergence of $\sum_{k=0}^{\infty} f^{(k)}(c)$ is replaced by the assumption that the power series

$$\sum_{k=0}^{\infty} \frac{f^{(k)}(c)}{(k + 1)^{1996}} z^k$$

has a positive radius of convergence?

28. Show that

$$\operatorname{Log} 2 = 1 - \frac{1}{2} + \frac{1}{3} - \cdots + (-1)^{k-1}\frac{1}{k} + \cdots,$$

by applying the Abel Continuity Theorem 2.6.4.

29. Reexpand the geometric series

$$1 + z + z^2 + \cdots + z^n + \cdots$$

around the points $a = \pm\frac{1}{2}$ and $a = \pm\frac{i}{2}$. What are the radii of convergence of each of the new series?

30. Show that an entire function that takes real values on the real axis and purely imaginary values on the imaginary axis must be an odd function:

$$f(-z) = -f(z) \quad \text{for all } z \in \mathbb{C}.$$

31. Prove, for z in the unit disc,

(a) $z + 2z^2 + 3z^3 + \cdots + nz^n + \cdots = \dfrac{z}{(1 - z)^2};$

(b) $1^2z + 2^2z^2 + 3^2z^3 + \cdots + n^2z^n + \cdots = \dfrac{z(1 + z)}{(1 - z)^3};$

(c) $1 + \binom{k+1}{1}z + \binom{k+2}{2}z^2 + \cdots + \binom{k+n}{n}z^n + \cdots = \dfrac{1}{(1 - z)^{k+1}}$ $(k = 0, 1, 2, \cdots),$

where the coefficients on the left are the binomial coefficients.

32. Find the sum of the infinite series

$$\frac{1^2}{3} - \frac{2^2}{3^2} + \frac{3^2}{3^3} - \frac{4^2}{3^4} + - \cdots + (-1)^{k-1}\frac{k^2}{3^k} + \cdots.$$

33. Show that

$$e^{i\pi} + 1 = 0.$$

Note that this equality ties together the five most important constants in mathematics: $0, 1, \pi, e, i$.

34. Show that

(a) $\displaystyle\int_{-\pi}^{\pi} e^{in\theta}\, d\theta = \begin{cases} 0 & (n \neq 0); \\ 2\pi & (n = 0). \end{cases}$

(b) $\displaystyle\int_{-\pi}^{\pi} \cos^4\theta\, d\theta = \int_{-\pi}^{\pi} \sin^4\theta\, d\theta = \frac{3\pi}{4}.$

Hint: $\displaystyle\cos^4\theta = \left(\frac{e^{i\theta} + e^{-i\theta}}{2}\right)^4$

$$= \frac{1}{2^4}\left(e^{4i\theta} + 4e^{2i\theta} + 6 + 4e^{-2i\theta} + e^{-4i\theta}\right).$$

(c) $\displaystyle\int_{0}^{\frac{\pi}{2}} \cos^{2n}\theta\, d\theta = \int_{0}^{\frac{\pi}{2}} \sin^{2n}\theta\, d\theta = \frac{1 \cdot 3 \cdot 5 \cdots (2n - 1)}{2 \cdot 4 \cdot 6 \cdots (2n)} \cdot \frac{\pi}{2}$

$(n \in \mathbb{N})$ (Wallis).

35. (a) Find the set of all z for which e^z is real.

(b) Same problem when e^z is purely imaginary.

36. Show that

$$e = 1 + \frac{1}{1!} + \frac{1}{2!} + \frac{1}{3!} + \cdots + \frac{1}{n!} + \cdots$$

is an irrational number.

37. Prove the addition formula for the exponential function:

$$e^{z+w} = e^z \cdot e^w \quad \text{for all } z, w \in \mathbb{C}$$

by using the Cauchy product.

38. Show that for any $w \neq 0$ there is a sequence $\{z_n\}_{n=1}^{\infty}$ converging to zero such that

$$\exp\left(\frac{1}{z_n}\right) = w \quad \text{for all } n \in \mathbb{N}.$$

39. (a) Show that for any $w_0 \neq 0$ and $\epsilon > 0$ there exist infinitely many $z \in \mathbb{C}$ satisfying

$$e^z = w_0 \quad \text{and} \quad \left| \arg z - \frac{\pi}{2} \right| < \epsilon.$$

(b) Can the argument $\frac{\pi}{2}$ be replaced by another angle?

40. Find the limits:

(a) $\lim\limits_{z \to 0} \dfrac{e^z - 1}{z}$;

(b) $\lim\limits_{z \to 0} \dfrac{\sin z}{z}$;

(c) $\lim\limits_{z \to 0} \dfrac{1 - \cos z}{z^2}$.

Hint: Use power-series expansions.

41. (a) Find the set of all $z \in \mathbb{C}$ for which $\sin z$ is real.

(b) Same problem for $\cos z$.

42. Find all the roots of

(a) $\cos z = 3$;

(b) $\sin z = \sqrt{2}$;

(c) $\sinh z = 0$;

(d) $\cosh z = 0$.

43. (a) Find the set of all $z \in \mathbb{C}$ for which $|\cos z| \leq 1$.

(b) Same problem for $|\sin z| \leq 1$.

44. (Abel transformation; summation by parts.)

(a) Suppose $B_n = \sum_{k=0}^{n} b_k$ $(n = 0, 1, 2, \cdots)$. Show that

$$\sum_{k=m}^{n} a_k b_k = -a_m B_{m-1} + \sum_{k=m}^{n-1} (a_k - a_{k+1}) B_k + a_n B_n \quad (n > m).$$

(b) Suppose $B_n = \sum_{k=0}^{n} b_k$ $(n = 0, 1, 2, \cdots)$ are bounded (where $b_k \in \mathbb{C}$), and $\{a_k\}_{k=0}^{\infty}$ is a sequence of positive numbers that are monotone nonincreasing to 0. Show that $\sum_{n=0}^{\infty} a_n b_n$ is convergent.

(c) Suppose $B_n = \sum_{k=0}^{n} b_k$ $(n = 0, 1, 2, \cdots)$ are bounded (where $b_k \in \mathbb{C}$), and $\sum_{k=0}^{\infty} a_k$ is an absolutely convergent series with complex terms. Show that $\sum_{n=0}^{\infty} a_n b_n$ is convergent.

(d) Show that the power series

$$z + \frac{z^2}{2} + \frac{z^3}{3} + \cdots + \frac{z^k}{k} + \cdots$$

converges uniformly on $D_\delta = \{z \in \bar{D}; |z - 1| \geq \delta\}$ for any $\delta > 0$. Note that this series does not converge absolutely at any of the points on the unit circle (hence the Weierstrass M-test 2.4.5 is not applicable).

45. (Pólya-Szegö) Construct a sequence $\{z_k\}_{k=1}^\infty$ for which all the series

$$\sum_{k=1}^\infty z_k^n \quad (n = 1, 2, 3, \ldots)$$

converge, and all the series

$$\sum_{k=1}^\infty |z_k|^n \quad (n = 1, 2, 3, \ldots)$$

diverge.

Hint: Try $z_k = \frac{e^{ik}}{\log(k+1)}$. Alternatively, let $\{p_m\}_{m=1}^\infty$ be a monotone increasing sequence of positive integers, and set the first p_1 terms to be

$$\frac{\omega_1}{\log 2}, \frac{\omega_1^2}{\log 2}, \frac{\omega_1^3}{\log 2}, \cdots, \frac{\omega_1^{p_1}}{\log 2}, \quad \left(\omega_1 = e^{\frac{2\pi i}{p_1}}\right);$$

the next p_2 terms to be

$$\frac{\omega_2}{\log 3}, \frac{\omega_2^2}{\log 3}, \frac{\omega_2^3}{\log 3}, \cdots, \frac{\omega_2^{p_2}}{\log 3}, \quad \left(\omega_2 = e^{\frac{2\pi i}{p_2}}\right); \quad \text{etc.}$$

46. (Pólya-Szegö) Construct a sequence $\{z_k\}_{k=1}^\infty$ for which the series $\sum_{k=1}^\infty z_k^n$ diverges for $n = 5$, but converges when n is any positive integer other than 5.

Hint: Let $\omega_0, \omega_1, \cdots, \omega_4$ be distinct roots of $z^5 = 1$. For $k = 5m + j$ $(0 \leq j < 5)$, set $z_k = \frac{\omega_j}{m^{\frac{1}{5}}}$.

47. (Takagi) Prove the addition formulas for the sine and cosine functions by expanding $\sin(z + w)$ and $\cos(z + w)$ into Taylor series around the point z, and then imitate the argument in Section 2.9.1(a), using

$$\frac{d}{dz}\sin z = \cos z, \qquad \frac{d}{dz}\cos z = -\sin z.$$

48. Find the most general power series (with two arbitrary constants) satisfying the differential equation

$$f''(z) + f(z) = 0.$$

49. Find a power series (centered at the origin) satisfying the Bessel differential equation

$$zf''(z) + f'(z) + zf(z) = 0$$

with the initial condition $f(0) = 1$. Show that this series converges for all $z \in \mathbb{C}$.

50. Show that both $\sum_{n=1}^{\infty} \dfrac{\cos nz}{n^p}$ and $\sum_{n=1}^{\infty} \dfrac{\sin nz}{n^p}$ diverge if z is not real ($p \in \mathbb{R}$).

51. Find all the values of

$$\cos i, \quad \sin\left(\frac{\pi}{4} + i\right), \quad \log e, \quad \log(-i), \quad i^i, \quad (-1)^i.$$

52. (a) Is the following equality valid?

$$\log(zw) = \log z + \log w.$$

(b) How about $z^\alpha \cdot z^\beta = z^{\alpha+\beta}$?

53. Find the fallacies in the following "proofs":

(a)
$$2\log(-1) = \log(-1)^2 = \log 1 = 0. \quad \therefore \ \log(-1) = 0.$$

It follows that

$$-1 = e^0 = 1. \quad \therefore \ 2 = 0.$$

(b) Let m and n be two arbitrary integers. Then

$$e^{2m\pi i} = 1 = e^{2n\pi i}.$$

$$\therefore \left(e^{2m\pi i}\right) = \left(e^{2n\pi i}\right)^i; \quad \text{i.e., } e^{-2m\pi} = e^{-2n\pi}.$$

Because e^x increases monotonically for $x \in \mathbb{R}$ and both $-2m\pi$ and $-2n\pi$ are real, we must have

$$-2m\pi = -2n\pi. \quad \therefore m = n.$$

54. Find the maximum modulus of the following functions on the closed unit disc:

(a) $z^2 - 1$;

(b) $z(z + i)(z + 2i)$;

(c) e^z.

55. As we have shown, the Minimum Modulus Principle 2.10.1 holds for power series, and this is the main ingredient in providing the fundamental theorem of algebra. Can we assert that every power series with infinite radius of convergence has a root?

56. (Schwarz) Let $f(z)$ be a function defined by a power series with radius of convergence R, and suppose

$$| f(z) | < M \quad \text{for } z \in D(0, R),$$

where M is a positive constant. If $f(0) - 0$, show that

$$|f'(0)| \leq \frac{M}{R} \quad \text{and} \quad | f(z) | \leq \frac{M}{R} \cdot | z | \quad \text{for all } z \in D'(0, R).$$

Discuss the possibility of equality.

Chapter 3

Analytic Functions

3.1 The Cauchy-Riemann Differential Equations

We defined the differentiability of a function in Chapter 2, and we have seen that any function defined by a convergent power series is infinitely differentiable inside its circle of convergence.

We remark that if the limit of the difference quotient obtained by approaching z_0 along one path is not equal to the limit obtained along a second path, then it is possible to construct a third path (as shown in Figure 3.1) along which the values oscillate, and hence along this path we get no limit at all.

In real analysis, if $f(x)$ is differentiable, its derivative $f'(x)$ may not even be continuous. We shall see that in complex analysis, by contrast, differentiability of $f(z)$ in some region implies that $f'(z)$ is also differentiable there, and hence $f(z)$ is infinitely differentiable. Recall that differentiability implies continuity, but the converse is not true.

Now, in defining differentiability of a complex-valued function $f(z)$ defined on a set Ω, the limit

$$f'(z_0) = \lim_{z \to z_0} \frac{f(z) - f(z_0)}{z - z_0}$$

is naturally taken with respect to values $z \in \Omega$. The limit will therefore have a different meaning depending on whether z_0 is an interior point or a boundary point of Ω. The best way to avoid this complication is simply to require that every differentiable function be defined on an *open* set. It is further advantageous if all functions are defined on a *connected* set. For, if a function is defined on a set consisting of a number of disjoint components, then the values of the function on the different components really have nothing to do with each other, and it is more reasonable to consider the function on each component separately.

Figure 3.1

DEFINITION 3.1.1 *A function $f(z)$ is said to be* **analytic** *in a region Ω if it is defined and differentiable at each point of Ω. A function $f(z)$ is* **analytic** *on an arbitrary point set E if it is analytic in some region that contains E. In particular, "$f(z)$ is analytic at a point z_0" means that $f(z)$ is defined and differentiable in some neighborhood of z_0.*

In general, the terms *analytic, regular, holomorphic,* and *monogenic* are equivalent, although slight niceties of difference exist as used by some authors.

A function defined by a convergent power series is analytic in its disc of convergence. The set of all functions analytic in Ω is denoted $H(\Omega)$.

Every complex-valued function $f(z)$ can be expressed as the sum of a real-valued function $u(x, y)$ and a real-valued function $v(x, y)$ multiplied by i. For example, suppose

$$f(z) = z^2 + 2z + 5\bar{z}.$$

Then

$$
\begin{aligned}
f(z) &= (x + iy)^2 + 2(x + iy) + 5(x - iy) \\
&= (x^2 - y^2 + 7x) + i(2xy - 3y) \\
&= u(x, y) + iv(x, y).
\end{aligned}
$$

It is traditional to denote the real part $u(x, y)$ and the imaginary part $v(x, y)$.

We raise the following question: Is there some way of testing u and v to determine if $f(z)$ is analytic? If we suppose $f(z)$ is differentiable at z_0, then the difference quotient $\frac{f(z)-f(z_0)}{z-z_0}$ approaches a limit as $z \to z_0 = x_0 + iy_0$ from any direction. In particular, let $z \to z_0$ horizontally. Then $y = y_0$ and $x \to x_0$, hence

$$\frac{f(z) - f(z_0)}{z - z_0} = \frac{f(x + iy_0) - f(x_0 + iy_0)}{x - x_0}.$$

If the limit exists, then

$$\left.\frac{df}{dz}\right|_{z=z_0} = f'(z_0) = \left.\frac{\partial f}{\partial x}\right|_{z=z_0} = \left.\left(\frac{\partial u}{\partial x} + i\frac{\partial v}{\partial x}\right)\right|_{z=z_0}.$$

Now let z approach z_0 vertically. Then $x = x_0$, $y \to y_0$, thus

$$\frac{f(z) - f(z_0)}{z - z_0} = \frac{f(x_0 + iy) - f(x_0 + iy_0)}{i(y - y_0)},$$

and in the limit we obtain

$$\left.\frac{df}{dz}\right|_{z=z_0} = f'(z_0) = \left.\frac{1}{i}\frac{\partial f}{\partial y}\right|_{z=z_0} = \left.\frac{1}{i}\left(\frac{\partial u}{\partial y} + i\frac{\partial v}{\partial y}\right)\right|_{z=z_0}.$$

Because the derivative exists, the two expressions for $f'(z_0)$ must be equal and we must have

$$\left.\frac{\partial f}{\partial x}\right|_{z=z_0} = -i\left.\frac{\partial f}{\partial y}\right|_{z=z_0}.$$

In terms of the real and imaginary parts, $u(x, y)$ and $v(x, y)$, of $f(z)$, this equation becomes

$$u_x + iv_x = -i(u_y + iv_y);$$

that is,

$$u_x(x_0, y_0) = v_y(x_0, y_0), \qquad u_y(x_0, y_0) = -v_x(x_0, y_0).$$

These are known as the *Cauchy-Riemann equations*. We have shown that differentiability at a point implies that the Cauchy-Riemann equations are satisfied there.

Let us return to our previous example. Here

$$\begin{aligned}
u(x, y) &= x^2 - y^2 + 7x, & v(x, y) &= 2xy - 3y, \\
u_x(x, y) &= 2x + 7, & v_x(x, y) &= 2y, \\
u_y(x, y) &= -2y, & v_y(x, y) &= 2x - 3.
\end{aligned}$$

We see that the partial derivatives exist everywhere, but the Cauchy-Riemann equations are not satisfied. Therefore, $f'(z)$ exists nowhere, for at no point does $2x + 7 = 2x - 3$.

Similarly, none of the functions $f_1(z) = \bar{z}$, $f_2(z) = \Re z$, $f_3(z) = \Im z$, $f_4(z) = |z|$, $f_5(z) = \arg z$ is analytic.

As a further example, let $f(z) = z^2 + 2z$.

$$f(z) = (x + iy)^2 + 2(x + iy) = (x^2 - y^2 + 2x) + i(2xy + 2y).$$

$$\begin{aligned}
u(x, y) &= x^2 - y^2 + 2x, & v(x, y) &= 2xy + 2y, \\
u_x(x, y) &= 2x + 2, & v_x(x, y) &= 2y, \\
u_y(x, y) &= -2y, & v_y(x, y) &= 2x + 2.
\end{aligned}$$

Here the Cauchy-Riemann equations are satisfied everywhere. Does this mean the function is analytic?

We have shown that if $f(z)$ is analytic on Ω, then the real and imaginary parts of $f(z)$ satisfy the Cauchy-Riemann equations there. The argument is not reversible. Even if the Cauchy-Riemann equations are satisfied, it could still happen that differentiability fails there.

EXAMPLE. (Menshoff) Let

$$f(z) = \begin{cases} \dfrac{z^5}{|z|^4}, & z \neq 0; \\ 0, & z = 0. \end{cases}$$

Then the limit

$$\lim_{h \to 0} \frac{f(h) - f(0)}{h} = \lim_{h \to 0} \left(\frac{h}{|h|}\right)^4$$

depends on the direction of approach to the origin, hence the function is not differentiable there, and so the function is not analytic at the origin. However, if h approaches the origin along either the real or imaginary axis, the limit exists and is 1 in both cases, and so the Cauchy-Riemann equations are satisfied at the origin.

Clearly, we need additional conditions for the analyticity of function $f(z)$. In the following theorem a useful sufficient condition is given, although weaker conditions are known. We shall not have occasion to become involved with such delicate problems.

THEOREM 3.1.2 *If the Cauchy-Riemann equations are satisfied in some neighborhood of a point* $(x_0,\ y_0)$, *and if* $u_x,\ u_y,\ v_x,\ v_y$ *are continuous in this neighborhood, then*

$$f(z) = u(x,\ y) + iv(x,\ y)$$

is analytic at $z_0 = x_0 + iy_0$.

Proof. We may assume the neighborhood of $z_0 = x_0 + iy_0$ to be a disc. Now let $z = x + iy$, $z + (h + ik) = (x + h) + i(y + k)$ $(h, k \in \mathbb{R})$ be two distinct points in this disc. Let

$$g(t) = u(x + th, y + tk) \quad (0 \le t \le 1).$$

Then, by the mean-value theorem, we have

$$g(1) - g(0) = g'(\theta) \quad \text{for some } 0 < \theta < 1.$$

Because

$$g'(t) = hu_x(x + th, y + tk) + ku_y(x + th, y + tk),$$

we obtain

$$\begin{aligned}
u(x + h, &y + k) - u(x, y) \\
&= hu_x(x + \theta h, y + \theta k) + ku_y(x + \theta h, y + \theta k) \quad (0 < \theta < 1).
\end{aligned}$$

Because we assumed u_x, u_y to be continuous, we may write

$$u(x + h, y + k) - u(x, y) = hu_x(x, y) + ku_y(x, y) + \epsilon,$$

where the error term, ϵ, tends to zero faster than $h + ik$ in the sense that $\frac{\epsilon}{h+ik} \longrightarrow 0$ as $h + ik \to 0$. A similar relation holds for the imaginary part, $v(x, y)$. Thus

$$\begin{aligned}
f(z + h + ik) - f(z) &= \{hu_x(x, y) + ku_y(x, y) + \epsilon\} \\
&\quad + i\{hv_x(x, y) + kv_y(x, y) + \epsilon'\} \\
&= \{u_x(x, y) + iv_x(x, y)\}(h + ik) + (\epsilon + i\epsilon'),
\end{aligned}$$

by the Cauchy-Riemann equations. Therefore,

$$\lim_{h+ik \to 0} \frac{f(z + h + ik) - f(z)}{h + ik} = u_x(x, y) + iv_x(x, y);$$

that is, the derivative exists at $z = x + iy$. Because $z = x + iy$ was an arbitrary point in the neighborhood of $z_0 = x_0 + iy_0$, we conclude that $f(z) = u(z) + iv(z)$ is analytic at z_0. (*Question*: Why did we assume our neighborhood to be a disc?)   □

3.2   Harmonic Functions

As a simple application of the Cauchy-Riemann equations, we prove:

COROLLARY 3.2.1 *Suppose* $f(z) = u(x, y) + iv(x, y) \in H(\Omega)$, *where* Ω *is a region, and* $u(x, y) \equiv$ *constant on* Ω; *then* $v(x, y) \equiv$ *constant on* Ω. *Thus,*

$$f(z) \equiv constant \quad and\ hence\ \ f'(z) \equiv 0\ on\ \Omega.$$

Proof. Because $u(x, y) \equiv$ constant in Ω, we have $u_x \equiv u_y \equiv 0$ in Ω. By the Cauchy-Riemann equations, this implies that $v_x \equiv v_y \equiv 0$ in Ω. "Therefore" v is constant in Ω, and the result follows immediately.   □

Remark. Here the connectedness of Ω is necessary. For example, let

$$f(z) = \begin{cases} 0, & \Re z > 0; \\ i, & \Re z < 0; \\ \text{undefined} & \text{for } \Re z = 0. \end{cases}$$

Now, $f(z)$ is analytic where it is defined, and $u(x, y) \equiv 0$ there, but $v(x, y)$ is *not* constant.

The problem in this case is that the domain of definition of the function $f(z)$ is not connected. In an open connected set any two distinct points can be joined by a polygonal path that consists only of horizontal and vertical segments. (Why?) The vanishing of the partial derivatives assures us that no change occurs in the function along any of these paths.

COROLLARY 3.2.2 *Suppose* $f(z), g(z) \in H(\Omega)$, *and* $\Re f(z) = \Re g(z)$ *for all* $z \in \Omega$. *Then* $f(z) = g(z) +$ constant. *A similar result holds if* $\Im f(z) = \Im g(z)$ *for all* $z \in \Omega$.

Restated, this result says: *The real part* $\Re f(z)$ *completely determines the analytic function* $f(z)$ *except for an additive imaginary constant, and a similar result is true for the imaginary part* $\Im f(z)$.

How, then, does one obtain $f(z)$ from $u(x, y)$ or $v(x, y)$? We give an example. Let $u(x, y) = x^2 - y^2$. Then

$$v_y = u_x = 2x, \quad v_x = -u_y = 2y,$$

$$\therefore v = 2xy + \varphi(x), \quad v = 2xy + \psi(y).$$

Hence,

$$2xy + \varphi(x) = 2xy + \psi(y) \quad \text{and so} \quad \varphi(x) = \psi(y).$$

It follows that $\varphi(x)$ and $\psi(y)$ are constant, and

$$f(z) = (x^2 - y^2) + i(2xy + c) = z^2 + ic \quad (c \in \mathbb{R}).$$

Does this mean that, given a real-valued function $u(x, y)$, we can always find a corresponding $v(x, y)$ such that $f(z) = u(x, y) + iv(x, y)$ is analytic? No! Consider $u(x, y) = x^2 + y^2$.

$$v_y = u_x = 2x, \quad v_x = -u_y = -2y,$$

$$v = 2xy + \varphi(x), \quad v = -2xy + \psi(y).$$

Therefore,

$$2xy + \varphi(x) = -2xy + \psi(y).$$

$$\therefore \varphi(x) = -4xy + \psi(y); \quad \text{i.e.,} \quad \varphi'(x) = -4y.$$

But this result is absurd, for $\varphi(x)$ was supposed to be independent of y. We conclude that no analytic function exists with $x^2 + y^2$ as its real part.

Have we some way of deciding which real-valued functions are real parts of analytic functions? Consider real-valued functions $u(x, y)$ and $v(x, y)$ satisfying the Cauchy-Riemann equations

$$u_x = v_y, \quad u_y = -v_x.$$

Then, formally,

$$u_{xx} = v_{yx}, \quad u_{yy} = -v_{xy}.$$

$$\therefore \nabla^2 u = u_{xx} + u_{yy} = v_{yx} - v_{xy} = 0.$$

There are two questions with the procedure employed above:

(a) Do the second derivatives exist?

(b) Does $v_{yx} = v_{xy}$?

We shall show later that the derivative of an analytic function is itself analytic, which implies that $u(x, y)$ and $v(x, y)$ will have continuous partial derivatives of all orders. In particular, the mixed partial derivatives will be equal. Accepting these facts temporarily, we get the necessary condition: *If $u(x, y)$ is the real part of an analytic function, then u_{xx}, u_{yy}*

exist and the Laplace equation, $\nabla^2 u = u_{xx} + u_{yy} = 0$, is satisfied at every point of the region.

Let us return to the example $u = x^2 + y^2$. Here

$$u_x = 2x, \quad u_y = 2y,$$

$$u_{xx} = 2, \quad u_{yy} = 2.$$

$$\therefore u_{xx} + u_{yy} = 4,$$

and the necessary condition is not fulfilled.

This result suggests the reverse question: Is the condition sufficient? Namely, is any solution of

$$\nabla^2 u = \frac{\partial^2 u}{\partial x^2} + \frac{\partial^2 u}{\partial y^2} = 0,$$

the real part of an analytic function? The answer is a qualified *yes*. The condition $u \in C^2$ (i.e., the second partial derivatives exist and are continuous) will do even though a weaker condition is known. A real-valued function $u(x, y) \in C^2$ satisfying the Laplace equation is called a *harmonic* function. Thus, a real-valued function $u(x, y)$ is harmonic if and only if it is "locally" the real part of an analytic function; that is, there exists another real-valued function $v(x, y)$ such that $u(x, y) + iv(x, y)$ is an analytic function of the complex variable $z = x + iy$. The harmonic function $v(x, y)$ is called the *harmonic conjugate* of $u(x, y)$; it is unique up to an additive constant, by Corollary 3.2.2. The existence of the harmonic conjugate will be discussed in Chapter 11.

In concluding this section, let us establish the *chain rule*. In calculus, the chain rule for differentiation of composite functions:

$$\frac{dy}{dt} = \frac{dy}{dx} \cdot \frac{dx}{dt}$$

is often proved by the following innocent argument:

$$\frac{\triangle y}{\triangle t} = \frac{\triangle y}{\triangle x} \cdot \frac{\triangle x}{\triangle t}.$$

As $\triangle t \to 0$, so too does $\triangle x$, and we have

$$\frac{\triangle x}{\triangle t} \longrightarrow \frac{dx}{dt}, \quad \frac{\triangle y}{\triangle x} \longrightarrow \frac{dy}{dx},$$

$$\therefore \; \frac{\triangle y}{\triangle t} \longrightarrow \frac{dy}{dx} \cdot \frac{dx}{dt}.$$

Here, the question of whether $\triangle x = 0$ for some $\triangle t$ is not considered. When x is an independent variable, $\triangle x$ is arbitrary and $\triangle x \neq 0$; however, if x is a function of t, then, depending on the value of $\triangle t$, it could happen that $\triangle x = 0$. Thus, in that case the expression $\frac{\triangle y}{\triangle x}$ is meaningless.[1] It is easier to give an alternate proof than to fix the argument above.

THEOREM 3.2.3 (Chain Rule) *Let* $f(z)$, $g(z) \in H(\mathbb{C})$ *and* $h(z) = (f \circ g)(z)$. *Then* $h(z) \in H(\mathbb{C})$, *and the derivative* $h'(z)$ *can be computed by the chain rule*

$$h'(z) = f'(g(z)) \cdot g'(z) \quad (z \in \mathbb{C}).$$

Proof. Fix $z_0 \in \mathbb{C}$, and set $w_0 = g(z_0)$. Then, from the definition of differentiability, we have

$$g(z) - g(z_0) = \{g'(z_0) + \epsilon(z)\} \cdot (z - z_0),$$

$$f(w) - f(w_0) = \{f'(w_0) + \delta(w)\} \cdot (w - w_0),$$

where $\epsilon(z) \to 0$ as $z \to z_0$ and $\delta(w) \to 0$ as $w \to w_0$. Setting $w = g(z)$ and substituting the first equation into the second, we obtain

$$\frac{h(z) - h(z_0)}{z - z_0} = \{f'(g(z_0)) + \delta(g(z_0))\} \cdot \{g'(z_0) + \epsilon(z)\} \quad (z \neq z_0).$$

Because the differentiability of $g(z)$ implies the continuity of $g(z)$ at z_0, we obtain the desired result.   □

In looking back at the proof, we note that the only reason for defining $f(z)$ and $g(z)$ in the whole plane $\mathbb{C}$ is to make sure the expression $h(z) = f(g(z))$ has meaning. Thus, we have

COROLLARY 3.2.4 *If* $f(z)$ *and* $g(z)$ *are analytic in some neighborhoods of* $g(z_0)$ *and* z_0, *respectively, and* $h(z) = f(g(z))$, *then* $h(z)$ *is analytic in a neighborhood of* z_0, *and*

$$h'(z_0) = f'(g(z_0)) \cdot g'(z_0).$$

---

[1]   It violates the *cardinal law of mathematics*: "Thou shalt not divide by zero."

EXAMPLE. Show that $u(x, y) = e^{x^2-y^2} \cdot \cos(2xy)$ satisfies the Laplace equation.

Solution. Because $g(z) = z^2$ is analytic, as is $f(z) = e^z$, we conclude that

$$h(z) = f(g(z)) = \exp(z^2)$$

is analytic, and so $u(x, y) = \Re h(z)$ satisfies the Laplace equation. (Readers are urged to check this result by computation.)

EXAMPLE. We have shown the addition formula for the cosine function:

$$\cos(z + w) = \cos z \cdot \cos w - \sin z \cdot \sin w.$$

Now, if we regard w as a constant, then $z + w$ is an analytic function of z, and so, differentiating the addition formula for the cosine function with respect to z, we obtain

$$\sin(z + w) = \sin z \cdot \cos w + \cos z \cdot \sin w,$$

which is the addition formula for the sine function.

3.3   Geometric Significance of the Derivative

Suppose a function $f(z)$ is analytic at z_0. This assumption implies that

$$\left| \frac{f(z) - f(z_0)}{z - z_0} - f'(z_0) \right|$$

is small when $|z - z_0|$ is small. More precisely, given any $\epsilon > 0$, there exists $\delta > 0$ such that

$$\left| \frac{f(z) - f(z_0)}{z - z_0} - f'(z_0) \right| < \epsilon \quad \text{whenever } z \in D'(z_0, \delta).$$

This result is equivalent to

$$|f(z) - f(z_0) - f'(z_0)(z - z_0)| < \epsilon |z - z_0| \quad \text{whenever } z \in D'(z_0, \delta).$$

Roughly speaking, this inequality says that

$$|f(z) - \{f(z_0) + f'(z_0)(z - z_0)\}|$$

is small in comparison with $|z - z_0|$. The quantity

$$f(z_0) + f'(z_0)(z - z_0)$$

is called the *linear approximation* or *linear part* of $f(z)$ at z_0. This is the complex analogue of the approximation of a function locally by a straight line in real variables.

Let us now consider how one draws graphs of complex functions. For real-valued functions of one real variable, we plot y ($= f(x)$) versus x in a rectangular coordinate system and speak of the tangent to the curve at a point $(x_0,\ f(x_0))$ as having slope $f'(x_0)$. However, we cannot perform this operation in the complex case, because we need a two-dimensional region to describe the path of the single variable z and we also need a two-dimensional region to describe the function values. For this reason we use two planes: the z-plane to describe the path of z and the w-plane to describe the path of $w = f(z)$ as z traverses its path in the z-plane.

Now, let $z = x + iy$, $w = f(z) = u(x,\ y) + iv(x,\ y)$. Using the Cauchy-Riemann equations, we see that the *Jacobian* of the mapping is

$$\frac{\partial(u,\ v)}{\partial(x,\ y)} = \begin{vmatrix} u_x & v_x \\ u_y & v_y \end{vmatrix} = u_x v_y - u_y v_x = u_x^2 + v_x^2 = |\ f'(z)\ |^2\ .$$

Therefore, if $f'(z)$ exists and is continuous in a neighborhood of a point z_0, where $f'(z_0) \neq 0$, we conclude (by the implicit function theorem) that, in some sufficiently small neighborhood of z_0, $f(z)$ provides a one-to-one correspondence between z and $w = f(z)$; that is, $f(z)$ is locally an injection.

Let us investigate this mapping in a little more detail. Suppose $\Gamma : z = z(t)$ $(a \leq t \leq b)$ is an arc. Its orientation is determined by the parameter t increasing from a to b, where $z(a)$ and $z(b)$ are called the *initial point* and the *terminal point* of the arc Γ, respectively. If Γ has a tangent at a point $z_0 = z(t_0)$; that is, if both

$$\lim_{t \to t_0+} \arg \{z(t) - z(t_0)\} \quad \text{and} \quad \lim_{t \to t_0-} \arg \{z(t_0) - z(t)\}$$

exist and are equal $(\text{mod } 2\pi)$, then its common value

$$\alpha = \lim_{t \to t_0+} \arg \{z(t) - z(t_0)\} = \lim_{t \to t_0-} \arg \{z(t_0) - z(t)\}$$

gives the angle between the directed tangent and the positive real axis. In particular, if the derivative $z'(t_0)$ exists and[2] $z'(t_0) \neq 0$, then the tangent

---

[2] Note that $z(t) = t(|\ t\ | + it)$ is differentiable everywhere for $t \in \mathbb{R}$, but $z'(0) = 0$ and it has no tangent at $t = 0$. On the other hand, $z(t) = t^3(1+i)$ has a tangent everywhere, yet $z'(0) = 0$.

exists, and we have

$$\alpha \equiv \arg z'(t_0) \qquad (\text{mod } 2\pi).$$

If two arcs Γ_1 and Γ_2 intersect at z_0 and each has a tangent at z_0, then the angle between the directed tangents is called the *angle between the arcs*.

Suppose $w = f(z) \in H(\Omega)$, where Ω is a region containing an arc $\Gamma : z = z(t)$ $(a \leq t \leq b)$. Then clearly

$$C : w = w(t) = f(z(t)) \quad (a \leq t \leq b)$$

is an arc in the w-plane that is the image of Γ under the mapping $f(z)$. Suppose Γ has a tangent at $z_0 = z(t_0)$ and that $f'(z_0) \neq 0$. Then

$$\lim_{z \to z_0} \frac{f(z) - f(z_0)}{z - z_0} = f'(z_0).$$

$$\therefore \ \lim_{t \to t_0} \arg \frac{f(z(t)) - f(z(t_0))}{z(t) - z(t_0)} \equiv \arg f'(z_0) \quad (\text{mod } 2\pi).$$

It follows that

$$\lim_{t \to t_0} \left[\arg\{w(t) - w(t_0)\} - \arg\{z(t) - z(t_0)\} \right] \equiv \arg f'(z_0) \quad (\text{mod } 2\pi).$$

Because we are assuming that Γ has a tangent at $z_0 = z(t_0)$, we have

$$\lim_{t \to t_0-} \arg\{z(t_0) - z(t)\} \equiv \lim_{t \to t_0+} \arg\{z(t) - z(t_0)\} \equiv \alpha \quad (\text{mod } 2\pi),$$

and this result implies

$$\lim_{t \to t_0-} \arg\{w(t_0) - w(t)\} \equiv \lim_{t \to t_0+} \arg\{w(t) - w(t_0)\} \equiv \beta \quad (\text{mod } 2\pi),$$

and the arc $C : w = w(t) = f(z(t))$ has a tangent at $t = t_0$. The angle β between the directed tangent and the positive real axis is given by

$$\beta \equiv \alpha + \arg f'(z_0) \quad (\text{mod } 2\pi).$$

This relation says that the tangent line at $z(t_0)$ is turned through an angle $\arg f'(z_0)$ under the transformation $w = f(z)$.

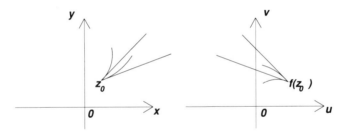

Figure 3.2

Now, consider two arcs Γ_1 and Γ_2 intersecting at z_0 that form angles α_1 and α_2, respectively, with the positive real axis in the z-plane (Figure 3.2). Then the corresponding angles β_1 and β_2 in the w-plane are given by

$$\beta_1 \equiv \alpha_1 + \arg f'(z_0), \quad \beta_2 \equiv \alpha_2 + \arg f'(z_0) \pmod{2\pi},$$

so that

$$\beta_2 - \beta_1 \equiv \alpha_2 - \alpha_1 \pmod{2\pi}.$$

The angle between Γ_1 and Γ_2 is preserved in sense as well as in size. Thus, at a point z_0 such that $f'(z_0) \neq 0$, the image arcs form the same angle as the original arcs.

In general, if a continuous mapping $w = f(z)$ satisfies the conditions that

(a) the property of arcs having tangents at z_0 is preserved, and

(b) the angle between two arcs at z_0 is also preserved in sense as well as in size,

then the mapping $w = f(z)$ is said to be *conformal* at z_0. We have shown that if $f(z)$ is analytic at z_0 and $f'(z_0) \neq 0$, then $w = f(z)$ is a conformal mapping at z_0. We emphasize that the condition $f'(z_0) \neq 0$ is necessary. For example, take $w = f(z) = z^2$ and let z approach the origin along the real and imaginary axes. Because $\arg w = 2 \arg z$, the angle is doubled. Exactly the same idea shows that if $f'(z_0) = 0$ then the angle is definitely not preserved, and the mapping is not injective in a neighborhood of z_0. (Why?) It follows that if an analytic function $f(z)$ gives a bijection between regions Ω and Ω', then $f'(z) \neq 0$ for every $z \in \Omega$, and so the mapping is conformal at every point. However, the condition that $f'(z) \neq 0$ everywhere in Ω does *not* guarantee that the

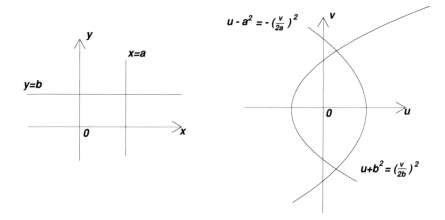

Figure 3.3

mapping $w = f(z)$ is a bijection. This fact is illustrated by the example $w = e^z$ below. We say an analytic function $f(z)$ maps Ω *conformally* onto Ω' if it gives a bijection of Ω to Ω'.

What about the geometric meaning of the modulus of the derivative? Because

$$\lim_{z \to z_0} \left| \frac{f(z) - f(z_0)}{z - z_0} \right| = |f'(z_0)|,$$

then, if z is close to z_0, we have

$$|f(z) - f(z_0)| \approx |f'(z_0)| \cdot |z - z_0|.$$

We see that the distance between z and z_0 is magnified (or contracted) by the factor $|f'(z_0)|$; this approximate statement becomes increasingly accurate as $z \to z_0$.

EXAMPLE. Consider $w = z^2$. The images of the lines $x = a$ ($a \neq 0$) and $y = b$ ($b \neq 0$) are parabolas

$$u - a^2 = -\left(\frac{v}{2a}\right)^2 \quad \text{and} \quad u + b^2 = \left(\frac{v}{2b}\right)^2,$$

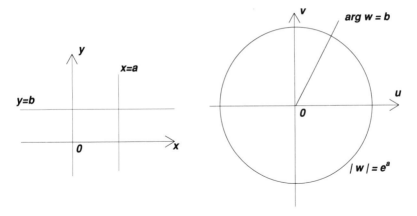

Figure 3.4

respectively (Figure 3.3). Note that as a and b vary, we obtain two families of parabolas intersecting orthogonally. Moreover, all the parabolas have their foci at the origin.

Question: What is the number of intersection points of the lines $x = a$ and $y = b$ in the z-plane? How about that of their images in the w-plane? Can you explain the phenomenon?

EXAMPLE.   Consider $w = e^z$. Line $x = a$ is mapped to the circle $\mid w \mid = e^a$, and line $y = b$ is mapped to the ray $\arg w = b$ (Figure 3.4). Again these two families of curves are orthogonal to each other.

3.4   Möbius Transformations

We devote this section to discussing a set of elementary, yet very useful, mappings called *Möbius transformations* (*linear fractional transformations, bilinear transformations, homographic transformations*, and so on). They are defined by the rational functions

$$w = Tz = \frac{\alpha z + \beta}{\gamma z + \delta}, \quad \begin{vmatrix} \alpha & \beta \\ \gamma & \delta \end{vmatrix} = \alpha\delta - \beta\gamma \neq 0.$$

The condition $\begin{vmatrix} \alpha & \beta \\ \gamma & \delta \end{vmatrix} \neq 0$ guarantees that T does not reduce to a constant.

It is convenient to regard this as a mapping of the Riemann sphere (the extended complex plane) to itself, with $-\frac{\delta}{\gamma}$ mapped to ∞, and ∞ to $\frac{\alpha}{\gamma}$ (if

$\gamma \neq 0$). It is then easy to see that each Möbius transformation is a bijection of the Riemann sphere onto itself.

If we map the z-plane onto the w_1-plane by

$$w_1 = Tz = \frac{\alpha z + \beta}{\gamma z + \delta}, \qquad \begin{vmatrix} \alpha & \beta \\ \gamma & \delta \end{vmatrix} \neq 0,$$

and follow this operation by mapping the w_1-plane onto the w-plane by

$$w = Sw_1 = \frac{aw_1 + b}{cw_1 + d}, \qquad \begin{vmatrix} a & b \\ c & d \end{vmatrix} \neq 0,$$

then the result is the Möbius transformation of the z-plane onto the w-plane given by

$$w = STz = \frac{(a\alpha + b\gamma)z + (a\beta + b\delta)}{(c\alpha + d\gamma)z + (c\beta + d\delta)},$$

with

$$\begin{vmatrix} a\alpha + b\gamma & a\beta + b\delta \\ c\alpha + d\gamma & c\beta + d\delta \end{vmatrix} = \begin{vmatrix} a & b \\ c & d \end{vmatrix} \cdot \begin{vmatrix} \alpha & \beta \\ \gamma & \delta \end{vmatrix} \neq 0.$$

Thus, if multiplication of transformations is defined as composition, the set of all Möbius transformations is closed under multiplication. Similarly, the associative law for multiplication can be verified. The identity mapping is clearly a Möbius transformation. Moreover, if we solve for z in terms of w, we obtain

$$z = T^{-1}w = \frac{\delta w - \beta}{-\gamma w + \alpha}, \qquad \begin{vmatrix} \delta & -\beta \\ -\gamma & \alpha \end{vmatrix} = \begin{vmatrix} \alpha & \beta \\ \gamma & \delta \end{vmatrix} \neq 0.$$

Hence, the inverse mapping of a Möbius transformation is a Möbius transformation.

We have established the following

THEOREM 3.4.1 *The set of all Möbius transformations is a group (with multiplication defined as composition).*

If we write the Möbius transformations T and S above as

$$T = \begin{pmatrix} \alpha & \beta \\ \gamma & \delta \end{pmatrix}, \quad S = \begin{pmatrix} a & b \\ c & d \end{pmatrix},$$

then

$$ST = \begin{pmatrix} a\alpha + b\gamma & a\beta + b\delta \\ c\alpha + d\gamma & c\beta + d\delta \end{pmatrix}.$$

It follows that *the group of all 2 × 2 invertible matrices is homomorphic to the group of all Möbius transformations*. (Note that a Möbius transformation is determined by the ratios among its coefficients.)

Now, let us express the rational function defining a Möbius transformation

$$w = Tz = \frac{\alpha z + \beta}{\gamma z + \delta}, \qquad \begin{vmatrix} \alpha & \beta \\ \gamma & \delta \end{vmatrix} \neq 0,$$

in terms of partial fractions.

(a) If $\gamma = 0$, then $\delta \neq 0$, and we have

$$w = \left(\frac{\alpha}{\delta}\right) z + \left(\frac{\beta}{\delta}\right).$$

(b) If $\gamma \neq 0$, then

$$w = Tz = \frac{\alpha z + \beta}{\gamma z + \delta} = \frac{\alpha}{\gamma} - \frac{(\alpha\delta - \beta\gamma)/\gamma}{\gamma z + \delta}.$$

It follows that a Möbius transformation is a composition of the following three types of Möbius transformations:

$$\begin{aligned}
&(1^0) \ \text{Dilation} & T_1 z &= az & (a \neq 0); \\
&(2^0) \ \text{Translation} & T_2 z &= z + b; \\
&(3^0) \ \text{Reciprocation} & T_3 z &= \frac{1}{z}.
\end{aligned}$$

Note that a dilation is a composition of a magnification (or contraction) $S_1 z = \mid a \mid z$ followed by a rotation $S_2 z = e^{i\theta} z$ of angle $\theta = \arg a$. (The order in which these two transformations are performed is inessential— they commute.)

Clearly, dilations and translations map lines to lines and circles to circles. This statement is not true for reciprocation. However, the operation of reciprocation *does* keep the family of all straight lines *and* circles invariant. For, let

$$A(x^2 + y^2) + Bx + Cy + D = 0 \quad (B^2 + C^2 > 4AD),$$

be an arbitrary circle if $A \neq 0$ or line if $A = 0$. In complex notation, this equation becomes

$$az\bar{z} + \bar{b}z + \bar{z} + c = 0 \quad (\mid b \mid^2 > ac),$$

where $a = A$ and $c = D$ are real and $b = \frac{1}{2}(B + iC)$, is an arbitrary circle (line if $a = 0$) in the plane. Performing the reciprocation $w = \frac{1}{z}$, we obtain

$$a + \bar{b}\bar{w} + bw + cw\bar{w} = 0,$$

which is an equation of the same type. We have shown:

THEOREM 3.4.2 *A Möbius transformation maps the family of all circles and straight lines in the plane to itself.*

It is convenient to treat straight lines as particular cases of circles when we are working with Möbius transformations.

Because a Möbius transformation is completely determined by the ratios among its coefficients, given three conditions we might expect to find a Möbius transformation that satisfies these conditions. In particular, it is natural to ask: Given three distinct points z_1, z_2, z_3, and three distinct points w_1, w_2, w_3 (where the value ∞ is allowed) can we always find a Möbius transformation that maps z_1, z_2, z_3 to w_1, w_2, w_3, respectively? To discuss this question, we start from the following observation.

A point z_0 is said to be a *fixed point* of a transformation T if $Tz_0 = z_0$. It is simple to verify that a Möbius transformation having 1, 0, and ∞ as fixed points must be the identity transformation.

Now, let us first consider the very special case in which $w_1 = 1$, $w_2 = 0$, $w_3 = \infty$; that is, given three distinct points z_1, z_2, z_3, we want to find a Möbius transformation that maps z_1, z_2, z_3 to $1, 0, \infty$, respectively. Such a Möbius transformation, if it exists, must be unique. For, if both T_1 and T_2 satisfy the required condition, then $T_2 T_1^{-1}$ has 1, 0, and ∞ as fixed points, and so $T_2 T_1^{-1}$ must be the identity. It follows that $T_1 = T_2$.

Now, regardless of the choice of the (nonzero) constant k, the Möbius transformation

$$Tz = k\frac{z - z_2}{z - z_3}$$

maps z_2, z_3 to 0 and ∞, respectively. It remains to choose the constant k in such a way that z_1 is mapped to 1. This is accomplished by choosing $k = \frac{z_1 - z_3}{z_1 - z_2}$. We obtain

$$Tz = \frac{z - z_2}{z - z_3} \bigg/ \frac{z_1 - z_2}{z_1 - z_3},$$

which has the desired property of mapping the three distinct points z_1, z_2, z_3 to $1, 0, \infty$, respectively.

Next, suppose S is a Möbius transformation that maps w_1, w_2, w_3 to 1, 0, ∞, respectively. Then, by the discussion above, S must be of the form

$$Sw = \frac{w - w_2}{w - w_3} \bigg/ \frac{w_1 - w_2}{w_1 - w_3},$$

and $S^{-1}T$ maps z_1, z_2, z_3 to w_1, w_2, w_3, respectively. Naturally, $S^{-1}T$ can be obtained from

$$\frac{w - w_2}{w - w_3} \bigg/ \frac{w_1 - w_2}{w_1 - w_3} = \frac{z - z_2}{z - z_3} \bigg/ \frac{z_1 - z_2}{z_1 - z_3}$$

by solving for w in terms of z. We have shown that *there exists a (unique) Möbius transformation that maps three arbitrary distinct points to three preassigned distinct points.*

In addition, if this Möbius transformation also maps a fourth distinct point z_0 to w_0, then we must have

$$\frac{w_0 - w_2}{w_0 - w_3} \bigg/ \frac{w_1 - w_2}{w_1 - w_3} = \frac{z_0 - z_2}{z_0 - z_3} \bigg/ \frac{z_1 - z_2}{z_1 - z_3}.$$

The expression $\frac{z_0 - z_2}{z_0 - z_3} \big/ \frac{z_1 - z_2}{z_1 - z_3}$ is called the *cross ratio* of the four distinct points z_0, z_1, z_2, z_3, and is denoted $(z_0, z_1; z_2, z_3)$:

$$(z_0, z_1; z_2, z_3) = \frac{z_0 - z_2}{z_0 - z_3} \bigg/ \frac{z_1 - z_2}{z_1 - z_3}.$$

We have established:

THEOREM 3.4.3 *Möbius transformations preserve cross ratios; that is,*

$$(Tz_0, Tz_1; Tz_2, Tz_3) = (z_0, z_1; z_2, z_3).$$

To render this theorem valid in full generality, we extend the definition of the cross ratio by continuity to the case in which one of the points is ∞. This extension is accomplished by simply dropping the factors that involve this point. For example,

$$(z, z_1; z_2, z_3) = \frac{z - z_2}{z - z_3} \bigg/ \frac{z_1 - z_2}{z_1 - z_3}$$

$$= \frac{1 - \dfrac{z_2}{z}}{1 - \dfrac{z_3}{z}} \cdot \frac{z_1 - z_3}{z_1 - z_2} \longrightarrow \frac{z_1 - z_3}{z_1 - z_2} \quad (\text{as } z \to \infty),$$

$$\therefore \ (\infty, z_1; z_2, z_3) = \frac{z_1 - z_3}{z_1 - z_2}.$$

Because a Möbius transformation maps a circle to a circle, and a circle is completely determined by three points on it, we can thus find Möbius transformations that map a given circle in the z-plane to a preassigned circle in the w-plane. Moreover, given three distinct points on the first circle, we can map them to three preassigned points on the second circle. Once this choice is made, the transformation is uniquely determined.

Remark. If three points on the first circle and three points on the second circle are of the same orientation (i.e., both counterclockwise or both clockwise), then the interior of the first circle is mapped to the interior of the second circle. On the other hand, if they are of the reverse orientation (one counterclockwise, the other clockwise), then the interior is mapped to the exterior. Note that the orientation has to be one way or the other.

COROLLARY 3.4.4 *Four points z_0, z_1, z_2, z_3 are cocyclic (i.e., they are on the same circle) if and only if their cross ratio $(z_0, z_1; z_2, z_3)$ is real.*

Proof. Four points are cocyclic if and only if there is a Möbius transformation that maps these points to points on the real axis. Because the cross ratio of four real numbers is obviously real, the theorem gives us our result.   □

This result can also be visualized in elementary geometry. Because

$$\arg(z_0, z_1; z_2, z_3) = \arg \frac{z_0 - z_2}{z_0 - z_3} - \arg \frac{z_1 - z_2}{z_1 - z_3},$$

then, if these four points are cocyclic, the right-hand side is 0 or $\pm \pi$ depending on whether z_0 and z_1 are on same side of the chord passing through z_2 and z_3 or not (Figure 3.5).

Because $w = Tz = \frac{\alpha z + \beta}{\gamma z + \delta}$ is analytic and

$$\frac{dw}{dz} = \frac{\alpha \delta - \beta \gamma}{(\gamma z + \delta)^2} \neq 0,$$

the mapping $w = Tz$ is conformal for $z \neq -\frac{\delta}{\gamma}, \infty$. It follows that a pair of circles that intersect at points other than $-\frac{\delta}{\gamma}, \infty$, are mapped to circles that intersect each other with the same angle (in size as well as sense).

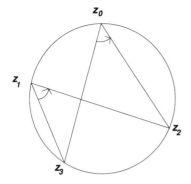
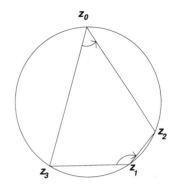

Figure 3.5

The exceptional points $z = -\frac{\delta}{\gamma}$ and ∞ can be treated by the following device: It is sufficient to consider the reciprocation $w = \frac{1}{z}$ on the Riemann sphere and the angle between two great circles passing through the origin 0 and the north pole N. The action of the reciprocation on the Riemann sphere is obtained by reflecting first with respect to the plane $\eta = 0$ (containing the north pole and the real axis) and then with respect to the plane $\zeta = \frac{1}{2}$ (passing through the center of the Riemann sphere parallel to the complex plane). These two reflections certainly preserve the size but reverse the sense of the angle between great circles that pass through the origin and the north pole. However, reversing the sense twice amounts to preserving the sense.

Two points P and Q are called *symmetric* (*reflections* of each other) with respect to a circle C with center at A and radius r, if P and Q are on the same ray from the center A, *and*

$$\overline{AP} \cdot \overline{AQ} = r^2.$$

For a straight line, a pair of points symmetric with respect to the line becomes a pair of points symmetric in the usual sense.

It is well known (and easy to prove), from elementary geometry, that any circle K that passes through a pair of points symmetric with respect to a circle C intersects C orthogonally; conversely, if a pair of points has the property that every circle K that passes through these two points is orthogonal to the circle C, then this pair of points is symmetric with respect to the circle C (Figure 3.6).

THEOREM 3.4.5 (Symmetry Principle) *Möbius transformations preserve symmetry; that is, if a Möbius transformation T maps a circle C to a circle*

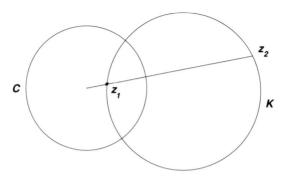

Figure 3.6

C', *then it maps any pair of points symmetric with respect to C to a pair of points symmetric with respect to C'.*

Proof. Let z_1 and z_2 be a pair of symmetric points with respect to the circle C. We want to show that their images Tz_1 and Tz_2 are symmetric with respect to the circle C' ($= TC$). Suppose K' is an arbitrary circle passing through the points Tz_1 and Tz_2. Then its preimage $T^{-1}K'$ passes through the pair of points z_1 and z_2 symmetric with respect to the circle C, hence is orthogonal to C. This result implies that the image, K', of $T^{-1}K'$ under T is orthogonal to the circle C', for T is conformal. We have shown that an arbitrary circle passing through the points Tz_1 and Tz_2 must be orthogonal to the circle C'. Therefore, Tz_1 and Tz_2 must be symmetric with respect to the circle C'.   □

EXAMPLE. Every Möbius transformation that maps the unit disc onto itself can be expressed in the form

$$w = k\frac{z - \alpha}{1 - \bar{\alpha}z} \quad (\mid k \mid = 1, \ \mid \alpha \mid < 1).$$

Solution. Suppose α ($\mid \alpha \mid < 1$) is the point mapped to $w = 0$. Because α and $\frac{1}{\bar{\alpha}}$ are symmetric with respect to the unit circle, $\frac{1}{\bar{\alpha}}$ must be mapped to $w = \infty$.

$$\therefore w = k'\frac{z - \alpha}{z - \dfrac{1}{\bar{\alpha}}} = k\frac{z - \alpha}{1 - \bar{\alpha}z} \quad (k = -\bar{\alpha}k').$$

Because we must have $\mid w \mid = 1$ when $\mid z \mid = 1$, we obtain, choosing $z = 1$,

$$1 = \mid k \mid \cdot \left|\frac{1 - \alpha}{1 - \bar{\alpha}}\right| = \mid k \mid.$$

Conversely, if $w = k\frac{z-\alpha}{1-\bar{\alpha}z}$ ($|\,k\,|=1,\;\;|\,\alpha\,|<1$), then, for $|\,z\,|=1$,

$$|\,w\,| = |\,k\,|\cdot\left|\frac{z-\alpha}{1-\bar{\alpha}z}\right| = \left|\frac{\bar{z}(z-\alpha)}{1-\bar{\alpha}z}\right| = \left|\frac{1-\alpha\bar{z}}{1-\bar{\alpha}z}\right| = 1.$$

Moreover, $|\,\alpha\,|<1$ implies that the interior of the unit circle in the z-plane is mapped to the interior of the unit circle in the w-plane.

Remark. It is obvious that if $|\,\alpha\,|>1$, then

$$w = k\frac{z-\alpha}{1-\bar{\alpha}z}\quad(|\,k\,|=1)$$

maps the unit disc to the exterior of the unit circle, and vice versa.

EXAMPLE. Any Möbius transformation that maps the upper half ($\Im z > 0$) of the z-plane to the unit disc in the w-plane can be expressed in the form

$$w = k\frac{z-\mu}{z-\bar{\mu}}\quad(|\,k\,|=1,\;\;\Im\mu>0).$$

Solution. The points on the z-plane that are mapped to $w = 0$ and $w = \infty$ have to be symmetric with respect to the real axis; that is, they are the complex conjugates of each other. Therefore,

$$w = k\frac{z-\mu}{z-\bar{\mu}}\quad(\Im\mu>0),$$

where k is some constant. When z is real, $\left|\frac{z-\mu}{z-\bar{\mu}}\right| = 1$ and w has to be on the unit circle; that is, $|\,w\,|=1$. It follows that $|\,k\,|=1$. Conversely, Möbius transformations of the above form clearly satisfy the stated condition.

Remark. If $\Im\mu<0$, then

$$w = k\frac{z-\mu}{z-\bar{\mu}}\quad(|\,k\,|=1)$$

maps the lower half-plane to the unit disc.

Given a Möbius transformation

$$w = \frac{\alpha z+\beta}{\gamma z+\delta}\quad\begin{vmatrix}\alpha&\beta\\\gamma&\delta\end{vmatrix}\neq 0,$$

we can write it in the form

$$w = k\frac{z-a}{z-b}\quad\left(k=\frac{\alpha}{\gamma},\;\;a=-\frac{\beta}{\alpha},\;\;b=-\frac{\delta}{\gamma},\;\;\text{if } \alpha\gamma\neq 0\right).$$

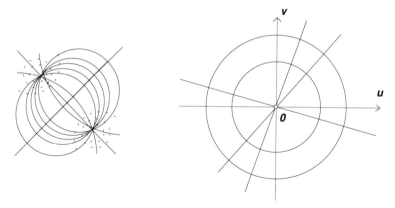

Figure 3.7

Each of the concentric circles $\mid w \mid = r$ (with common center at the origin) is the image of the points z in the z-plane satisfying the equations

$$\left| \frac{z-a}{z-b} \right| = \frac{r}{\mid k \mid}.$$

Each of these equations gives the locus of the points whose distances from the points a and b have a constant ratio. Clearly, this locus is a circle. The circles obtained for $r > 0$ are known as *Apollonius circles* (Figure 3.7).

On the other hand, $z = a$ and $z = b$ are mapped to $w = 0$ and $w = \infty$, respectively. Therefore, the straight lines through the origin in the w-plane are images of the circles passing through the points a and b. More precisely, the ray $\arg w = \theta$ is the image of the arc $\angle bza = \theta - \arg k$, and the ray $\arg w = \theta + \pi$ that of the arc $\angle bza = \pi + \theta - \arg k$. The union of these two rays is a straight line and the union of these two arcs is a full circle which intersects the Apollonius circles orthogonally.

In general, if two points a and b in the z-plane correspond to two points α and β in the w-plane, respectively, then we have

$$\frac{w-\alpha}{w-\beta} = k \frac{z-a}{z-b}.$$

Therefore, the pencil of circles C passing through the points a and b corresponds to the pencil of circles C' passing through the points α and β; and the pencil of the Apollonius circles K with limiting points a and b corresponds to the pencil of Apollonius circles K' with limiting points α and β. These two pairs of pencils of circles intersect each other orthogonally.

It follows from the symmetry principle that reflection with respect to a circle C maps an Apollonius circle K to itself, and a reflection with respect to an Apollonius circle K maps a circle C to another circle in the same pencil.

Remark. For each point in the plane there exist exactly one member of C and one member of K that pass through the point, with the limiting points a and b as the only exceptions.

Exercises

1. Let $p(z)$ be a polynomial of degree $n \geq 1$, and $z_1, z_2, \ldots, z_n$ its roots (each repeated as often as its multiplicity). Show that

   (a) $\dfrac{p'(z)}{p(z)} = \displaystyle\sum_{k=1}^{n} \dfrac{1}{z - z_k}$ $(z \neq z_k, \ k = 1, 2, \ldots, n);$

   (b) all the roots of the polynomial $p'(z)$ are in the convex hull spanned by the roots of $p(z)$; that is, in the smallest convex set containing the roots $z_1, z_2, \ldots, z_n$ of $p(z)$.

2. Verify the Cauchy-Riemann equations for the functions

$$z^3, \quad e^z, \quad \sin z, \quad \text{and} \quad \cos z.$$

3. Verify that the function

$$f(z) = \begin{cases} \dfrac{xy}{x^2 + y^2}, & z = x + iy \neq 0; \\ 0, & z = 0, \end{cases}$$

has partial derivatives with respect to x and y equal to zero at the origin, and hence it satisfies the Cauchy-Riemann equations there; show, however, that it is not even continuous there.

4. (a) Let $f(z)$ be analytic; show that

$$\left(\frac{\partial^2}{\partial x^2} + \frac{\partial^2}{\partial y^2} \right) |f(z)|^2 = 4 |f'(z)|^2 .$$

   (b) Let $f_1(z), f_2(z), \ldots, f_n(z) \in H(\Omega)$. Show that

$$| f_1(z) |^2 + | f_2(z) |^2 + \cdots + | f_n(z) |^2$$

   is harmonic on Ω only if all the functions $f_k(z)$ $(k = 1, 2, \ldots, n)$ are constant.

5. Let $f(z) \in H(\Omega)$. Show that if $| f(z) | \equiv$ constant on Ω then $f(z) \equiv$ constant on Ω.

6. Suppose $f(z)$ is analytic (say, in the upper half-plane). Show that $\overline{f(\bar{z})}$ is also analytic. Where is it defined?

7. Suppose $f(z) \in H(\Omega)$ and $|f(z)| < 1$ for $z \in \Omega$. Show that $g(z) \in H(\Omega)$, where

$$g(z) = \sum_{n=1}^{\infty} n\{f(z)\}^n.$$

8. Show that in polar coordinates the Cauchy-Riemann equations assume the form (for $r \neq 0$)

$$\frac{\partial u}{\partial r} = \frac{1}{r} \frac{\partial v}{\partial \theta}, \qquad \frac{\partial v}{\partial r} = -\frac{1}{r} \frac{\partial u}{\partial \theta}$$

and the Laplace equation becomes

$$\frac{\partial^2 u}{\partial r^2} + \frac{1}{r} \frac{\partial u}{\partial r} + \frac{1}{r^2} \frac{\partial^2 u}{\partial \theta^2} = 0.$$

9. Find the most general harmonic function of the form

$$u(x, y) = ax^3 + bx^2 y + cxy^2 + dy^3,$$

where a, b, c, d are real coefficients. What is its harmonic conjugate?

10. Show that, for $0 \le r < 1$,

(a) $1 + 2\sum_{k=1}^{\infty} r^k \cos k\theta = \dfrac{1 - r^2}{1 - 2r \cos \theta + r^2}$;

(b) $\dfrac{1}{2\pi} \displaystyle\int_{-\pi}^{\pi} \dfrac{1 - r^2}{1 - 2r \cos \theta + r^2} d\theta = 1$;

(c) the *Poisson kernel*

$$P_r(\theta) = \frac{1 - r^2}{1 - 2r \cos \theta + r^2}$$

is a harmonic function in D. What is its harmonic conjugate?

11. Suppose $u(x, y)$ and $v(x, y)$ are harmonic in a region Ω. Under what condition is the product $u(x, y) \cdot v(x, y)$ harmonic? In particular, when is $u^2(x, y)$ harmonic?

12. Suppose $f(z) \in H(\Omega)$, where Ω is a convex region, and a and b are two distinct points in Ω.

(a) (Mean-Value Theorem) Show that there exist points c_1 and c_2 on the line segment joining a and b such that

$$\frac{f(b) - f(a)}{b - a} = \Re f'(c_1) + i\Im f'(c_2).$$

(b) Does there necessarily exist a point c on the line segment joining a and b such that

$$\frac{f(b) - f(a)}{b - a} = f'(c) ?$$

13. (l'Hôpital's Rule) Let $f(z)$, $g(z) \in H(\Omega)$ and $z_0 \in \Omega$. Suppose $f(z_0) = g(z_0) = 0$, but $g'(z_0) \neq 0$. Show that

$$\lim_{z \to z_0} \frac{f(z)}{g(z)} = \frac{f'(z_0)}{g'(z_0)}.$$

14. Find the preimages of lines $u = a$ and $v = b$ under the mapping $w = z^2$ (where $w = u + iv$).

15. Show that the function $w = \frac{1}{2}\left(z + \frac{1}{z}\right)$ maps circles $| z | =$ constant to ellipses and rays $\arg z =$ constant to hyperbolas, where the foci of all these conics are at $w = \pm 1$.

16. Suppose a function $w = f(z) \in H(\Omega)$ maps a region Ω conformally onto a region Ω'; show that the area of Ω' is given by

$$\int \int_\Omega |f'(z)|^2 \, dx \, dy.$$

17. Suppose a function $w = f(z) \in H(D)$ maps the unit disc D conformally onto a region of area A. Show that

$$A \geq \pi |f'(0)|^2.$$

When does equality hold?

18. Suppose $z = x + iy$ moves in direction θ. Show that the corresponding direction ψ of the image under mapping $w = u(x, y) + iv(x, y) \in C^1$ is given by

$$\tan \psi = \frac{\dfrac{\partial v}{\partial x} + \dfrac{\partial v}{\partial y} \tan \theta}{\dfrac{\partial u}{\partial x} + \dfrac{\partial u}{\partial y} \tan \theta};$$

then deduce the Cauchy-Riemann equations under the assumption that $w = u + iv$ preserves both the size and the sense of every angle.

19. Let
$$\frac{\partial}{\partial z} = \frac{1}{2}\left(\frac{\partial}{\partial x} - i\frac{\partial}{\partial y}\right), \quad \frac{\partial}{\partial \bar{z}} = \frac{1}{2}\left(\frac{\partial}{\partial x} + i\frac{\partial}{\partial y}\right).$$

(a) Show that the Cauchy-Riemann equations can be expressed as
$$\frac{\partial}{\partial \bar{z}}f(z) = 0.$$

(b) Show that
$$\frac{\partial}{\partial z}z = 1, \quad \frac{\partial}{\partial \bar{z}}z = 0,$$
$$\frac{\partial}{\partial z}\bar{z} = 0, \quad \frac{\partial}{\partial \bar{z}}\bar{z} = 1.$$

(c) Show that, if $f(z)$ is analytic, then
$$f'(z) = \frac{\partial f}{\partial z}.$$

(d) Show that the *Laplace operator*
$$\nabla^2 = \frac{\partial^2}{\partial x^2} + \frac{\partial^2}{\partial y^2}$$

can be expressed as
$$\nabla^2 = 4\frac{\partial}{\partial z}\frac{\partial}{\partial \bar{z}} = 4\frac{\partial}{\partial \bar{z}}\frac{\partial}{\partial z}.$$

(e) Suppose $f(z)$ is analytic in a region Ω and never vanishes there; show that
$$\nabla^2\left(\log\{|f(z)|\}^2\right) = 0 \quad \text{for } z \in \Omega.$$

(f) (Chain rule) Suppose the partial derivatives (with respect to x and y) of $f(z)$ and $g(z)$ exist and $f \circ g$ is well defined. Show that
$$\frac{\partial}{\partial z}(f \circ g)(z) = \frac{\partial f}{\partial z}(g(z))\frac{\partial g}{\partial z}(z) + \frac{\partial f}{\partial \bar{z}}(g(z))\frac{\partial \bar{g}}{\partial z}(z),$$
$$\frac{\partial}{\partial \bar{z}}(f \circ g)(z) = \frac{\partial f}{\partial z}(g(z))\frac{\partial g}{\partial \bar{z}}(z) + \frac{\partial f}{\partial \bar{z}}(g(z))\frac{\partial \bar{g}}{\partial \bar{z}}(z).$$

(g) Show that, if either $f(z)$ or $g(z)$ is analytic, then

$$\frac{\partial}{\partial z}(f \circ g)(z) = \frac{\partial f}{\partial z}(g(z))\frac{\partial g}{\partial z}(z).$$

20. Show that the following six Möbius transformations form a group under composition:

$$z, \ \frac{1}{z}, \ 1 - z, \ \frac{1}{1-z}, \ \frac{z}{z-1}, \ \frac{z-1}{z}.$$

21. Show that the cross ratios corresponding to the 24 permutations of four points z_0, z_1, z_2, z_3 can have only the following six values:

$$\lambda = (z_0, z_1; z_2, z_3), \ \frac{1}{\lambda}, \ 1 - \lambda, \ \frac{1}{1-\lambda}, \ \frac{\lambda}{\lambda-1}, \ \frac{\lambda-1}{\lambda}.$$

Discuss the situation when at least two of these six values coincide.

22. (a) Show that a Möbius transformation that has three distinct fixed points must be the identity transformation.

(b) Suppose Möbius transformations S and T have the property that $Sz_i = Tz_i$ for three distinct points z_i $(i = 1, 2, 3)$; show that $S = T$.

23. Show that a Möbius transformation maps the family of all circles and straight lines *onto* itself.

24. Given four points A, B, C, D, let A', B', C', D' denote the centroids of the triangles BCD, ACD, ABD, ABC, respectively. Show that A', B', C', D' are cocyclic if and only if A, B, C, D are cocyclic.

25. Show that the four roots of

$$z^2 + 2a_1z + a_2 = 0 \quad \text{and} \quad z^2 + 2b_1z + b_2 = 0$$

are cocyclic if $2a_1b_1 = a_2 + b_2$.

26. Find the Möbius transformation that maps $-1, 0, 1$ to $-i, 1, i$, respectively.

27. Show that any four distinct cocyclic points can be mapped to ± 1 and $\pm k$ for some suitable k $(0 < k < 1)$ by a Möbius transformation.

28. (a) Verify the invariance of cross ratio for translations, dilations, and reciprocations by computation.

(b) Verify the symmetry principle for translations, dilations, and reciprocations by computation.

(c) Using the property that a Möbius transformation preserves cross ratios, show that a Möbius transformation maps a circle to a circle.

29. (a) Show that z_0 and z_0^* are symmetric with respect to the circle $C(a,\, r)$ if and only if

$$(\bar{z}_0 - \bar{a})(z_0^* - a) = r^2.$$

(b) Show that z_0 and z_0^* are symmetric with respect to the circumcircle of $\triangle z_1 z_2 z_3$ if and only if

$$(z_0^*,\, z_1;\, z_2,\, z_3) = \overline{(z_0,\, z_1;\, z_2,\, z_3)}.$$

(c) Use (b) to prove the symmetry principle.

30. If z_0 and z_0^* are symmetric with respect to a circle C and z_1, z_2 are two arbitrary points on C, show that the cross ratio $(z_0,\, z_0^*;\, z_1,\, z_2)$ has absolute value 1.

31. (a) Find the most general Möbius transformation that maps the right half-plane to the unit disc carrying the point 17 to the origin.

(b) Find a Möbius transformation that maps the right half-plane to the upper half-plane carrying the point $7 + 5i$ to $3i$.

32. (a) Show that any Möbius transformation that maps the real axis to itself can be written with real coefficients.

(b) Show that any Möbius transformation that maps the upper half-plane of the z-plane to the unit disc of the w-plane can be expressed in the form

$$w = \frac{\alpha z - \beta}{\bar{\alpha} z - \bar{\beta}},$$

where $\Im\left(\frac{\beta}{\alpha}\right) > 0$.

(c) Show that any Möbius transformation that maps the unit disc to itself can be expressed in the form

$$w = \frac{\alpha z - \beta}{\bar{\beta} z - \bar{\alpha}},$$

where $|\,\alpha\,| > |\,\beta\,|$.

33. Is there a Möbius transformation mapping the upper half-plane onto itself that interchanges two preassigned points in the upper half-plane? If so, how many such Möbius transformations are there?

34. (Poincaré)[3] Let $f(z)$ be analytic in the unit disc. Define $g(w) = f(z)$, where $w = Tz$ is a Möbius transformation mapping the unit disc conformally onto itself. Show that

$$\left(1 - |w|^2\right)\left|\frac{dg}{dw}\right| = \left(1 - |z|^2\right)\left|\frac{df}{dz}\right|.$$

35. Let $Tz = \frac{\alpha z + \beta}{\gamma z + \delta}$ be a Möbius transformation having two distinct fixed points a and b.

   (a) Show that the equation $w = Tz$ can also be written in the form

   $$\frac{w - a}{w - b} = k\frac{z - a}{z - b},$$

   where $k = \frac{\alpha - \gamma a}{\alpha - \gamma b}$ is a (nonzero) constant.

   (b) Show that the values of $\frac{dw}{dz}$ at the two fixed points are reciprocals of each other.

36. (a) Suppose z_1, z_2 are points in the complex plane whose stereographic images on the Riemann sphere are at opposite ends of a diameter. Show that $z_1 \bar{z}_2 = -1$.

   (b) Show that a Möbius transformation that defines a rotation of the Riemann sphere is given by

   $$\frac{w - \alpha}{1 + \bar{\alpha}w} = k\frac{z - \alpha}{1 + \bar{\alpha}z}, \quad |k| = 1.$$

   Show that they form a group.

37. Suppose a, b, c, d are integers and $\begin{vmatrix} a & b \\ c & d \end{vmatrix} = 1$. Show that the corresponding Möbius transformations form a group (known as the *unimodular group*). Show also that every Möbius transformation in this group maps the upper half-plane onto itself.

38. (a) Show that two circles tangent to each other can always be mapped by a suitable Möbius transformation $w = Tz$ to the pair of parallel lines $\Im w = \pm 1$.

---

[3] This equality is the starting point for Poincaré geometry.

(b) Show that two intersecting circles can always be mapped by a suitable Möbius transformation to two lines which intersect at the origin and which together are symmetric to the coordinate axes.

(c) Show that two nonintersecting circles can always be mapped by a suitable Möbius transformation to two concentric circles.

39. (a) Find a Möbius transformation that maps the pair of circles $C(1, 1)$ and $C(-1, 1)$ to the pair of circles $C(1, 1)$ and $C(2, 2)$.

(b) Find a Möbius transformation that maps the pair of circles $C(1, 2)$ and $C(-1, 2)$ to the pair of circles $C(0, 1)$ and $C(\sqrt{3}, 2)$.

(c) Find a Möbius transformation that maps the pair of circles $C(0, 1)$ and $C(4, 2)$ to a pair of concentric circles.

Chapter 4

The Cauchy Theorem

4.1 Some Remarks on Curves

In this chapter we establish, among other things, that analytic functions are exactly those which have local power series expansions. *The* royal road toward this objective uses integrals. However, a fully rigorous development of complex integration would take too much space and time in an introductory course, and so we give a somewhat intuitive account.

In Chapter 3, we defined an arc as a continuous mapping of a closed bounded interval into $\mathbb{C}$. This definition turned out to be too general. In 1890, Peano showed that it is possible to map an interval continuously onto a whole square. This discovery led to some consternation among those who felt that a curve is an intuitively obvious object. His example is not a homeomorphism; that is, some pairs of distinct numbers in the interval are mapped into the same points of $\mathbb{C}$. On the other hand, Brouwer proved that there is no *one-to-one* continuous mapping of an interval onto a square. Thus we define a *simple* arc as a continuous mapping of a closed bounded interval into $\mathbb{C}$ that is one-to-one, except that the endpoints of the interval *may* be mapped to the same point; when this occurs we speak of a *simple closed curve* (or *Jordan curve*).

EXAMPLE. A line segment $\Gamma : z(t) = \alpha + \beta t$ $(0 \le t \le 1)$, where α, $\beta \in \mathbb{C}$, is a simple arc (if $\beta \ne 0$).

EXAMPLE. A circle $z(t) = a + re^{it}$ $(0 \le t \le 2\pi)$, where $a \in \mathbb{C}$ and $r \in \mathbb{R}_+$, is a simple closed arc.

Several remarks are in order:

(a) The same subset of $\mathbb{C}$ can be parameterized as an arc in many ways:

EXAMPLE.

1^0 $z_1(t) = (1 + 2i) + (1 + i\sqrt{3})t,$ $0 \le t \le 4$;

2^0 $z_2(t) = (1 + 2i) + 4(1 + i\sqrt{3})t,$ $0 \le t \le 1$;

3^0 $z_3(t) = (1 + 2i) + 2t^2 e^{i\frac{\pi}{3}},$ $0 \le t \le 2$.

In all three the same set of points is traversed. Thus, strictly speaking, an arc is an equivalence class of continuous mappings; that is, $z = z_1(t)$ $(a \le t \le b)$ and $z = z_2(s)$ $(c \le s \le d)$ represent the same arc if there is a continuous, strictly monotone increasing function $\varphi(t)$, mapping the interval $[a, b]$ onto $[c, d]$, such that $z_1(t) = z_2(\varphi(t))$ $(a \le t \le b)$. In particular, for closed arcs we may shift the parameter to get rid of the distinguished position of the initial point. For example, circle $C(a, r)$ may be written $z(t) = a + re^{it}$ $(0 \le t \le 2\pi)$, or $z(t) = a + re^{it}$ $(-\pi \le t \le \pi)$. Although it is not quite correct, we tend to use the words *arc* and *trace* $\{z(t)\,;\, a \le t \le b\}$ interchangeably when there is no danger of confusion.

(b) An arc has an orientation. For example, in (1^0), the initial point is $z_0 = 1 + 2i$ and the terminal point is $z_1 = 5 + 2(1 + 2\sqrt{3})i$. But if we change t to $-t$, and let t run from -4 to 0, we obtain the same trace in reverse order. We denote the arc with opposite sense $-\Gamma$. For example, if $\Gamma : z = z(t)$ $(a \le t \le b)$, then $-\Gamma : z = z(-t)$, $(-b \le t \le -a)$.

(c) Any finite number of arcs can be *added* to get a new arc if the terminal point of the kth arc coincides with the initial point of the $(k + 1)$st.

The *length* of an arc is defined as the supremum of all sums

$$\sum_{k=1}^{n} |\, z(t_k) - z(t_{k-1})\,|,$$

where $a = t_0 < t_1 < t_2 < \cdots < t_{n-1} < t_n = b$, $n \in \mathbb{N}$. If this supremum is finite, then we say that the arc is *rectifiable*. Clearly, the definition of the arclength is independent of the parameterization. The *sum* of a finite number of rectifiable arcs is rectifiable, and the length of the sum is the sum of the lengths of the pieces. Note that there exist infinitely long arcs (Figure 4.1):

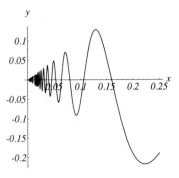

Figure 4.1

EXAMPLE.

$$\Gamma : z(t) = \begin{cases} t\left(1 + i \sin \dfrac{1}{t}\right), & (0 < t \le 1); \\ 0, & (t = 0). \end{cases}$$

An arc $z = z(t)$ $(a \le t \le b)$ is *smooth* if it can be parameterized by a function $z(t)$ which is *continuously differentiable* and which satisfies the condition that $z'(t) \ne 0$ $(a \le t \le b)$. If these conditions are satisfied, then the arc will have a tangent at all points. An arc is *piecewise smooth* if it is the sum of a finite number of smooth arcs.

A (piecewise) smooth arc is rectifiable, and its length is given by

$$\int_a^b |\, z'(t)\, |\, dt.$$

The chain rule implies that this integral is independent of the parameterization employed.

EXAMPLE. The length of circle $C(a, r) : z(t) = a + re^{it}$ $(0 \le t \le 2\pi)$ is

$$\int_0^{2\pi} |\, z'(t)\, |\, dt = \int_0^{2\pi} |\, ire^{it}\, |\, dt = r \int_0^{2\pi} dt = 2\pi r.$$

4.2   Line Integrals

For the rest of this chapter, *all arcs that we consider are rectifiable*. Let $f(z)$ be a function defined on an arc Γ (or in a set containing Γ). Further, let $[\alpha = z_0, z_1, \ldots, z_n = \beta]$ be $n + 1$ distinct points on Γ (z_k is situated on the section of Γ that joins z_{k-1} and z_{k+1}). Consider a *Riemann sum*

$$\sum_{k=1}^{n} f(\zeta_k)(z_k - z_{k-1}),$$

where ζ_k is an arbitrary point on the section Γ_k of Γ that joins z_{k-1} and z_k. If this sum tends to a limit independent of the choice of ζ's as the maximum of the lengths of subarcs tends to zero then this limit is called the *integral* of $f(z)$ along Γ and is denoted

$$\int_{\Gamma} f(z)\,dz.$$

Note that the definition of the integral does not depend on the parameterization of the arc Γ. The existence of the integral for every *continuous* integrand $f(z)$ can be shown by separating the Riemann sums into real and imaginary parts and appealing to the real case. From now on, *all integrands are assumed to be continuous*.

Below are some elementary properties of line integrals that follow immediately from the definition:

(a) $\displaystyle\int_{\Gamma} \{f(z) + g(z)\}\,dz = \int_{\Gamma} f(z)\,dz + \int_{\Gamma} g(z)\,dz.$

(b) $\displaystyle\int_{\Gamma} k f(z)\,dz = k \int_{\Gamma} f(z)\,dz \quad (k \in \mathbb{C}).$

(c) Reversing the sense of the arc reverses the sign of the integral:

$$\int_{-\Gamma} f(z)\,dz = - \int_{\Gamma} f(z)\,dz.$$

(d) If Γ is the sum of a set of arcs Γ_k $(k = 1, 2, \ldots, n)$, then

$$\int_{\Gamma} f(z)\,dz = \sum_{k=1}^{n} \int_{\Gamma_k} f(z)\,dz.$$

Consequently, if Γ is a closed arc then $\int_{\Gamma}$ is the same for any initial point.

(e) If L is the length of an arc Γ and $M = \sup\{|f(z)| \, ; \, z \in \Gamma\}$, then

$$\left| \int_{\Gamma} f(z)\,dz \right| \leq ML.$$

The proof is similar to that of the real case. Indeed, by the triangle inequality, we have

$$\left| \sum_{k=1}^{n} f(\zeta_k)(z_k - z_{k-1}) \right| \leq \sum_{k=1}^{n} |f(\zeta_k)| \cdot |z_k - z_{k-1}|$$
$$\leq M \sum_{k=1}^{n} |z_k - z_{k-1}| \leq ML,$$

because $\sum_{k=1}^{n} |z_k - z_{k-1}|$ is the length of the inscribed polygon. Now pass to the limit.

(f) More precisely,

$$\left| \int_{\Gamma} f(z)\, dz \right| \leq \int_{\Gamma} |f(z)| \cdot |dz|,$$

where the second integral is with respect to arc length.

Proof.

$$\left| \sum_{k=1}^{n} f(\zeta_k)(z_k - z_{k-1}) \right| \leq \sum_{k=1}^{n} |f(\zeta_k)| \cdot |z_k - z_{k-1}|$$
$$\leq \sum_{k=1}^{n} |f(\zeta_k)| \cdot (s_k - s_{k-1}),$$

where $|z_k - z_{k-1}|$ is the length of the chord of the subarc Γ_k and $(s_k - s_{k-1})$ is the arclength of Γ_k. Taking the limit, we obtain the desired result. □

(g) If $\{f_n(z)\}_{n=1}^{\infty}$ is a sequence of continuous functions converging uniformly to $f(z)$ on Γ, then

$$\int_{\Gamma} f(z)\, dz = \lim_{n \to \infty} \int_{\Gamma} f_n(z)\, dz.$$

Proof. Uniform convergence means that for arbitrary $\epsilon > 0$ there exists an integer N such that

$$|f(z) - f_n(z)| < \epsilon \quad \text{for all } n \geq N \text{ and all } z \in \Gamma.$$

It follows from (e) that

$$\left| \int_{\Gamma} f(z)\, dz - \int_{\Gamma} f_n(z)\, dz \right| = \left| \int_{\Gamma} \{f(z) - f_n(z)\}\, dz \right| < \epsilon \int_{\Gamma} |dz|$$
$$= \epsilon L,$$

where L is the length of Γ. □

As an immediate consequence, we obtain

(h) If $\sum_{n=1}^{\infty} f_n(z)$ converges uniformly to $f(z)$ on Γ, then termwise integration is permissible:

$$\int_{\Gamma} f(z)\, dz = \sum_{n=1}^{\infty} \int_{\Gamma} f_n(z)\, dz.$$

(i) If $f(z) = u(z) + iv(z)$, where u and v are real-valued functions, then

$$\int_{\Gamma} f(z)\, dz = \int_{\Gamma} \{u(z)\, dx - v(z)\, dy\} + i \int_{\Gamma} \{v(z)\, dx + u(z)\, dy\}$$

$$(z = x + iy).$$

This follows from the analogous relation for Riemann sums:

$$\sum_{k=1}^{n} f(\zeta_k)(z_k - z_{k-1})$$

$$= \sum_{k=1}^{n} \{u(\zeta_k) + iv(\zeta_k)\} \cdot (x_k + iy_k - x_{k-1} - iy_{k-1})$$

$$= \sum_{k=1}^{n} \{u(\zeta_k)(x_k - x_{k-1}) - v(\zeta_k)(y_k - y_{k-1})\}$$

$$+ i \sum_{k=1}^{n} \{v(\zeta_k)(x_k - x_{k-1}) + u(\zeta_k)(y_k - y_{k-1})\}.$$

(j) In particular, if $\Gamma : z(t) = x(t) + iy(t)$ $(a \le t \le b)$ is a smooth arc, then the Riemann sum above becomes

$$\sum_{k=1}^{n} \{u(\zeta_k)x'(t_k^{(1)}) - v(\zeta_k)y'(t_k^{(2)})\}(t_k - t_{k-1})$$

$$+ i \sum_{k=1}^{n} \{v(\zeta_k)x'(t_k^{(3)}) + u(\zeta_k)y'(t_k^{(4)})\}(t_k - t_{k-1}),$$

where the values $t_k^{(1)}, \ldots, t_k^{(4)}$ are obtained from the mean-value theorem. Thus, taking the limit of the Riemann sum above as four separate terms (or applying the *Duhamel principle* in calculus), we obtain

$$\int_\Gamma f(z)\,dz = \int_a^b \{u(z)x'(t) - v(z)y'(t)\}\,dt$$
$$+ i \int_a^b \{v(z)x'(t) + u(z)y'(t)\}\,dt$$
$$= \int_a^b f(z(t)) \cdot z'(t)\,dt.$$

We shall need the following

LEMMA 4.2.1 *Let $f(z)$ be a continuous function on a region Ω, and $\Gamma : z = z(t)$ $(a \le t \le b)$ be a rectifiable arc with its trace in Ω. (We denote the initial and the terminal points α and β, respectively.) Then for any $\epsilon > 0$ there exists $\delta > 0$ such that, for any polygon $\Gamma' : [\alpha = z_0, z_1, \ldots, z_n = \beta]$ with vertices z_k on Γ and $\eta = \max\{|z_k - z_{k-1}|; k = 1, 2, \ldots, n\} < \delta$, the relation*

$$\left| \int_\Gamma f(z)\,dz - \int_{\Gamma'} f(z)\,dz \right| < \epsilon$$

holds. Namely, $\int_\Gamma f(z)\,dz$ can be approximated arbitrary closely by integration over a polygon Γ' sufficiently close to Γ. (Note that for η sufficiently small Γ' is entirely in Ω.)

   Proof. Let d_0 denote the distance between the trace of Γ and the complement of Ω:

$$d_0 = \inf\{|z(t) - w|; \ a \le t \le b, \ w \notin \Omega\},$$

and K denote a closed strip around Γ that is in Ω:

$$K = \{z' \in \mathbb{C}; \ |z' - z(t)| \le \frac{d_0}{2} \quad \text{for some } t \in [a, b]\}.$$

Clearly, $d_0 > 0$ (Exercises 2.9–2.10) and K is compact; therefore, $f(z)$ is uniformly continuous on K. It follows that, given $\epsilon > 0$, there exists $\delta > 0$ such that

$$|f(z) - f(z')| < \frac{\epsilon}{2L} \quad \text{whenever } z, z' \in K \text{ and } |z - z'| < \delta,$$

where L is the length of the arc Γ. We may assume that $\delta < \frac{d_0}{2}$.

   Now, take partition points $\alpha = z_0, z_1, \ldots, z_n = \beta$ on Γ such that the maximum, σ, of the lengths σ_k of the subarcs $\Gamma_k = (z_{k-1}, z_k)$ is less than δ. Then the polygon $\Gamma' : [\alpha = z_0, z_1, \ldots, z_n = \beta]$ is in K, and each chord Γ'_k corresponding to Γ_k has length $|z_k - z_{k-1}| \leq \sigma < \delta$. Therefore, for an arbitrary point z on Γ_k we have $|z - z_k| < \delta$, and hence $|f(z) - f(z_k)| < \frac{\epsilon}{2L}$ for $z \in \Gamma_k$. Thus, using (f), we obtain

$$\left| \int_{\Gamma_k} f(z)\,dz - \int_{\Gamma_k} f(z_k)\,dz \right| \leq \int_{\Gamma_k} |f(z) - f(z_k)| \cdot |dz| < \frac{\epsilon}{2L}\sigma_k.$$

From the definition of the integral,

$$\int_{\Gamma_k} f(z_k)\,dz = f(z_k) \int_{\Gamma_k} dz = f(z_k)(z_k - z_{k-1}).$$

$$\therefore \ \left| \int_{\Gamma_k} f(z)\,dz - f(z_k)(z_k - z_{k-1}) \right| < \frac{\epsilon}{2L}\sigma_k.$$

Thus

$$\left| \int_{\Gamma} f(z)\,dz - \sum_{k=1}^{n} f(z_k)(z_k - z_{k-1}) \right|$$

$$= \left| \sum_{k=1}^{n} \int_{\Gamma_k} f(z)\,dz - \sum_{k=1}^{n} f(z_k)(z_k - z_{k-1}) \right|$$

$$\leq \sum_{k=1}^{n} \left| \int_{\Gamma_k} f(z)\,dz - f(z_k)(z_k - z_{k-1}) \right|$$

$$< \frac{\epsilon}{2L} \cdot \sum_{k=1}^{n} \sigma_k = \frac{\epsilon}{2}.$$

Similarly, integrating over Γ' instead of Γ, we obtain

$$\left| \int_{\Gamma'} f(z)\,dz - \sum_{k=1}^{n} f(z_k)(z_k - z_{k-1}) \right| < \frac{\epsilon}{2}.$$

We have used the fact that the length of the polygon Γ' is not greater than that of Γ. Hence

$$\left| \int_{\Gamma} f(z)\,dz - \int_{\Gamma'} f(z)\,dz \right| \leq \left| \int_{\Gamma} f(z)\,dz - \sum_{k=1}^{n} f(z_k)(z_k - z_{k-1}) \right|$$

$$+ \left| \sum_{k=1}^{n} f(z_k)(z_k - z_{k-1}) - \int_{\Gamma'} f(z)\,dz \right|$$

$$< \frac{\epsilon}{2} + \frac{\epsilon}{2} = \epsilon. \qquad \qquad \square$$

EXAMPLE. Let us compute $\int_\Gamma z^n dz$ $(n \geq 0)$, where $\Gamma = [\alpha, \beta]$ is the line segment joining α to β; that is,

$$\Gamma : z = \alpha + t(\beta - \alpha), \quad (0 \leq t \leq 1).$$

Then

$$\int_\Gamma z^n dz = \int_0^1 \{\alpha + t(\beta - \alpha)\}^n \cdot (\beta - \alpha)\, dt$$

$$= \frac{1}{n+1} \{\alpha + t(\beta - \alpha)\}^{n+1} \Big|_0^1 = \frac{\beta^{n+1} - \alpha^{n+1}}{n+1}.$$

Consequently, if $\triangle$ is the triangle with vertices at α, β, γ, then, integrating along its boundary $\partial\triangle$, we obtain

$$\int_{\partial\triangle} z^n dz = \frac{\beta^{n+1} - \alpha^{n+1}}{n+1} + \frac{\gamma^{n+1} - \beta^{n+1}}{n+1} + \frac{\alpha^{n+1} - \gamma^{n+1}}{n+1} = 0.$$

More generally, if $p(z)$ is a polynomial, then

$$\int_{\partial\triangle} p(z)\, dz = 0,$$

where $\partial\triangle$ is the boundary of an arbitrary triangle $\triangle$. The same is true if $\partial\triangle$ is replaced by the boundary of a (simple) closed polygon. This result suggests a number of questions: How far can we extend this result? Can we replace the polygon by a piecewise smooth arc, or even a rectifiable arc? Can we replace the polynomial by a power series? It turns out that answers to all these questions are affirmative, even if asked together; this will be the topic of Section 4.3.

4.3   The Cauchy-Goursat Theorem

THEOREM 4.3.1 (Cauchy-Goursat) *Let* $f(z) \in H(\Omega)$. *If a simple closed arc* Γ *together with its interior is in region* Ω, *then*

$$\int_\Gamma f(z)\, dz = 0.$$

In conjunction with the Cauchy theorem, Jordan raised the question: What is the interior of a closed curve? His answer (for *simple* closed curves) was: Remove the simple closed curve from the plane. The remainder (which is easily shown to be open) possesses two components, of which one is bounded and the other is unbounded. By definition, the bounded component is the *interior* and the unbounded component is the *exterior*.

Cauchy proved the theorem under additional assumptions. He began with

$$\int_{\Gamma} f(z)\,dz = \int_{\Gamma}(u + iv)(dx + i\,dy)$$
$$= \int_{\Gamma}(u\,dx - v\,dy) + i\int_{\Gamma}(v\,dx + u\,dy).$$

Now two applications of the Green theorem

$$\int_{\Gamma} P\,dx + Q\,dy = \int\int_{(\Gamma)}\left(\frac{\partial Q}{\partial x} - \frac{\partial P}{\partial y}\right)dx\,dy$$

enable us to write

$$\int_{\Gamma} f(z)\,dz = -\int\int_{(\Gamma)}\left(\frac{\partial v}{\partial x} + \frac{\partial u}{\partial y}\right)dx\,dy + i\int\int_{(\Gamma)}\left(\frac{\partial u}{\partial x} - \frac{\partial v}{\partial y}\right)dx\,dy,$$

and the Cauchy-Riemann equations are equivalent to the identical vanishing of the integrands over the region enclosed by Γ (denoted (Γ) here). Thus

$$\int_{\Gamma} f(z)\,dz = 0.$$

However, the applicability of the Green theorem requires certain smoothness conditions on Γ, u, and v; in particular, the continuity of the partial derivatives of u and v is needed. About 1900, Goursat showed that this assumption can be completely avoided; indeed, it becomes a *consequence* of his version of the Cauchy theorem!

Before proving Theorem 4.3.1, we perform the preliminary simplification of confining the proof to the case that Γ is a simple closed polygon. Thanks to Lemma 4.2.1, $|\int_{\Gamma} - \int_{\Gamma'}|$ can be made arbitrarily small when Γ' is a suitably chosen polygon, and so if we can prove that $\int_{\Gamma'} = 0$ we shall actually have proven the theorem for the more general case. Of course Γ' has to be in Ω, but this can be achieved by taking the partition points sufficiently close. Also, it may happen that polygon Γ' has "double points" and/or "double segments"; however, for a polygon, the number of double points is at most finite, and double segments are traversed twice in opposite directions, and so the integrals over double segments cancel themselves out. In any case, Γ' can be assumed to be the union of a finite number of simple closed polygons (Figure 4.2).

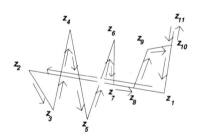
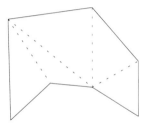

Figure 4.2

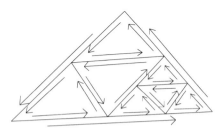

Figure 4.3

Thus it is sufficient to prove the theorem for an arbitrary simple closed polygon. Now if this polygon is convex we can obviously divide it into a finite number of triangles. If the polygon is concave we first divide it into a finite number of convex polygons, and then proceed to divide the convex polygons into a finite number of triangles.

Thus, we may assume that Γ is the boundary of a triangle and it is sufficient to prove

$$\int_{\partial \triangle} f(z)\, dz = 0,$$

where $f(z)$ is analytic on the triangle $\triangle$.

Let

$$M = \left| \int_{\partial \triangle} f(z)\, dz \right|.$$

We want to show that $M = 0$. The proof is based on the Cantor Intersection Theorem 2.3.2. First, divide the triangle $\triangle$ into four congruent triangles $\triangle_1^{(1)}, \triangle_1^{(2)}, \triangle_1^{(3)}, \triangle_1^{(4)}$, by joining the midpoints of the sides (Figure 4.3). Then the sum of the integrals of $f(z)$ along the boundaries of these four triangles, traversed in the same sense, is the same as the integral along the boundary of the original triangle. We have

$$\int_{\partial\triangle} = \int_{\partial\triangle_1^{(1)}} + \int_{\partial\triangle_1^{(2)}} + \int_{\partial\triangle_1^{(3)}} + \int_{\partial\triangle_1^{(4)}},$$

because the integrals over the common sides cancel each other. Choose one of these four integrals over the small triangles with the largest absolute value, and denote this triangle $\triangle_1$. Then

$$M = \left|\int_{\partial\triangle}\right| \leq 4 \left|\int_{\partial\triangle_1}\right|; \quad \text{i.e.,} \quad \left|\int_{\partial\triangle_1}\right| \geq \frac{M}{4}.$$

Now, dividing the triangle $\triangle_1$ in the same manner, and choosing $\triangle_2$ as $\triangle_1$ was chosen, we have $\left|\int_{\partial\triangle_2}\right| \geq \frac{M}{4^2}$. Continuing this procedure, we obtain

$$\left|\int_{\partial\triangle_n}\right| \geq \frac{M}{4^n} \quad \text{(for all } n \in \mathbb{N}).$$

Each triangle $\triangle_n$ (including its interior) is a compact set contained in the preceding triangle $\triangle_{n-1}$, and its diameter is less than $\frac{L}{2^{n+1}}$, where L is the length of the perimeter of the original triangle $\triangle$. Therefore, the Cantor Intersection Theorem 2.3.2 is applicable, and the triangles $\triangle_n$ converge to a point z_0, $\{z_0\} = \bigcap_{n=1}^{\infty} \triangle_n$, where z_0 belongs to $\triangle$ and, *a fortiori*, to Ω. By assumption, $f(z)$ is analytic at z_0, and therefore, given $\epsilon > 0$, there exists $\delta > 0$ such that

$$\left|f(z) - \{f(z_0) + f'(z_0)(z - z_0)\}\right| \leq \epsilon|z - z_0| \quad \text{if } z \in D(z_0, \delta).$$

Now, for sufficiently large n, $\triangle_n$ is entirely in $D(z_0, \delta)$, and therefore, for such n,

$$\left|\int_{\partial\triangle_n} [f(z) - \{f(z_0) + f'(z_0)(z - z_0)\}]\, dz\right| \leq \epsilon \int_{\partial\triangle_n} |z - z_0| \cdot |dz|.$$

Because

$$\int_{\partial\triangle_n} \{f(z_0) + f'(z_0)(z - z_0)\}\, dz = 0,$$

by the example at the end of Section 4.2, and as

$$|z - z_0| < L_n = \frac{L}{2^n}$$

for z on the boundary $\partial \triangle_n$ of the triangle $\triangle_n$ (L_n is the length of the perimeter of $\triangle_n$), we have

$$\left| \int_{\partial \triangle_n} f(z)\, dz \right| < \frac{\epsilon L}{2^n} \int_{\partial \triangle_n} |dz| = \frac{\epsilon L^2}{2^{2n}}.$$

Combining these results all together we obtain the estimates

$$\frac{M}{4^n} \leq \left| \int_{\partial \triangle_n} f(z)\, dz \right| < \frac{\epsilon L^2}{2^{2n}},$$

for large values of n. Hence $M < \epsilon L^2$, and because $\epsilon > 0$ is arbitrary, we conclude that $M = 0$.

EXAMPLE. Let $f(z) = \frac{1}{(z-a)^n}$ ($n \in \mathbb{N}$, $a \in \mathbb{C}$). Then $f(z)$ is analytic except at $z = a$. Let Γ be the circle $C(a, r)$. Because $C(a, r)$ can be represented parametrically as $z = a + re^{i\theta}$ ($0 \leq \theta \leq 2\pi$), we have

$$z - a = re^{i\theta}, \quad dz = ire^{i\theta}\, d\theta. \quad \therefore \quad \frac{dz}{z - a} = i\, d\theta.$$

(a) If $n = 1$, then

$$\int_{C(a,r)} \frac{dz}{z - a} = i \int_0^{2\pi} d\theta = 2\pi i,$$

which is not zero; note that $f(z) = \frac{1}{z-a}$ is not analytic at the point $z = a$ interior to $C(a, r)$.

(b) If $n > 1$, then

$$\int_{C(a,r)} \frac{dz}{(z - a)^n} = \int_{C(a,r)} \frac{1}{(z - a)^{n-1}} \cdot \frac{dz}{z - a}$$

$$= i \int_0^{2\pi} \frac{d\theta}{r^{n-1} e^{i(n-1)\theta}}$$

$$= \frac{i}{r^{n-1}} \int_0^{2\pi} e^{-i(n-1)\theta}\, d\theta$$

$$= 0.$$

Although $f(z)$ is not analytic at $z = a$, we still obtain (curiously)

$$\int_{C(a,r)} f(z)\, dz = 0.$$

Of course this result does not contradict the theorem, for the theorem says nothing about the situation if $f(z)$ is not analytic at even one point inside Γ.

Note that in all cases the integral does not depend on the radius r.

4.4   The Cauchy Integral Formula

Let Γ be a simple closed arc. Extending the notations used for real intervals, we define by (Γ) the region enclosed by Γ, and by $[\Gamma]$ its closure; that is, $[\Gamma] = (\Gamma) \cup \Gamma$. The boundary of a region is said to be traversed in the *positive sense* if an observer traveling in this direction has the region to the left.

LEMMA 4.4.1 *Let Γ_1 and Γ_2 be two simple closed arcs with the same sense, and $\Gamma_1 \subset (\Gamma_2)$. If $f(z)$ is analytic on the union of the curvilinear annulus between Γ_1 and Γ_2 and on the boundaries, Γ_1 and Γ_2, then*

$$\int_{\Gamma_1} f(z)\, dz = \int_{\Gamma_2} f(z)\, dz.$$

Proof. If we join Γ_1 and Γ_2 by two mutually nonintersecting arcs C_1 and C_2, then the curvilinear annulus will be divided into two regions, Ω_1 and Ω_2, say.

Using the notation in Figure 4.4, the Cauchy-Goursat Theorem 4.3.1 gives

$$\int_{ABC} + \int_{CE} + \int_{EFG} + \int_{GA} = \int_{\partial\Omega_1} = 0,$$

and

$$\int_{CDA} + \int_{AG} + \int_{GHE} + \int_{EC} = \int_{\partial\Omega_2} = 0.$$

Adding, we obtain

$$\left(\int_{ABC} + \int_{CDA}\right) + \left(\int_{CE} + \int_{EC}\right) + \left(\int_{EFG} + \int_{GHE}\right)$$

$$+ \left(\int_{GA} + \int_{AG}\right) = 0.$$

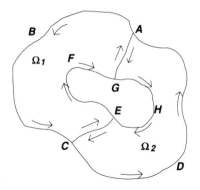
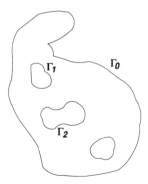

Figure 4.4

Because $\int_{CE} + \int_{EC} = 0$ and $\int_{GA} + \int_{AG} = 0$, we obtain the desired result,

$$\int_{\Gamma_1} - \int_{\Gamma_2} = 0. \qquad\qquad \square$$

COROLLARY 4.4.2 *Let $\Gamma_0, \Gamma_1, \ldots, \Gamma_n$ be simple closed arcs with positive orientation. Suppose Γ_j is in the interior of Γ_0 but in the exterior of Γ_k for $j, k = 1, 2, \ldots, n, j \neq k$. If $f(z)$ is analytic on the region bounded by these arcs and the boundaries, then*

$$\int_{\Gamma_0} f(z)\, dz = \sum_{k=1}^{n} \int_{\Gamma_k} f(z)\, dz.$$

THEOREM 4.4.3 (Cauchy Integral Formula) *Suppose $f(z)$ is analytic in $[\Gamma]$, where Γ is a simple closed arc. Then*

$$f(a) = \frac{1}{2\pi i} \int_{\Gamma} \frac{f(z)}{z - a}\, dz \quad a \in (\Gamma),$$

where the integral is taken in the positive sense.

Proof. Although the integrand $\frac{f(z)}{z-a}$ is not continuous at a, it is analytic in the curvilinear annulus between Γ and the circle $C(a, r)$, where $0 < r < \mathrm{dist}(a, \Gamma) = \inf\{|z - a| \, ; \, z \in \Gamma\}$. Therefore, by the corollary, we have

$$\int_{\Gamma} \frac{f(z)}{z - a}\, dz = \int_{C(a,r)} \frac{f(z)}{z - a}\, dz.$$

In particular, $\int_{C(a,r)} \frac{f(z)}{z-a} dz$ is independent of radius r (as long as r is small enough so that $\bar{D}(a, r) \subset (\Gamma)$). Choose a constant M such that $|f'(a)| < M$. Then, for z sufficiently close to a, $\left| \frac{f(z)-f(a)}{z-a} \right| < M$. By the example at the end of Section 4.3, we have[1]

$$f(a) = \frac{f(a)}{2\pi i} \int_{C(a,r)} \frac{dz}{z-a}.$$

It now follows that

$$\left| \frac{1}{2\pi i} \int_{\Gamma} \frac{f(z)}{z-a} dz - f(a) \right| = \left| \frac{1}{2\pi i} \int_{C(a,r)} \frac{f(z)-f(a)}{z-a} dz \right|$$

$$\leq \frac{1}{2\pi} \int_{C(a,r)} \left| \frac{f(z)-f(a)}{z-a} \right| \cdot |dz|$$

$$< \frac{M}{2\pi} \int_{C(a,r)} |dz| = \frac{M}{2\pi} \cdot 2\pi r = Mr.$$

Because r can be arbitrary small, we obtain the desired result. $\qquad\square$

The Cauchy Integral Formula 4.4.3 shows a truly remarkable feature of analytic functions. If $f(z)$ is analytic inside and on a simple closed arc then the value of $f(z)$ at any point a inside this arc is completely determined by the values of $f(z)$ *on* the arc.

We can now tie together the Cauchy-Riemann approach and the power-series approach of Weierstrass.

THEOREM 4.4.4 *Suppose* $f(z) \in H(\Omega)$, *and let* a *be an arbitrary point in the region* Ω. *Then there exists a power series about the point* a *that converges to* $f(z)$ *for all* $z \in D(a, R)$, *where* $R = dist(a, \partial\Omega)$.

Proof. Let $|z - a| = \rho < r < R$. Then $z \in D(a, r)$ and $\bar{D}(a, r) \subset \Omega$. If ζ is a point on the circle $C(a, r)$, then $\left| \frac{z-a}{\zeta-a} \right| = \frac{\rho}{r} < 1$. Hence

$$\frac{1}{\zeta - z} = \frac{1}{(\zeta - a) - (z - a)} = \frac{1}{\zeta - a} \cdot \frac{1}{1 - \dfrac{z-a}{\zeta-a}}$$

---

[1] One of the most frequently used tricks in mathematics is to express 1 in a fancy way. In our case, we have

$$1 = \frac{1}{2\pi i} \int_{C(a,r)} \frac{dz}{z-a}.$$

$$= \frac{1}{\zeta - a} + \frac{z - a}{(\zeta - a)^2} + \frac{(z - a)^2}{(\zeta - a)^3} + \cdots + \frac{(z - a)^k}{(\zeta - a)^{k+1}} + \cdots.$$

Therefore,

$$\frac{f(\zeta)}{\zeta - z} = \frac{f(\zeta)}{\zeta - a} + (z - a)\frac{f(\zeta)}{(\zeta - a)^2} + (z - a)^2\frac{f(\zeta)}{(\zeta - a)^3} + \cdots$$
$$+ (z - a)^k\frac{f(\zeta)}{(\zeta - a)^{k+1}} + \cdots,$$

where the convergence is uniform for ζ on $C(a, r)$. Indeed, if we denote

$$M = \max\{|f(\zeta)| \, ; \, \zeta \in C(a, r)\},$$

then

$$\left|(z - a)^k\frac{f(\zeta)}{(\zeta - a)^{k+1}}\right| \le \frac{M}{r}\left(\frac{\rho}{r}\right)^k, \quad \text{and} \quad \sum_{k=0}^{\infty}\left\{\frac{M}{r}\left(\frac{\rho}{r}\right)^k\right\} < \infty.$$

Hence, by the Weierstrass M-test 2.4.5, the convergence is uniform for ζ on $C(a, r)$, as stated above. Therefore, by Property (h) of Section 4.2, we may integrate termwise (with respect to ζ), obtaining

$$f(z) = \frac{1}{2\pi i}\int_{C(a,r)}\frac{f(\zeta)}{\zeta - z}d\zeta$$
$$= \sum_{k=0}^{\infty}\left\{\frac{(z - a)^k}{2\pi i}\int_{C(a,r)}\frac{f(\zeta)}{(\zeta - a)^{k+1}}d\zeta\right\} = \sum_{k=0}^{\infty}A_k(z - a)^k,$$

where

$$A_k = \frac{1}{2\pi i}\int_{C(a,r)}\frac{f(\zeta)}{(\zeta - a)^{k+1}}d\zeta \quad (k = 0, 1, 2, \ldots).$$

Thus, we obtain a Taylor-series representation of $f(z)$ around a. $\square$

Remark. As a consequence, results such as the Uniqueness Theorem 2.7.3 and the Maximum and Minimum Modulus Principles 2.10.1, which we proved for power series, are now valid for all analytic functions.

COROLLARY 4.4.5 *On the circle of convergence of a power series, there exists at least one point that does not belong to the region of analyticity of the function defined by the power series.*

Combining this theorem with Corollary 2.8.3 (and the Uniqueness Theorem 2.7.3 for power series), we obtain

THEOREM 4.4.6 (Goursat 1900) *If $f(z) \in H(\Omega)$, then $f(z)$ has derivatives of all orders, and $f^{(k)}(z) \in H(\Omega)$. Moreover,*

$$f^{(k)}(a) = \frac{k!}{2\pi i} \int_{\Gamma} \frac{f(\zeta)}{(\zeta - a)^{k+1}} d\zeta \quad (k = 0, 1, 2, \ldots),$$

where Γ is a simple closed positively oriented arc such that $a \in (\Gamma)$ and $[\Gamma] \subset \Omega$.

This extension of the Cauchy integral formula to $f^{(k)}(z)$ is a remarkable result. We started out by assuming only that the first derivative exists, and this assumption is now shown to imply that the derivatives of *all* orders exist. In calculus, the existence of the derivative does not guarantee the continuity, much less the differentiability, of the derivative.

As an immediate corollary to the Goursat Theorem 4.4.6, we establish the useful *Cauchy estimates.*

THEOREM 4.4.7 (Cauchy) *Suppose $f(z)$ is analytic on the closed disc $\bar{D}(a, r)$ and that there exists a constant M such that $|f(z)| \leq M$ for $z \in C(a, r)$. Then*

$$|f^{(k)}(a)| \leq \frac{k! M}{r^k} \quad (k = 0, 1, 2, \ldots).$$

Proof.

$$\begin{aligned}
|f^{(k)}(a)| &= \left| \frac{k!}{2\pi i} \int_{C(a,r)} \frac{f(\zeta)}{(\zeta - a)^{k+1}} d\zeta \right| \\
&\leq \frac{k!}{2\pi} \int_{C(a,r)} \left| \frac{f(\zeta)}{(\zeta - a)^{k+1}} \right| \cdot |d\zeta| \\
&\leq \frac{k!}{2\pi} \cdot \frac{M}{r^{k+1}} \cdot 2\pi r = \frac{k! M}{r^k}. \qquad \square
\end{aligned}$$

Remark. The Cauchy estimates reveal yet another contrast with real analysis. Given *any* sequence $\{m_k\}_{k=0}^{\infty}$ of real numbers, we can always construct a function $f(x) \in C^{\infty}(\mathbb{R})$ (i.e., $f(x)$ is infinitely differentiable on $\mathbb{R}$) satisfying

$$f^{(k)}(0) = m_k \quad (k = 0, 1, 2, \ldots),$$

but the Cauchy estimates show that the corresponding statement for analytic functions is false.

DEFINITION 4.4.8 *A function that is analytic in the whole plane $\mathbb{C}$ is called an* **entire** *function.*

EXAMPLE. Polynomial functions, the exponential function, the sine and cosine functions, and the hyperbolic sine and cosine functions are all entire functions.

THEOREM 4.4.9 (Liouville) *A bounded entire function must be constant.*

Proof. Because $f(z)$ is bounded, there exists a constant M such that

$$|f(z)| < M \quad \text{for all } z \in \mathbb{C}.$$

Let a be an arbitrary point in the complex plane $\mathbb{C}$ and r an arbitrary positive number. By the Cauchy estimates,

$$|f^{(k)}(a)| \leq \frac{k!M}{r^k}.$$

Because $f(z)$ is entire, we may let r tend to ∞ (with $k = 1$ fixed), and so we obtain

$$f'(a) = 0.$$

Because a is an arbitrary point in the plane $\mathbb{C}$, this result implies that $f(z)$ is a constant function.   □

The Fundamental Theorem of Algebra 2.10.3 is an immediate consequence of the Liouville Theorem 4.4.9. Let $p(z)$ be a polynomial of degree ≥ 1 and suppose $p(z)$ is never zero. Then $\frac{1}{p(z)}$ is an entire function. To show that $\frac{1}{p(z)}$ is bounded, recall that as $|z|$ increases the leading term of $p(z)$ dominates the rest of the terms in the polynomial (see the proof of Theorem 2.10.3 and the remark after it). Hence there exists some circle outside which $|p(z)|$ is larger than any preassigned positive number, and so outside this circle $\left|\frac{1}{p(z)}\right|$ is small. The interior of the circle along with its boundary forms a compact set, and because the function $\left|\frac{1}{p(z)}\right|$ is continuous, it is bounded inside and on the circle. Hence $\frac{1}{p(z)}$ is a bounded entire function. Thus, by the Liouville Theorem 4.4.9, $\frac{1}{p(z)}$ must be constant. This implies that $p(z)$ must be constant, and thus contradicts the assumption that $\deg p(z) \geq 1$. Therefore, $p(z)$ must be zero somewhere.

We could have used the Maximum Modulus Principle 2.10.1 instead of the Liouville Theorem 4.4.9. Suppose

$$p(z) = a_n z^n + a_{n-1} z^{n-1} + \cdots + a_1 z + a_0 \quad (a_n \neq 0, \, n \geq 1).$$

If $a_0 = 0$, then $z = 0$ is a root and we have nothing left to prove. If $a_0 \neq 0$, choose R so large that

$$|p(z)| > |a_0| \quad \text{for } z \in C(0, R).$$

Then

$$\left|\frac{1}{p(z)}\right| < \frac{1}{|a_0|} = \left|\frac{1}{p(0)}\right| \quad \text{for } z \in C(0, R),$$

and so $\frac{1}{p(z)}$ does not attain its maximum modulus on the boundary circle $C(0, R)$. If $p(z)$ does not vanish at all, then $\frac{1}{p(z)}$ is entire, and so, by the Maximum Modulus Principle 2.10.1, it cannot attain its maximum modulus in $D(0, R)$. But $\left|\frac{1}{p(z)}\right|$ is a real-valued continuous function and $\bar{D}(0, R)$ is compact, and so $\left|\frac{1}{p(z)}\right|$ must attain its maximum value somewhere in $\bar{D}(0, R)$. The contradiction shows that our assumption that $p(z)$ never vanishes is not tenable.

4.5   The Morera Theorem

DEFINITION 4.5.1 *A* **simply connected** *region Ω is a region with the property that if a simple closed arc lies in Ω, then the interior of the arc does so as well; that is, if Γ is a simple closed arc and $\Gamma \subset \Omega$, then $[\Gamma] \subset \Omega$. Otherwise, we say that the region Ω is* **multiply connected.**

LEMMA 4.5.2 *Let Ω be a simply connected region. Then the following are equivalent:*

(a) $f(z) \in H(\Omega)$;

(b) $f(z) \in C(\Omega)$ (i.e., $f(z)$ is continuous on Ω), and $\int_\Gamma f(z)\,dz = 0$ for every simple closed arc Γ in Ω;

(c) $f(z) \in C(\Omega)$, and if z_0 and z_1 are points in Ω, then $\int_\Gamma f(\zeta)\,d\zeta$ is independent of the arc Γ (in Ω) joining z_0 and z_1.

Proof. (a) $\Longrightarrow$ (b). The Cauchy-Goursat Theorem 4.3.1.
(b) $\Longrightarrow$ (c). If Γ_1 and Γ_2 are two simple arcs joining points z_0 and z_1 in Ω, and they have only the points z_0 and z_1 in common, then $\Gamma_1 - \Gamma_2$ is a simple closed arc, and

$$\therefore \int_{\Gamma_1} - \int_{\Gamma_2} = \int_{\Gamma_1 - \Gamma_2} = 0.$$

In the general case, given any $\epsilon > 0$, by Lemma 4.2.1, we can approximate $\int_{\Gamma_1}$ and $\int_{\Gamma_2}$ within ϵ by $\int_{\Gamma_1'}$ and $\int_{\Gamma_2'}$, respectively, where Γ_1' and Γ_2' are suitable polygons. Then $\Gamma_1' - \Gamma_2'$ is a closed polygon, and hence it can be expressed as a finite union of simple closed polygons. The assumption (b) implies that the integral along a simple closed polygon is zero.

$$\therefore \int_{\Gamma_1'} - \int_{\Gamma_2'} = \int_{\Gamma_1' - \Gamma_2'} = 0.$$

It follows that

$$\left| \int_{\Gamma_1} - \int_{\Gamma_2} \right| \leq \left| \int_{\Gamma_1} - \int_{\Gamma_1'} \right| + \left| \int_{\Gamma_1'} - \int_{\Gamma_2'} \right| + \left| \int_{\Gamma_2'} - \int_{\Gamma_2} \right|$$
$$\leq \epsilon + 0 + \epsilon = 2\epsilon.$$

Because $\epsilon > 0$ is arbitrary, we have $\int_{\Gamma_1} = \int_{\Gamma_2}$.

(c) $\Longrightarrow$ (a). Because $\int_\Gamma f(\zeta)\,d\zeta$ depends on the initial and terminal points z_0 and z, but not on the particular arc Γ joining them, we may write it as $\int_{z_0}^z f(\zeta)\,d\zeta$. If we fix z_0, then $\int_{z_0}^z f(\zeta)\,d\zeta$ is a function of z. Let

$$F(z) = \int_{z_0}^z f(\zeta)\,d\zeta.$$

Now, in

$$F(z+h) = \int_{z_0}^{z+h} f(\zeta)\,d\zeta$$

we may choose a path of integration passing through point z and then joining z to $z + h$ by a line segment. Because Ω is open, this line segment is in Ω when $|h|$ is sufficiently small. We have

$$
\begin{aligned}
F(z + h) - F(z) &= \int_z^{z+h} f(\zeta)\, d\zeta \\
&= \int_z^{z+h} f(z)\, d\zeta + \int_z^{z+h} \{f(\zeta) - f(z)\}\, d\zeta \\
&= h f(z) + \int_z^{z+h} \{f(\zeta) - f(z)\}\, d\zeta.
\end{aligned}
$$

By the continuity of $f(z)$, for any given $\epsilon > 0$ there corresponds $\delta > 0$ such that

$$
|f(\zeta) - f(z)| < \epsilon \quad \text{whenever } \zeta \in D(z, \delta).
$$

Then

$$
\begin{aligned}
|F(z + h) &- F(z) - h f(z)| \\
&= \left| \int_z^{z+h} \{f(\zeta) - f(z)\}\, d\zeta \right| < \epsilon \cdot |h| \quad (|h| < \delta).
\end{aligned}
$$

$$
\therefore \left| \frac{F(z + h) - F(z)}{h} - f(z) \right| < \epsilon \quad (|h| < \delta).
$$

Because $\epsilon > 0$ is arbitrary, this inequality means

$$
\lim_{h \to 0} \frac{F(z + h) - F(z)}{h} = f(z);
$$

that is, $F'(z)$ exists and equals $f(z)$. Therefore, $F(z)$ is analytic in Ω, which in turn implies that $F'(z)$, which equals $f(z)$, is also analytic there.   □

COROLLARY 4.5.3 *A function analytic in a simply connected region has a primitive.*

Although simple connectedness is not required in the proof of (b) $\Longrightarrow$ (a) (via (c)), the hypothesis of (b) is too stringent to be useful. If the region is not simply connected, then it usually occurs that the integral along a closed arc in the region does not vanish, and so the lemma cannot be applied, even though the function turns out to be analytic in the region. (For example, consider $f(z) = \frac{1}{z}$ in the region $\mathbb{C} \setminus \{0\}$.) To remedy this situation, we present the following extension of (b) $\Longrightarrow$ (a), due to Morera:

THEOREM 4.5.4 (Morera) *Let $f(z)$ be a continuous function on a region Ω (not necessarily simply connected) with the property that*

$$\int_\Gamma f(z)\,dz = 0$$

for every simple closed arc Γ such that $[\Gamma] \subset \Omega$. Then $f(z) \in H(\Omega)$.

Proof. Let z_0 be an arbitrary point in Ω and $D(z_0, r) \subset \Omega$. Because $D(z_0, r)$ is simply connected, the hypothesis of this theorem, together with Lemma 4.5.2, implies the existence of an analytic function $F(z)$ such that

$$F'(z) = f(z) \qquad z \in D(z_0, r).$$

Thus $f(z)$, being the derivative of an analytic function, is analytic in $D(z_0, r)$. Because z_0 is an arbitrary point in Ω, this implies $f(z)$ is differentiable everywhere in Ω; that is, $f(z) \in H(\Omega)$. $\qquad\square$

Recalling the proof of the Cauchy-Goursat Theorem 4.3.1, we see that the hypothesis in the Morera Theorem 4.5.4 can be weakened to require $\int_\Gamma f(z)\,dz = 0$ only when Γ is the boundary of an arbitrary triangle in Ω.

Remark à la Takagi. We defined the analyticity of a function in a region as differentiability everywhere in the region. If we agree, for a moment, to call a function *integrable* if it has a *local* primitive, then the Cauchy-Goursat Theorem 4.3.1 asserts that *differentiability* implies *integrability*, and the Morera Theorem 4.5.4 asserts that *integrability* implies *differentiability*. Hence *differentiability* and *integrability* are synonymous in complex analysis. In real analysis, we have

$$\textit{differentiability} \Longrightarrow \textit{continuity} \Longrightarrow \textit{integrability},$$

but none of these implications is reversible.

As a consequence of the Morera Theorem 4.5.4, we have the following

COROLLARY 4.5.5 *Let $\{f_n(z)\}_{n=1}^\infty$ be a sequence of analytic functions in a region Ω that converges locally uniformly to $f(z)$ there. Then*

(a) $f(z) \in H(\Omega)$;

(b) $f^{(k)}(z) = \lim_{n \to \infty} f_n^{(k)}(z)$ *locally uniformly in* Ω $(k = 0, 1, 2, \ldots)$.

Proof. (a) Let Γ be an arbitrary simple closed arc such that $[\Gamma] \subset \Omega$. Then, by the Cauchy-Goursat Theorem 4.3.1,

$$\int_\Gamma f_n(z)\, dz = 0 \quad \text{for all } n \in \mathbb{N}.$$

Therefore, because the convergence is uniform on the compact set Γ, we can apply Property (g) in Section 4.2 to obtain

$$\int_\Gamma f(z)\, dz = \lim_{n \to \infty} \int_\Gamma f_n(z)\, dz = 0.$$

The continuity of the limit function $f(z)$ follows from the local uniform convergence and the Morera Theorem 4.5.4 implies that $f(z) \in H(\Omega)$.

(b) Let z_0 be an arbitrary point in Ω and choose $r > 0$ satisfying $\bar{D}(z_0, 2r) \subset \Omega$. By the uniform convergence of $\{f_n(\zeta)\}_{n=1}^\infty$ to $f(\zeta)$ for $\zeta \in C(z_0, 2r)$, for any $\epsilon > 0$ there exists a positive integer N with the property that

$$|f_n(\zeta) - f(\zeta)| < \epsilon \quad \text{for } \zeta \in C(z_0, 2r) \text{ and all } n \geq N.$$

Therefore, for $z \in D(z_0, r)$, we have

$$
\begin{aligned}
|f_n^{(k)}(z) - f^{(k)}(z)| &= \left| \frac{k!}{2\pi i} \int_{C(z_0, 2r)} \frac{f_n(\zeta)}{(\zeta - z)^{k+1}} d\zeta \right. \\
&\quad \left. - \frac{k!}{2\pi i} \int_{C(z_0, 2r)} \frac{f(\zeta)}{(\zeta - z)^{k+1}} d\zeta \right| \\
&\leq \frac{k!}{2\pi} \int_{C(z_0, 2r)} \left| \frac{f_n(\zeta) - f(\zeta)}{(\zeta - z)^{k+1}} \right| \cdot |d\zeta| \\
&\leq \frac{k!}{2\pi} \cdot \frac{\epsilon}{r^{k+1}} \cdot 4\pi r = \frac{2k! \cdot \epsilon}{r^k} \quad \text{for } n \geq N.
\end{aligned}
$$

This shows uniform convergence on $D(z_0, r)$, and now, by Lemma 2.4.4, we have local uniform convergence. $\qquad\square$

Question: If we replace "*local uniform convergence*" by "*uniform convergence*" in the hypotheses, can we conclude uniform convergence in (b)?

Naturally, the series version of this corollary is also true. In particular, termwise differentiation of a power series within its disc of convergence is again justified.

Remark. Compare this result with its real counterpart. There the uniform convergence of $\{f_n^{(k)}(x)\}_{n=1}^{\infty}$ in an (open) interval of $\mathbb{R}$ is usually assumed in order to establish the equality in (b), while in complex analysis we have local uniform convergence (which is often an adequate substitute for uniform convergence) as a *consequence*. In fact, in the real case a sequence of infinitely differentiable functions can converge uniformly to a nowhere differentiable function.

As an application of the above theorem, let us prove the following theorem, known as the *Weierstrass double series theorem*, which is frequently used in computing Taylor series.

THEOREM 4.5.6 (Weierstrass) *Suppose the disc of convergence of each of the power series*

$$f_n(z) = \sum_{k=0}^{\infty} c_k^{(n)}(z-a)^k \quad (n = 0, 1, 2, \ldots)$$

contains $D(a, r)$ and that

$$F(z) = \sum_{n=0}^{\infty} f_n(z)$$

converges locally uniformly in $D(a, r)$. Then each of the series

$$\sum_{n=0}^{\infty} c_k^{(n)} = A_k \quad (k = 0, 1, 2, \ldots)$$

converges, and

$$F(z) = \sum_{k=0}^{\infty} A_k(z-a)^k \quad in \ D(a, r).$$

Proof. By the preceding corollary, $F(z) = \sum_{n=0}^{\infty} f_n(z)$ is analytic in $D(a, r)$ and we may differentiate $F(z) = \sum_{n=0}^{\infty} f_n(z)$ termwise k times, obtaining

$$F^{(k)}(a) = \sum_{n=0}^{\infty} f_n^{(k)}(a) = k! \sum_{n=0}^{\infty} c_k^{(n)} = k! A_k. \quad (k = 0, 1, 2, \ldots).$$

$\square$

Note that we obtain this result without appealing to absolute convergence, which is the tool usually employed to justify rearrangement of series. Similar reasoning may be employed to show locally uniform convergence of the Cauchy product of two Taylor series within the smaller of the two concentric discs of convergence.

One final remark: As illustrated by the above proofs, the usefulness of the Morera Theorem 4.5.4 illustrates the general principle: *"Integration is easier to handle than differentiation."*

Exercises

1. Evaluate

   (a) $\int_{[\alpha, \beta]} \Re z \, dz$,

   (b) $\int_{C(0,2)} \frac{dz}{1+z^2}$,

   (c) $\int_{C(0,1)} \bar{z}^n \, dz$,

   (d) $\int_{C(0,1)} \frac{e^z}{z} dz$,

   (e) $\int_{C(0,1)} |z - 1| \cdot |dz|$,

   where $[\alpha, \beta]$ in (a) denotes the line segment joining α to β.

2. Evaluate the integral $\int_\Gamma \bar{z} \, dz$ along the arc Γ of the parabola $y = x^2$ from $(0, 0)$ to $(3, 9)$.

3. Evaluate $\int_\Gamma |z|^2 dz$ along the ellipse $z = a \cos\theta + ib \sin\theta, (0 \le \theta \le 2\pi, a > 0, b > 0)$.

4. Evaluate the line integral $\int_{-i}^{i} |z| \, dz$ along

   (a) the imaginary axis,

   (b) the right half of the unit circle,

   (c) the left half of the unit circle.

5. Evaluate the integral

$$\int_1^z \frac{dt}{t} \quad (z \in \mathbb{C} \setminus \{0\})$$

   (a) along an arc on the unit circle from 1 to sgn z $\left(= \frac{z}{|z|}\right)$, and then from sgn z to z along the line segment joining them;

   (b) along the complement of the arc on the unit circle used in (a) from 1 to sgn z, and then from sgn z to z along the line segment.

(c) Let the value obtained in (a) be ℓ_1 and that obtained in (b) ℓ_2. Show that $e^{\ell_1} = e^{\ell_2} = z$.

6. Let $0 < |a| < |b|$. Evaluate

$$\int_{C(0,r)} \frac{dz}{(z-a)(z-b)},$$

where

$$(a)\ 0 < r < |a|, \quad (b)\ |a| < r < |b|, \quad (c)\ |b| < r.$$

7. (a) Let $f(z)$ be analytic on $[\Gamma]$, except possibly at a finite number of points where $f(z)$ is bounded in some neighborhood of each of these exceptional points. Show that

$$\int_\Gamma f(z)\,dz = 0.$$

(b) Apply (a) to prove the Cauchy integral formula.
   Hint: Apply (a) to

$$\varphi(z) = \begin{cases} \dfrac{f(z) - f(a)}{z - a}, & (z \neq a); \\ f'(a), & (z = a). \end{cases}$$

8. (a) Evaluate $\displaystyle \int_{C(0,1)} z^n\,dz \quad (n \in \mathbb{Z})$.

   (b) Evaluate $\displaystyle \int_{C(0,1)} \left(z + \frac{1}{z}\right)^{2n} \frac{dz}{z}$.

   (c) Show that

$$\int_0^{2\pi} \sin^{2n}\theta\,d\theta = \int_0^{2\pi} \cos^{2n}\theta\,d\theta = 2\pi \cdot \frac{1\cdot 3\cdot 5\cdots(2n-1)}{2\cdot 4\cdot 6\cdots(2n)}.$$

(Wallis)

9. Evaluate

$$\int_\Gamma \frac{e^z}{(z-2)(z+4)^2}\,dz$$

where Γ is $(a)\ C(0, 1),\quad (b)\ C(i, 3),\quad (c)\ C(-1, 5),$ all in the positive sense.

10. Let

$$f(z) = \int_{C(0,1)} \frac{d\zeta}{\zeta(\zeta + z)}.$$

Find the limit as $z \to i$ from the interior of the unit circle. What if $z \to i$ from the exterior of the unit circle?

11. Let $p(z) = a_0 + a_1 z + a_2 z^2 + \cdots + a_n z^n$ ($a_n \neq 0$). Show that for sufficiently large r,

$$\frac{1}{2\pi i} \int_{C(0,r)} \frac{p'(z)}{p(z)} dz = n.$$

What can be said for smaller r?

12. If Γ is a simple closed arc, show that the area of the region enclosed by Γ is given by the line integral

$$A = \frac{1}{2i} \int_\Gamma \bar{z} \, dz \left(= \frac{1}{2} \int_\Gamma r^2 d\theta \right).$$

13. Suppose f and g are analytic functions on $[\Gamma]$, where Γ is a simple closed piecewise smooth arc. Show that

$$\int_\Gamma \overline{f(z)} \cdot g(z) \, dz = 2i \int\int_{[\Gamma]} \overline{f'(z)} \cdot g(z) \, dx \, dy.$$

Hint: Apply the Gauss-Green theorem.

14. Prove the Cauchy-Goursat Theorem 4.3.1 for the case that $f(z) = \sum_{n=0}^\infty a_n z^n$ in $D(0, R)$ and Γ is a simple closed arc in the disc $D(0, R)$, without appealing to the Cantor Intersection Theorem 2.3.2.

15. (Gauss) Let $f(z)$ be continuous in the closed disc $\bar{D}(a, R)$ and analytic in the disc $D(a, R)$. Show that

$$f(a) = \frac{1}{2\pi} \int_0^{2\pi} f(a + Re^{i\theta}) \, d\theta.$$

More generally, show that

$$f^{(n)}(a) = \frac{n!}{2\pi R^n} \int_0^{2\pi} f(a + Re^{i\theta}) e^{-in\theta} \, d\theta \quad (n \geq 0).$$

16. Let $f(z)$ be continuous in the closed unit disc and analytic in the unit disc. Prove the *Poisson integral formula*

$$f(a) = \frac{1}{2\pi} \int_0^{2\pi} f(e^{i\theta}) \frac{1 - r^2}{1 - 2r\cos(t - \theta) + r^2} d\theta$$

$$(a = re^{it}, \ 0 \le r < 1).$$

Hint: Note that because $\displaystyle\int_{C(0,1)} \frac{f(z)}{z - \dfrac{1}{\bar{a}}} dz = 0$, it follows that

$$f(a) = \frac{1}{2\pi i} \int_{C(0,1)} f(z) \left(\frac{1}{z - a} - \frac{1}{z - \dfrac{1}{\bar{a}}} \right) dz.$$

17. Derive

$$f^{(k)}(z) = \frac{k!}{2\pi i} \int_\Gamma \frac{f(\zeta)}{(\zeta - z)^{k+1}} d\zeta \quad (k = 1, 2, 3, \ldots)$$

by induction from the Cauchy Integral Formula 4.4.3, taking the limit of the difference quotient of $f^{(k-1)}(z)$.

18. Let $f(z) \in H(D)$ and suppose that

$$|f(z)| \le \frac{1}{1 - |z|} \quad \text{for all } z \in D.$$

Show that

$$|f^{(n)}(0)| \le (n + 1)!e.$$

Hint: Use the Cauchy integral formula with a path of integration suitably chosen for each n.

19. Let $\varphi(\zeta)$ be a continuous function on a simple arc Γ. Show that

$$f(z) = \int_\Gamma \frac{\varphi(\zeta)}{\zeta - z} d\zeta$$

is analytic in $\mathbb{C} \setminus \Gamma$. (This set is not necessarily connected.)

20. Prove the Liouville Theorem 4.4.9 by evaluating the following integral:

$$\int_{C(0,R)} f(z) \left(\frac{1}{z-a} - \frac{1}{z-b} \right) dz.$$

21. (a) Suppose that for some positive integer n and some positive constant M an entire function $f(z)$ satisfies the inequality

$$|f(z)| \leq M \cdot |z|^n$$

for all $z \in \mathbb{C}$ with sufficiently large modulus. Show that $f(z)$ must be a polynomial whose degree does not exceed n.

(b) Find all entire functions that are uniformly continuous on the complex plane.

22. Suppose $f(z) \in H(D)$.

(a) Is it possible to have

$$|f^{(n)}(0)| \geq e^n \cdot n! \quad \text{for infinitely many } n?$$

(b) Suppose there is a constant M such that

$$|f^{(n)}(0)| \leq M^n \quad \text{for all } n \text{ sufficiently large.}$$

Show that $f(z)$ can be extended to all of $\mathbb{C}$ as an entire function.

23. Prove the Fundamental Theorem of Algebra 2.10.3 using the following ideas of Boas: Suppose $p(z) = a_0 + a_1 z + \cdots + a_n z^n$, $a_n \neq 0$, $n \geq 1$, never vanishes. Show that $q(z) = \bar{a}_0 + \bar{a}_1 z + \cdots + \bar{a}_n z^n$ also never vanishes and apply the Cauchy-Goursat Theorem 4.3.1 for the perimeter of a semicircular disc in the upper half-plane to $\frac{1}{p(z)q(z)}$. Let the radius tend to ∞ to obtain

$$\int_{-\infty}^{\infty} |p(x)|^{-2} dx = 0.$$

What contradiction do we obtain?

24. (a) Show that an entire function whose real part is positive throughout the complex plane must be a constant.

(b) Show that the range of a nonconstant entire function is dense in the complex plane.

25. Give an example of an entire function that is not a polynomial and has

    (a) no root;

    (b) exactly n roots;

    (c) infinitely many roots.

    (d) Does there exist an entire function (not identically zero) with uncountably many roots?

26. Prove or disprove: Suppose $f(z), g(z) \in H(\Omega)$ and

$$f(z) \cdot g(z) \equiv 0 \quad \text{in the region } \Omega;$$

then either $f(z) \equiv 0$ or $g(z) \equiv 0$ in Ω.

27. Let $f(z)$ be an entire function satisfying

$$\left| f\left(\frac{1}{\log(n+2)} \right) \right| < \frac{1}{n} \quad (n \in \mathbb{N}).$$

Show that $f(z) \equiv 0$.

28. Expand the following functions in Taylor series around the origin and find the radius of convergence:

    (a) $\dfrac{1}{z^2 - 3z + 2}$;

    (b) $\dfrac{\sin z}{1 + z^2}$;

    (c) $\dfrac{z}{1 + e^z}$.

29. Find the first 3 nonvanishing terms of the Taylor-series expansion around the origin of the function $\tan z = \frac{\sin z}{\cos z}$. What is the radius of convergence?

30. Let $f(z) = \sum_{n=0}^{\infty} c_n (z - a)^n$, $z \in D(a, R)$.

    (a) Prove the *Parseval equality*

$$\frac{1}{2\pi} \int_0^{2\pi} |f(a + re^{i\theta})|^2 d\theta = \sum_{n=0}^{\infty} |c_n|^2 r^{2n} \quad (0 \le r < R).$$

    (b) Use this equality to derive the Maximum Modulus Principle 2.10.1, the Cauchy Estimates 4.4.7, and the Liouville Theorem 4.4.9.

(c) Show that

$$\frac{1}{\pi} \int \int_{\bar{D}(a,r)} |f(z)|^2 dx \, dy = \sum_{n=0}^{\infty} \frac{|c_n|^2}{n+1} r^{2n+2} \quad (0 \le r < R).$$

Can we also derive the Maximum Modulus Principle 2.10.1, the Cauchy Estimates 4.4.7, and the Liouville Theorem 4.4.9 from this equality?

31. Let $f(t)$ be a (complex-valued) continuous function defined on a real interval $[a, b]$. Show that the *Laplace transform*

$$F(z) = \int_a^b e^{-zt} f(t) \, dt$$

is an entire function and that

$$F'(z) = -\int_a^b te^{-zt} f(t) \, dt \quad (z \in \mathbb{C}).$$

32. Does a function that is analytic in a multiply connected region have a (global) primitive?

33. Find the fallacy: For any simple closed arc Γ (whether it contains the origin in its interior or not), we have

$$\int_\Gamma \frac{dz}{z^n} = 0 \quad (n \ge 2).$$

Therefore, by the Morera Theorem 4.5.4, $f(z) = \frac{1}{z^n}$ $(n \ge 2)$ are entire functions.

34. Show that $f(z) = \sum_{n=1}^{\infty} \frac{z^n}{1-z^n}$ converges locally uniformly in the unit disc. Find the values of $f'''(0)$ and $f^{(17)}(0)$.

35. Let $f_n(z) = \frac{1}{n} \cos(nz)$ $(n \in \mathbb{N})$. Show that $f_n(x) \longrightarrow 0$ uniformly on $\mathbb{R}$, yet $\{f_n'(z)\}_{n=1}^{\infty}$ converges only at the points $k\pi$ $(k \in \mathbb{Z})$. Does $\{f_n(z)\}_{n=1}^{\infty}$ converge locally uniformly in some region of the plane?

36. Show that the set of all polynomial functions is not dense in $C(\bar{D})$, the Banach space of all functions continuous on the closed unit disc $\bar{D}$ with maximum norm:

$$\|f\| = \max\{|f(z)| \, ; \, z \in \bar{D}\}.$$

37. Let $C(\Omega)$ be the vector space of all continuous functions in a region Ω. We introduce a metric ρ in $C(\Omega)$ as follows: Let

$$D_n = \left\{ z \in \Omega \cap \bar{D}(0,\, n)\,;\; \mathrm{dist}(z, \partial\Omega) \geq \frac{1}{n} \right\} \quad (n \in \mathbb{N}),$$

and

$$\rho(f,\, g) = \sum_{n=1}^{\infty} \frac{1}{2^n} \cdot \frac{\|f - g\|_n}{1 + \|f - g\|_n}, \quad f,\, g \in C(\Omega),$$

where $\|f - g\|_n = \max\{|f(z) - g(z)|\,;\; z \in D_n\}$. Show that

(a) the function ρ defined above is a metric;

(b) the space $C(\Omega)$ is complete (with ρ as its metric);

(c) convergence in this metric is equivalent to local uniform convergence in Ω; that is, $\rho(f_k,\, f) \longrightarrow 0$ (as $k \to \infty$) if and only if $\{f_k(z)\}_{k=1}^{\infty}$ converges to $f(z)$ locally uniformly in Ω.

(d) the space $H(\Omega)$, as a subspace of $C(\Omega)$, is complete.

Chapter 5

Singularities and Residues

5.1 Laurent Series

Suppose $f(z)$ is analytic in the annulus $\Omega = \{z \in \mathbb{C};\ r < |z - a| < R\}$ bounded by the concentric circles $C(a, r)$ and $C(a, R)$ $(0 \le r < R \le \infty)$. Let z be a point in this annulus and choose r' and R' such that $r < r' < |z - a| < R' < R$ (Figure 5.1). By the Cauchy Integral Formula 4.4.3 and Corollary 4.4.2, we have

$$
\begin{aligned}
f(z) &= \frac{1}{2\pi i} \int_{C(z,d)} \frac{f(\zeta)}{\zeta - z} d\zeta \\
&= \frac{1}{2\pi i} \int_{C(a,R')} \frac{f(\zeta)}{\zeta - z} d\zeta - \frac{1}{2\pi i} \int_{C(a,r')} \frac{f(\zeta)}{\zeta - z} d\zeta,
\end{aligned}
$$

where $0 < d < \min\{R' - |z - a|,\ |z - a| - r'\}$, so that $C(z, d)$ is inside the annulus $\{z \in \mathbb{C};\ r' < |z - a| < R'\}$; all three integrals are taken in the positive sense.

Proceeding as we did with Taylor series, we expand the first integral on the right:

$$
\begin{aligned}
\frac{1}{2\pi i} \int_{C(a,R')} \frac{f(\zeta)}{\zeta - z} d\zeta &= \frac{1}{2\pi i} \int_{C(a,R')} \frac{f(\zeta)}{(\zeta - a) - (z - a)} d\zeta \\
&= \frac{1}{2\pi i} \int_{C(a,R')} \frac{f(\zeta)}{\zeta - a} \cdot \frac{1}{1 - \dfrac{z - a}{\zeta - a}} d\zeta \\
&= \frac{1}{2\pi i} \int_{C(a,R')} f(\zeta) \cdot \sum_{k=0}^{\infty} \frac{(z - a)^k}{(\zeta - a)^{k+1}} d\zeta \\
&= \sum_{k=0}^{\infty} A_k (z - a)^k,
\end{aligned}
$$

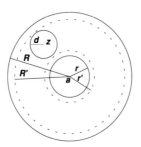

Figure 5.1

where

$$A_k = \frac{1}{2\pi i} \int_{C(a,R')} \frac{f(\zeta)}{(\zeta - a)^{k+1}} d\zeta \quad (k = 0, 1, 2, \dots).$$

The termwise integration is legitimate because the series under the integral sign converges uniformly for ζ on $C(a, R')$, since $\left|\frac{z-a}{\zeta-a}\right| \leq \frac{|z-a|}{R'} < 1$ for $\zeta \in C(a, R')$. The Taylor series converges locally uniformly for z inside $C(a, R')$.

Now, in the second integral, because

$$\left|\frac{\zeta - a}{z - a}\right| \leq \frac{r'}{|z - a|} < 1 \quad \text{for } \zeta \text{ on } C(a, r'),$$

the geometric series

$$-\frac{1}{\zeta - z} = \frac{1}{(z - a) - (\zeta - a)} = \frac{1}{(z - a)\left(1 - \dfrac{\zeta - a}{z - a}\right)}$$

$$= \frac{1}{z - a} \cdot \sum_{k=0}^{\infty} \left(\frac{\zeta - a}{z - a}\right)^k$$

converges uniformly for ζ on $C(a, r')$. Therefore, we have

$$-\frac{1}{2\pi i} \int_{C(a,r')} \frac{f(\zeta)}{\zeta - z} d\zeta$$

$$= \sum_{k=0}^{\infty} \left\{ (z - a)^{-k-1} \cdot \frac{1}{2\pi i} \int_{C(a,r')} f(\zeta)(\zeta - a)^k d\zeta \right\}$$

$$= \sum_{k=1}^{\infty} A_{-k}(z - a)^{-k},$$

where

$$A_{-k} = \frac{1}{2\pi i} \int_{C(a,r')} f(\zeta)(\zeta - a)^{k-1} d\zeta \quad (k \in \mathbb{N}).$$

This is a power series in $(z - a)^{-1}$ that converges (locally uniformly) for z outside the circle $C(a, r')$.

Therefore, for z in the annulus Ω, we have

$$f(z) = \sum_{k=-\infty}^{\infty} A_k (z - a)^k,$$

where the coefficients A_k are obtained by integration over the circle $C(a, R')$ or $C(a, r')$, depending on whether k is nonnegative or negative.

However, the integrands in both cases are analytic in the annulus Ω, hence the paths can be replaced by any simple closed positively oriented rectifiable arc Γ in the annulus Ω that contains the circle $C(a, r)$ in its interior. We then can express A_k as

$$A_k = \frac{1}{2\pi i} \int_{\Gamma} \frac{f(\zeta)}{(\zeta - a)^{k+1}} d\zeta \quad (k \in \mathbb{Z}).$$

We have thus proved the following

THEOREM 5.1.1 (Laurent) *If* $f(z) \in H(\Omega)$, *where* Ω *is an annulus* $\{z \in \mathbb{C}; r < |z - a| < R\}$, *then it can be expanded into a series of the form*

$$f(z) = \sum_{k=-\infty}^{\infty} A_k (z - a)^k,$$

where

$$A_k = \frac{1}{2\pi i} \int_{C(a,\rho)} \frac{f(\zeta)}{(\zeta - a)^{k+1}} d\zeta \quad (r < \rho < R, \ k \in \mathbb{Z}).$$

The series converges absolutely and locally uniformly in the annulus Ω.

Remark. The convergence of a two-sided series $\sum_{n=-\infty}^{\infty} c_n$ means the existence of

$$\lim_{M,N \to \infty} \sum_{n=-M}^{N} c_n$$

where M and N tend to ∞ *independently*. In other words, a two-sided series $\sum_{n=-\infty}^{\infty} c_n$ is convergent if and only if both $\sum_{n=-\infty}^{-1} c_n$ and $\sum_{n=0}^{\infty} c_n$ are convergent.

The Laurent expansion of a function in a given annulus is unique: For, if

$$f(z) = \sum_{k=-\infty}^{\infty} A_k(z-a)^k \quad (r < |z-a| < R),$$

then, multiplying by $(z-a)^{-n-1}$ and integrating along a circle $C(a, \rho)$ $(r < \rho < R)$, we obtain

$$\int_{C(a,\rho)} f(z)(z-a)^{-n-1}dz = \sum_{k=-\infty}^{\infty} A_k \int_{C(a,\rho)} (z-a)^{k-n-1}dz$$
$$= 2\pi i A_n \quad (n \in \mathbb{Z}).$$

Note carefully that this result does *not* imply that there exists at most one Laurent expansion centered at a of a given function. This is indicated by the following example.

EXAMPLE. Find all Laurent series of $f(z) = \frac{z+5}{z^2-2z-3}$ centered at the origin, and indicate the annulus of convergence of each series.
    Solution. $f(z) = \frac{2}{z-3} - \frac{1}{z+1}$.
(a) For $|z| < 1$, we have

$$f(z) = -\frac{2}{3\left(1 - \dfrac{z}{3}\right)} - \frac{1}{1 - (-z)}$$
$$= -\frac{2}{3}\sum_{k=0}^{\infty} \left(\frac{z}{3}\right)^k - \sum_{k=0}^{\infty}(-z)^k$$
$$= -\sum_{k=0}^{\infty} \left\{\frac{2}{3^{k+1}} + (-1)^k\right\} z^k.$$

Thus, for z in the unit disc, we obtain an ordinary Taylor series, for $f(z)$ is analytic there.

(b) For $1 < |z| < 3$, we have

$$f(z) = -\frac{2}{3} \sum_{k=0}^{\infty} \left(\frac{z}{3}\right)^k - \frac{1}{z\left(1 + \dfrac{1}{z}\right)}$$

$$= -\frac{2}{3} \sum_{k=0}^{\infty} \left(\frac{z}{3}\right)^k - \frac{1}{z} \sum_{k=0}^{\infty} \left(-\frac{1}{z}\right)^k$$

$$= -\sum_{k=0}^{\infty} \frac{2}{3^{k+1}} z^k + \sum_{k=1}^{\infty} \frac{(-1)^k}{z^k}.$$

(c) For $|z| > 3$, we have

$$f(z) = \frac{2}{z\left(1 - \dfrac{3}{z}\right)} + \sum_{k=1}^{\infty} \frac{(-1)^k}{z^k}$$

$$= \frac{2}{z} \sum_{k=0}^{\infty} \left(\frac{3}{z}\right)^k + \sum_{k=1}^{\infty} \frac{(-1)^k}{z^k}$$

$$= \sum_{k=1}^{\infty} \left\{2 \cdot 3^{k-1} + (-1)^k\right\} \frac{1}{z^k}.$$

Moreover, for functions such as $\cot z = \frac{\cos z}{\sin z}$, there exist infinitely many Laurent series centered at the origin, one associated with each of the annuli

$$\Omega_k = \{z \in \mathbb{C}; \, k\pi < |z| < (k+1)\pi\}, \quad (k = 0, 1, 2, \ldots),$$

because $\cot z$ is analytic in each of them.

5.2   Isolated Singularities

If $f(z)$ is analytic in some punctured disc $D'(a, r) = \{z \in \mathbb{C}; 0 < |z - a| < r\}, r > 0$ (but may even be undefined at $z = a$), then the point $z = a$ is called an *isolated singular point* of $f(z)$, and the function $f(z)$ is said to have an *isolated singularity* at $z = a$.

EXAMPLE.

(a) $f(z) = \begin{cases} \sin z & (z \neq 0); \\ 17 & (z = 0). \end{cases}$

(b) $f(z) = \dfrac{\sin z}{z}$ $(z \neq 0)$; $f(0)$ not defined.

(c) $f(z) = \dfrac{1}{z^2}$ $(z \neq 0)$.

(d) $f(z) = e^{\frac{1}{z}} = 1 + \dfrac{1}{z} + \dfrac{1}{2!z^2} + \dfrac{1}{3!z^3} + \cdots + \dfrac{1}{k!z^k} + \cdots$ $(z \neq 0)$.

For each of the functions defined above, the point $z = 0$ is an isolated singularity. However, nonisolated, as well as isolated, singularities exist.

EXAMPLE. Consider the function $f(z) = \dfrac{1}{\sin\frac{1}{z}}$. The origin $z = 0$ is an accumulation point of zeros of $\sin \frac{1}{z}$, hence is a nonisolated singular point of $f(z)$.

The Laurent series allows us to analyze what happens at an isolated singularity. Let $f(z)$ be analytic in a region Ω except possibly at the point $a \in \Omega$. Then in the formula for the coefficients of the Laurent series

$$A_k = \frac{1}{2\pi i} \int_\Gamma \frac{f(z)}{(\zeta - a)^{k+1}} d\zeta \quad (k \in \mathbb{Z}),$$

the path Γ of integration can be an arbitrary simple closed positively oriented rectifiable arc (to save words, such an arc is called a *curve* from now on) such that $a \in (\Gamma)$ and $[\Gamma] \subset \Omega$. Moreover, in the Laurent expansion

$$f(z) = \sum_{k=-\infty}^{\infty} A_k(z - a)^k,$$

the series of positive powers converges at least inside the largest circle contained in Ω, and its sum is also analytic at a. On the other hand, the inner circle $C(a, r)$ can be arbitrarily small, hence the series of negative powers

$$\frac{A_{-1}}{(z - a)} + \frac{A_{-2}}{(z - a)^2} + \cdots + \frac{A_{-k}}{(z - a)^k} + \cdots$$

converges for all z except at a. Near a, the series of these negative powers determines the "bad" behavior of the function $f(z)$; thus the series of these negative powers

$$\sum_{-\infty}^{-1} A_k(z - a)^k$$

is called the *singular part* (or the *principal part*) of $f(z)$ at the singular point a.

The singular part may assume three possible forms:

(a) No singular part; in this case the Laurent series simply reduces to a Taylor series

$$f(z) = A_0 + A_1(z - a) + A_2(z - a)^2 + \cdots + A_k(z - a)^k + \cdots$$

for $z \in D'(a, R)$. Though the point a is excluded, $f(z)$ becomes analytic at a if we define $f(a) = A_0$. Thus, if $f(a)$ is not A_0, or if $f(a)$ is undefined, then $f(z)$ is not analytic at a. By defining or redefining $f(z)$ at the point a, we can make $f(z)$ analytic there; then, following Riemann, we say that $f(z)$ has a *removable singularity* at a. (The first two examples at the beginning of this section are removable singularities.)

THEOREM 5.2.1 (Riemann) *If $f(z)$ is analytic and bounded in a punctured neighborhood $D'(a, R)$ of an isolated singular point a, then the singularity at a is removable.*

Proof. Let $f(z)$ be bounded by M in $D'(a, R)$; that is, $|f(z)| < M$ ($z \in D'(a, R)$), and let k be an arbitrary positive integer, then

$$A_{-k} = \frac{1}{2\pi i} \int_{C(a,\rho)} f(z)(z - a)^{k-1} dz \quad (0 < \rho < R).$$

$$\therefore |A_{-k}| \leq \frac{1}{2\pi} \cdot M \cdot \rho^{k-1} \cdot 2\pi\rho = M\rho^k.$$

Because the singularity is isolated, we can make ρ as small as we please; hence

$$|A_{-k}| \leq M\rho^k \longrightarrow 0 \quad (\text{as } \rho \to 0).$$

$$\therefore A_{-k} = 0 \quad (k \in \mathbb{N}).$$

It follows that the Laurent series at the point a has no negative powers and reduces to an ordinary Taylor series.   □

In the sequel, it will be understood that *if a function has a removable singularity, we shall always remove it by defining $f(a) = A_0$.*

(b) The singular part consists of a finite number of terms; that is,

$$f(z) = \frac{A_{-m}}{(z - a)^m} + \cdots + \frac{A_{-1}}{(z - a)} + \sum_{k=0}^{\infty} A_k(z-a)^k \quad (A_{-m} \neq 0, \ m \geq 1).$$

The point a is now called a *pole of order* m (or *multiplicity* m). In this case,

$$\lim_{z \to a}(z - a)^m f(z) = A_{-m} \neq 0.$$

Therefore, $f(z)$ behaves like $\frac{A_{-m}}{(z-a)^m}$ when z is near a; in particular, $f(z) \longrightarrow \infty$ (as $z \to a$). Moreover, because

$$(z-a)^m f(z) = A_{-m} + A_{-m+1}(z-a) + \cdots + A_{-1}(z-a)^{m-1}$$
$$+ \sum_{k=0}^{\infty} A_k(z-a)^{k+m}$$

and $A_{-m} \neq 0$, $(z-a)^m f(z)$ is analytic and does not vanish in a (sufficiently small) neighborhood of a. It follows that the reciprocal $\{(z-a)^m f(z)\}^{-1}$ has a Taylor series expansion

$$\frac{1}{(z-a)^m f(z)} = B_0 + B_1(z-a) + B_2(z-a)^2 + \cdots \quad (B_0 = \frac{1}{A_{-m}} \neq 0),$$

$$\therefore \frac{1}{f(z)} = B_0(z-a)^m + B_1(z-a)^{m+1} + B_2(z-a)^{m+2} + \cdots \quad (B_0 \neq 0).$$

Therefore, *if $f(z)$ has a pole of order m at a point a, then $\frac{1}{f(z)}$ is analytic at a and has a root of order m there.* The reasoning above is reversible, and thus the converse is also true: *if an analytic function has a root of order m at a point a, then its reciprocal has a pole of order m there.* Note that a pole is always an isolated singularity, because the roots of a function analytic in a region are isolated (unless the function is identically zero).

(c) The singular part consists of infinitely many terms; that is,

$$f(z) = \sum_{k=-\infty}^{\infty} A_k(z-a)^k,$$

where $A_k \neq 0$ for infinitely many negative integers k. We now say that $f(z)$ has an *essential singularity* at the point a. The behavior near an essential singularity is strange. To see what may happen, let us consider the function

$$e^{\frac{1}{z}} = 1 + \frac{1}{z} + \frac{1}{2!z^2} + \cdots + \frac{1}{k!z^k} + \cdots .$$

The point $z = 0$ is an essential singularity. If z approaches 0 along the positive real axis, then $e^{\frac{1}{z}} \longrightarrow \infty$; but for $z \to 0$ along the negative real axis we have $e^{\frac{1}{z}} \longrightarrow 0$. If $z \to 0$ along the imaginary axis, we find that $e^{\frac{1}{z}}$ describes the unit circle infinitely many times. If $c \neq 0$, then the roots of $e^{\frac{1}{z}} = c$ are $z_n = (\text{Log}\,c + 2n\pi i)^{-1}$ $(n \in \mathbb{Z})$, which accumulate at $z = 0$.

This peculiar behavior is characteristic of an essential singularity. In fact, we have the following result, known as the *"great"* Picard theorem:

THEOREM 5.2.2 (Picard) *If $f(z)$ has an isolated essential singularity at a point a, then the image of any punctured neighborhood of a under $f(z)$ is either the whole plane $\mathbb{C}$ or the plane $\mathbb{C}$ with one point deleted. Equivalently, for every complex number c, with perhaps one exception, the function $f(z)$ assumes the value c infinitely often in any punctured neighborhood of a.*

Naturally, there must be a theorem called the *"little"* Picard theorem:

THEOREM 5.2.3 (Picard) *The image of the complex plane $\mathbb{C}$ under a nonconstant entire function is either the whole complex plane $\mathbb{C}$ or the plane $\mathbb{C}$ with one point deleted.*

We do not yet have the tools needed to prove these theorems (see Chapter 12); at this point, however, we can prove the following weaker version.

THEOREM 5.2.4 (Casorati-Weierstrass) *If $f(z)$ has an isolated essential singularity at a point a, then the image of any punctured neighborhood of a under the function $f(z)$ is dense in the plane $\mathbb{C}$.*

Proof. Assume that the conclusion is false; that is, there exist a complex number c and positive numbers r and ϵ such that

$$|f(z) - c| \geq \epsilon \quad \text{for } z \in D'(a, r).$$

Set $g(z) = \frac{1}{f(z)-c}$. Then we have

$$|g(z)| \leq \frac{1}{\epsilon} \quad \text{for } z \in D'(a, r),$$

and $g(z)$ is analytic in the punctured neighborhood $D'(a, r)$ of the point a. Because $g(z)$ is bounded, $g(z)$ can be extended, by the Riemann Theorem 5.2.1, to an analytic function in $D(a, r)$. But then $f(z) = c + \frac{1}{g(z)}$ is analytic at a if $g(a) \neq 0$. If $g(a) = 0$, then $g(z)$ has a root at $z = a$, hence its reciprocal $f(z) - c$ has a pole at the point a. Either way, $z = a$ cannot be an essential singularity of $f(z)$. □

Summing up: Suppose $f(z)$ is analytic in a punctured neighborhood $D'(a, r)$ of a point a.

(a) If $f(z)$ is continuous at a point a or even only bounded near a, then it is also analytic there;

(b) If $f(z) \longrightarrow \infty$ (as $z \to a$), then the point a is a pole;

(c) If $\lim_{z \to a} f(z)$ does not exist, then the point a is an essential singular point.

So far we have been discussing the case $r = 0$ in a Laurent expansion

$$f(z) = \sum_{k=-\infty}^{\infty} A_k (z-a)^k \quad (r < |z-a| < R).$$

Let us now turn to the other extreme case, $R = \infty$. In this case, the series of negative powers

$$\frac{A_{-1}}{(z-a)} + \frac{A_{-2}}{(z-a)^2} + \cdots + \frac{A_{-k}}{(z-a)^k} + \cdots$$

converges outside the circle $C(a, r)$ and tends to zero as $z \to \infty$. On the other hand, because $R = \infty$, the series of the positive powers

$$A_1(z-a) + A_2(z-a)^2 + \cdots + A_k(z-a)^k + \cdots$$

converges for all $z \in \mathbb{C}$. Because the behavior of $f(z)$ (when $z \to \infty$) is determined by the series of the positive powers, we call $\sum_{k=1}^{\infty} A_k (z-a)^k$ the *singular part* (or *principal part*) of $f(z)$ at $z = \infty$.

To study the behavior of $f(z)$ at the point at infinity, we introduce a function $g(\zeta)$ defined by $g(\zeta) = g\left(\frac{1}{z-a}\right) = f(z)$ and consider the behavior of $g(\zeta)$ near the origin, $\zeta = 0$. Again we have three possibilities:

(a') No singular part; that is,

$$f(z) = A_0 + \frac{A_{-1}}{z-a} + \frac{A_{-2}}{(z-a)^2} + \cdots \quad z \notin \bar{D}(a, r).$$

In this case, $g(\zeta) = A_0 + A_{-1}\zeta + A_{-2}\zeta^2 + \cdots$ is the Taylor-series expansion of $g(\zeta)$ around the origin $\zeta = 0$. Because $g(\zeta)$ is analytic at $\zeta = 0$ and $f(z) = g(\zeta) \longrightarrow A_0$ (as $z \to \infty$), we say that $f(z)$ is *analytic at $z = \infty$*.

(b') The singular part consists of a finite number of terms; that is,

$$f(z) = A_m(z-a)^m + A_{m-1}(z-a)^{m-1} + \cdots + A_1(z-a)$$
$$+ \sum_{k=0}^{\infty} A_{-k}(z-a)^{-k},$$

with $A_m \neq 0$, $m \geq 1$. Because $g(\zeta) = \frac{A_m}{\zeta^m} + \frac{A_{m-1}}{\zeta^{m-1}} + \cdots + \frac{A_1}{\zeta} + \sum_{k=0}^{\infty} A_{-k}\zeta^k$ has a pole of order m at $\zeta = 0$, we say that $f(z)$ *has a pole of order m at $z = \infty$.* Note that in this case $f(z) \longrightarrow \infty$ (as $z \to \infty$), and $\frac{f(z)}{(z-a)^m}$ is analytic at $z = \infty$.

(c′) The singular part consists of infinitely many terms; that is,

$$f(z) = \sum_{k=-\infty}^{\infty} A_k(z-a)^k,$$

where $A_k \neq 0$ for infinitely many positive integers k. In this case, the function $g(\zeta)$ has an isolated essential singularity at the origin $\zeta = 0$. Applying the Casorati-Weierstrass Theorem 5.2.4 to the function $g(\zeta)$ at $\zeta = 0$, we see that the image of any neighborhood of $z = \infty$ (i.e., the complement of a compact set in the plane $\mathbb{C}$) is dense in the plane $\mathbb{C}$. We say that $f(z)$ *has an essential singularity at $z = \infty$.*

Observe that the nature of the singularity of $f(z)$ at $z = \infty$ does not depend on a particular choice of the point a around which we expand $f(z)$ in Laurent series. For, if

$$f(z) = \sum_{k=-\infty}^{\infty} A_k(z-a)^k$$

is analytic for $r < |z-a| < \infty$, then $f(z)$ can also be expanded in a Laurent series centered at the origin; using the Weierstrass Double Series Theorem 4.5.6,

$$f(z) = \sum_{k=-\infty}^{\infty} B_k z^k \quad \text{for } r + |a| < |z| < \infty.$$

During this reexpansion, if $f(z)$ has a pole at $z = \infty$ then the order of the pole at $z = \infty$ and the coefficient of the highest positive power are unchanged; moreover, by the contrapositive, if $A_k \neq 0$ for infinitely many positive integers k, then the coefficients B_k have the same property. Thus, the singularity of $f(z)$ at $z = \infty$ is the same as that of $g(\zeta) = f\left(\frac{1}{\zeta}\right)$ at $\zeta = 0$.

As a byproduct of the above discussion, we see that if $f(z) = \sum_{k=0}^{\infty} A_k z^k$ is an entire function, then we have three possibilities:

(a'') $f(z) = A_0$ (constant).

(b'') $f(z) = A_0 + A_1 z + \cdots + A_n z^n$ $(A_n \neq 0, n \geq 1)$.
  In this case, $\lim_{z \to \infty} f(z) = \infty$.

(c'') $f(z) = \sum_{k=0}^{\infty} A_k z^k$ (where $A_k \neq 0$ for infinitely many $k \in \mathbb{N}$) is an infinite series with radius of convergence $R = \infty$.
  In this case, $\lim_{z \to \infty} f(z)$ does not exist.

  In cases (a'') and (b'') $f(z)$ is a polynomial, or, as it is sometimes called, a *rational entire function*; in the case (c'') we say $f(z)$ is a *transcendental entire function*.

  From this classification the Liouville Theorem 4.4.9 follows immediately, for in cases (b'') and (c'') the function $f(z)$ cannot remain bounded. Note also that nonconstant polynomials are the only entire functions $f(z)$ with the property $\lim_{z \to \infty} f(z) = \infty$. This property is used in every proof of the Fundamental Theorem of Algebra 2.10.3.

5.3   Rational Functions

A rational function $R(z)$ is the quotient of two polynomials:

$$R(z) = \frac{p(z)}{q(z)}.$$

It is understood that the quotient is reduced to lowest terms. Thus the polynomials $p(z)$ and $q(z)$ do not both become zero at the same point. Under this assumption, the roots of $q(z)$ are the poles of $R(z)$; that is, if $q(z)$ has a root of order m at a point a, then

$$p(a) \neq 0, \quad q(z) = (z - a)^m q_0(z), \quad q_0(a) \neq 0.$$

Therefore,

$$(z - a)^m R(z) = \frac{p(z)}{q_0(z)}$$

is analytic at a, and $R(z)$ has a pole of order m at a.

  Let the singular part of $R(z)$ at a be

$$P(z; a) = \frac{A_{-m}}{(z - a)^m} + \cdots + \frac{A_{-1}}{z - a} \quad \left(A_{-m} = \frac{p(a)}{q_0(a)} \neq 0 \right);$$

this result can be obtained by dividing the first m terms of the Taylor-series expansion of $\frac{p(z)}{q_0(z)}$ by $(z - a)^m$.

Suppose the *distinct* roots of $q(z)$ are $a_1, a_2, \ldots, a_n$ $(n \leq \deg q)$, and let $P(z; a_j)$ be the singular part of $R(z)$ at a_j $(j = 1, 2, \ldots, n)$. Note that if $q(z)$ is not a constant, then $R(z)$ has at least one singular point and may have as many singular parts as the degree of $q(z)$. Consider the function

$$\varphi(z) = R(z) - \sum_{j=1}^{n} P(z; a_j).$$

Near a_k,

$$R(z) = P(z; a_k) + (\text{a Taylor series}).$$

Thus, near a_k,

$$\varphi(z) = (\text{a Taylor series}) - \sum_{j \neq k} P(z; a_j),$$

and this is analytic at a_k $(k = 1, 2, \ldots, n)$. Because $P(z; a_j)$ is analytic everywhere except at a_j, and so is analytic at a_k $(j \neq k)$, therefore $\varphi(z)$ is analytic everywhere; that is, $\varphi(z)$ is an entire function.

Suppose $\deg p < \deg q$; then

$$R(z) \longrightarrow 0 \quad \text{as } |z| \to \infty.$$

Similarly, the singular parts

$$P(z; a_j) \longrightarrow 0 \quad \text{as } |z| \to \infty \quad (j = 1, 2, \ldots, n).$$

Thus the entire function $\varphi(z) = R(z) - \sum_{j=1}^{n} P(z; a_j)$ is bounded and tends to zero (as $|z| \to \infty$). By the Liouville Theorem 4.4.9, we conclude that

$$\varphi(z) \equiv 0.$$

Thus we have

$$R(z) = \varphi(z) + \sum_{j=1}^{n} P(z; a_j) = \sum_{j=1}^{n} P(z; a_j).$$

A proper rational function $R(z) = \frac{p(z)}{q(z)}$ (deg $p <$ deg q) is the sum of its singular parts.

If $\deg p \geq \deg q$, then we may write

$$R(z) = \frac{p(z)}{q(z)} = \psi(z) + \frac{p_0(z)}{q(z)},$$

where $\psi(z)$ and $p_0(z)$ are polynomials and $\deg p_0 < \deg q$. Thus the singular parts of $R(z)$ and $\frac{p_0(z)}{q(z)}$ at the poles in $\mathbb{C}$ are identical (because $\psi(z)$ is entire) and we obtain

$$R(z) = \psi(z) + \sum_{j=1}^{n} P(z; a_j).$$

If $\deg \psi \geq 1$, then $z = \infty$ is a pole of $R(z)$, and the singular part of $R(z)$ is $\psi(z)$ with the constant term deleted.

   We have just derived the *partial fraction* development of a rational function.

THEOREM 5.3.1 *A rational function is the sum of all its singular parts (including the one at the point at infinity) and a constant. A nonconstant rational function $R(z) = \frac{p(z)}{q(z)}$ takes all values c in the extended complex plane $\hat{\mathbb{C}}$ exactly the same number of times (counting the multiplicities) in the extended complex plane $\hat{\mathbb{C}}$.*

In particular, the number of roots and the number of poles are the same. This common number is called the *order* of the rational function $R(z) = \frac{p(z)}{q(z)}$, and is given by

$$\max\{\deg p(z),\ \deg q(z)\}.$$

(Remember that we are assuming that the polynomials $p(z)$ and $q(z)$ have no common roots.)

5.4   Residues

We have seen that if $f(z)$ has an isolated singularity at a point a, then we may expand $f(z)$ in a Laurent series

$$f(z) = \sum_{k=-\infty}^{\infty} A_k(z - a)^k \quad (0 < |z - a| < R),$$

where

$$A_k = \frac{1}{2\pi i} \int_{\Gamma} \frac{f(\zeta)}{(\zeta - a)^{k+1}} d\zeta \quad (k \in \mathbb{Z})$$

and Γ is a curve such that $a \in (\Gamma)$, $\Gamma \subset D'(a, R)$. In particular,

$$A_{-1} = \frac{1}{2\pi i} \int_{\Gamma} f(\zeta)\, d\zeta.$$

Therefore, if by hook or by crook we can find A_{-1}, then we have determined the value of the integral of $f(z)$ around any curve that encloses no singular point other than a; A_{-1} is called the *residue* of $f(z)$ at the point a and is denoted Res $(a; f)$, or simply Res (a):

$$\text{Res } (a; f) = A_{-1} = \frac{1}{2\pi i} \int_{\Gamma} f(z)\, dz,$$

where we may choose $C(a, r)$ $(0 < r < R)$ as Γ.

If $a = \infty$ (i.e., if $f(z)$ is analytic for $r < |z| < \infty$) and

$$f(z) = \sum_{k=-\infty}^{\infty} B_k z^k$$

is its Laurent series expansion around the point at infinity, then we define the *residue* at $z = \infty$ as

$$\text{Res } (\infty; f) = -B_{-1} = -\frac{1}{2\pi i} \int_{C(0,\rho)} f(z)\, dz \quad (\rho > r).$$

The reason for choosing the negative sign will become clear in the sequel. Note that for $a \in \mathbb{C}$, if Res $(a; f) \neq 0$ then a is a singular point of $f(z)$; but $f(z)$ being analytic at ∞ does *not* imply that Res $(\infty; f) = 0$. Moreover, Res $(\infty; f)$ is *not* equal to the residue of

$$f\left(\frac{1}{\zeta}\right) = \sum_{k=-\infty}^{\infty} B_{-k}\zeta^k$$

at $\zeta = 0$. However, if $\lim_{z\to\infty} z f(z)$ $\left(= \lim_{\zeta\to 0} \frac{1}{\zeta} f\left(\frac{1}{\zeta}\right)\right)$ exists and is not equal to 0 or ∞, then $f(z)$ has a simple zero at $z = \infty$, and

$$\text{Res } (\infty; f) = -\lim_{z\to\infty} z f(z).$$

In some cases we have easy formulas for computing the residue. For example, if a is a pole of order m, then $(z - a)^m f(z)$ is analytic at $z = a$, and

$$(z - a)^m f(z) = A_{-m} + A_{-m+1}(z - a) + \cdots + A_{-1}(z - a)^{m-1}$$
$$+ \sum_{k=0}^{\infty} A_k(z - a)^{k+m}.$$

$$\therefore A_{-1} = \frac{1}{(m-1)!} \lim_{z \to a} \frac{d^{m-1}}{dz^{m-1}} \{(z-a)^m \cdot f(z)\}.$$

In particular, for a *simple pole* (i.e., when $m = 1$), we have

$$\text{Res} \,(a; \, f) = A_{-1} = \lim_{z \to a} \{(z-a) \cdot f(z)\}.$$

Thus, if $f(z) = \frac{h(z)}{g(z)}$, where $g(z)$ has a simple root at $z = a$ and $h(a) \neq 0$, then

$$\begin{aligned}
\text{Res} \,(a; \, f) &= \lim_{z \to a} (z-a) \cdot f(z) \\
&= \lim_{z \to a} \frac{h(z)}{\dfrac{g(z) - g(a)}{z - a}} = \frac{h(a)}{g'(a)}.
\end{aligned}$$

(Note that $g'(a) \neq 0$.)

EXAMPLE. Let $f(z) = \cot z = \frac{\cos z}{\sin z}$. Here $g(z) = \sin z$ and $h(z) = \cos z$. If $z = k\pi$ is any root of $\sin z$, then the residue there is

$$\text{Res} \,(k\pi; \cot z) = \frac{h(k\pi)}{g'(k\pi)} = \frac{\cos k\pi}{\cos k\pi} = 1;$$

that is, all residues of $\cot z$ are 1. Therefore,

$$\int_\Gamma \cot z \, dz = 2\pi i N,$$

where N is the number of poles of $\cot z$ enclosed in Γ (i.e., the number of multiples of π enclosed in Γ).

This example suggests the following

THEOREM 5.4.1 (Cauchy Residue Theorem) *Suppose $f(z)$ is analytic on $[\Gamma]$ except at a finite number of (isolated) singular points $a_1, a_2, \ldots, a_n$ in (Γ) (where Γ is a curve). Then*

$$\int_\Gamma f(z) \, dz = 2\pi i \sum_{j=1}^n \text{Res} \,(a_j; \, f).$$

Proof. Choose small circles C_j in (Γ) centered at each of the singular points a_j in (Γ) and excluding all the rest of the singular points. Then, by Corollary 4.4.2, we have

$$\int_\Gamma f(z)\,dz = \sum_{j=1}^n \int_{C_j} f(z)\,dz = 2\pi i \sum_{j=1}^n \text{Res}\,(a_j;\,f). \qquad \square$$

COROLLARY 5.4.2 *If $f(z)$ is analytic in the extended complex plane $\hat{\mathbb{C}}$ except at a finite number of points, then the sum of all its residues (including the one at the point at infinity) is 0.*

Proof. Choose $R > 0$ so large that all the singular points $a_1, a_2, \ldots,$ a_n in the complex plane $\mathbb{C}$ are in $D(0,\,R)$; then

$$\int_{C(0,R)} f(z)\,dz = 2\pi i \sum_{j=1}^n \text{Res}\,(a_j;\,f).$$

On the other hand, the integral on the left-hand side is $-2\pi i\,\text{Res}\,(\infty;\,f)$.

$$\therefore\ \text{Res}\,(\infty;\,f) + \sum_{j=1}^n \text{Res}\,(a_j;\,f) = 0. \qquad \square$$

COROLLARY 5.4.3 *The sum of all residues of a rational function is 0.*

5.5   Evaluation of Real Integrals

It is possible in some cases to evaluate a real definite integral by using complex integration. More specifically, if the integrand in a given integral is the restriction to the real axis or to the unit circle of an analytic function, then it may be possible to exploit properties of analytic functions to evaluate the original integral. (For example, the Wallis formula in Exercise 4.8.) We illustrate some techniques below.

EXAMPLE.

$$\int_{-\infty}^{\infty} \frac{dx}{1+x^2} = \pi.$$

Of course the primitive of this integrand is well known, but we choose this integral as our first example to demonstrate the technique of using the Cauchy Residue Theorem 5.4.1 to compute real integrals. The integrand $\frac{1}{1+x^2}$ suggests using the complex function $\frac{1}{1+z^2}$; it has a unique simple pole at $z = i$ in the semicircle Γ_R $(R > 1)$ shown in Figure 5.2.

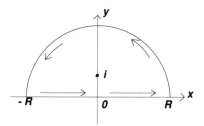

Figure 5.2

We have

$$\int_{\Gamma_R} \frac{dz}{1 + z^2} = \int_{\Gamma_R} \left(\frac{1}{z + i} \right) \frac{dz}{z - i}.$$

The function $f(z) = \frac{1}{z+i}$ is analytic everywhere on the closed semidisc; therefore, by the Cauchy Integral Formula 4.4.3,

$$\int_{\Gamma_R} \frac{dz}{1 + z^2} = 2\pi i \cdot f(i) = \frac{2\pi i}{2i} = \pi.$$

Alternatively,

$$\int_{\Gamma_R} \frac{dz}{1 + z^2} = \frac{1}{2i} \int_{\Gamma_R} \left(\frac{1}{z - i} - \frac{1}{z + i} \right) dz = \frac{1}{2i} \int_{\Gamma_R} \frac{dz}{z - i} = \pi.$$

(Because $\frac{1}{z+i}$ is analytic on the closed semidisc $[\Gamma_R]$, its integral is zero, and the last equality follows from the example at the end of Section 4.3.) Now,

$$\int_{\Gamma_R} \frac{dz}{1 + z^2} = \int_{-R}^{R} \frac{dx}{1 + x^2} + \int_{C^+(0,R)} \frac{dz}{1 + z^2},$$

where

$$C^+(0, R) = \{z \in \mathbb{C}; |z| = R, \ \Im z \geq 0\}.$$

On $C^+(0, R)$, we have $|1 + z^2| \geq R^2 - 1$, and so

$$\left| \int_{C^+(0,R)} \frac{dz}{1 + z^2} \right| \leq \frac{\pi R}{R^2 - 1} \longrightarrow 0 \quad (R \to \infty),$$

and we obtain the desired result.

   Remark. In calculus, an improper integral $\int_{-\infty}^{\infty} f(x)\, dx$ is defined as

$$\int_{-\infty}^{\infty} f(x)\, dx = \lim_{M, N \to \infty} \int_{-M}^{N} f(x)\, dx,$$

where M and N tend to infinity *independently*. Equivalently, the improper integral $\int_{-\infty}^{\infty} f(x)\,dx$ converges if and only if both $\int_{-\infty}^{0} f(x)\,dx$ and $\int_{0}^{\infty} f(x)\,dx$ converge. But what we have computed here is the *Cauchy principal value* of the integral; viz.,

$$\lim_{R \to \infty} \int_{-R}^{R} f(x)\,dx.$$

However, in this case the improper integral converges (absolutely), and so the Cauchy principal value *is* the value of the integral. In fact, if the integrand $f(x)$ is an even function, then the improper integral converges if and only if the Cauchy principal value exists.

EXAMPLE.

$$\int_{-\infty}^{\infty} \frac{dx}{(1+x^2)^{n+1}} = \pi \cdot \frac{1 \cdot 3 \cdot 5 \cdots (2n-1)}{2 \cdot 4 \cdot 6 \cdots (2n)} \quad (n \in \mathbb{N}).$$

Let $f(z) = \frac{1}{(1+z^2)^{n+1}}$ and use the same contour as in the preceding example. Then $f(z)$ has a single pole in the upper semicircle, $z = i$, which is of order $n + 1$. The residue is given by

$$\frac{1}{n!} \left\{ \frac{d^n}{dz^n} \frac{(z-i)^{n+1}}{(1+z^2)^{n+1}} \right\} \bigg|_{z=i} = \frac{1}{n!} \left\{ \frac{d^n}{dz^n} \frac{1}{(z+i)^{n+1}} \right\} \bigg|_{z=i}$$

$$= (-1)^n \frac{(n+1)(n+2)\cdots(2n)}{n!(2i)^{2n+1}}$$

$$= \frac{(2n)!}{2^{2n}(n!)^2} \cdot \frac{1}{2i}.$$

$$\therefore \int_{\Gamma_R} \frac{dz}{(1+z^2)^{n+1}} = \frac{\pi(2n)!}{2^{2n}(n!)^2}.$$

As before,

$$\left| \int_{C^+(0,R)} \right| < \frac{\pi R}{(R^2-1)^{n+1}} \longrightarrow 0 \quad (R \to \infty),$$

and we obtain the desired result.

Remarks. (a) If $f(z) = \frac{p(z)}{q(z)}$ is a rational function with $\deg p(z) \le \deg q(z) - 2$, then the integral of $f(z)$ over the upper semicircle $C^+(0, R)$ tends to 0 (as $R \to \infty$).

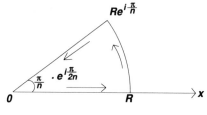

Figure 5.3

(b) Differentiate both sides of

$$\int_{-\infty}^{\infty} \frac{dx}{a + x^2} = \frac{\pi}{\sqrt{a}} \quad (a > 0)$$

n times with respect to a (why is this differentiation under the integral sign legitimate?), then setting $a = 1$, we obtain the same result.

EXAMPLE.

$$\int_0^{\infty} \frac{dx}{1 + x^{2n}} = \frac{\pi}{2n \sin \frac{\pi}{2n}} \quad (n \in \mathbb{N}).$$

We could integrate the function $f(z) = \frac{1}{1+z^{2n}}$ over the same contour as in the preceding examples, but then we would have to sum the residues at n poles in the upper semicircle. (The reader is advised to carry out this computation for comparison.) Instead we choose the boundary contour of the sector of angle $\frac{\pi}{n}$ as in Figure 5.3. Then inside the path there exists only one pole, a simple one at $z = e^{i\frac{\pi}{2n}}$, where the residue is

$$\frac{1}{2nz^{2n-1}}\Bigg|_{z=e^{i\frac{\pi}{2n}}} = -\frac{e^{i\frac{\pi}{2n}}}{2n}.$$

$$\therefore \int_0^R \frac{dx}{1 + x^{2n}} + \int_{C_R} \frac{dz}{1 + z^{2n}} + \int_{\Gamma_R} \frac{dz}{1 + z^{2n}} = 2\pi i \cdot \left(-\frac{e^{i\frac{\pi}{2n}}}{2n}\right)$$

$$= -i\frac{\pi e^{i\frac{\pi}{2n}}}{n}.$$

On C_R, $z = Re^{i\theta}$ ($0 \leq \theta \leq \frac{\pi}{n}$), and we have

$$\left|\int_{C_R} \frac{dz}{1 + z^{2n}}\right| \leq \frac{\pi R}{n(R^{2n} - 1)} \longrightarrow 0 \quad (R \to \infty).$$

On Γ_R, $z = re^{i\frac{\pi}{n}}$ ($0 \le r \le R$), and we have

$$\int_{\Gamma_R} \frac{dz}{1 + z^{2n}} = \int_R^0 \frac{e^{i\frac{\pi}{n}} dr}{1 + (re^{i\frac{\pi}{n}})^{2n}} = -e^{i\frac{\pi}{n}} \int_0^R \frac{dr}{1 + r^{2n}}.$$

Letting $R \to \infty$, we obtain

$$\left(1 - e^{i\frac{\pi}{n}}\right) \int_0^\infty \frac{dx}{1 + x^{2n}} = -i \frac{\pi e^{i\frac{\pi}{2n}}}{n}.$$

$$\int_0^\infty \frac{dx}{1 + x^{2n}} = -i \frac{e^{i\frac{\pi}{2n}}}{1 - e^{i\frac{\pi}{n}}} \cdot \frac{\pi}{n} = \frac{2i}{e^{i\frac{\pi}{2n}} - e^{-i\frac{\pi}{2n}}} \cdot \frac{\pi}{2n} = \frac{\pi}{2n \sin \dfrac{\pi}{2n}}.$$

EXAMPLE.

$$\int_0^\infty \frac{\cos x}{x^2 + a^2} dx = \frac{\pi e^{-a}}{2a} \quad (a > 0).$$

Because

$$\begin{aligned}
2 \int_0^R \frac{\cos x}{x^2 + a^2} dx &= \int_0^R \frac{e^{ix} + e^{-ix}}{x^2 + a^2} dx \\
&= \int_0^R \frac{e^{ix}}{x^2 + a^2} dx + \int_0^R \frac{e^{-ix}}{x^2 + a^2} dx \\
&= \int_0^R \frac{e^{ix}}{x^2 + a^2} dx + \int_{-R}^0 \frac{e^{ix}}{x^2 + a^2} dx \\
&= \int_{-R}^R \frac{e^{ix}}{x^2 + a^2} dx,
\end{aligned}$$

we integrate the function $f(z) = \frac{e^{iz}}{z^2 + a^2}$ over the upper semicircle with center at the origin and radius R (Figure 5.4). Our function $f(z)$ is analytic everywhere in the upper half-plane except at ai, where it has a simple pole. The residue there is

$$\lim_{z \to ai} (z - ai) \frac{e^{iz}}{z^2 + a^2} = \frac{e^{-a}}{2ai}.$$

$$\therefore \int_{-R}^R \frac{e^{ix}}{x^2 + a^2} dx + \int_{C^+(0,R)} \frac{e^{iz}}{z^2 + a^2} dz = 2\pi i \cdot \frac{e^{-a}}{2ai} = \frac{\pi e^{-a}}{a}.$$

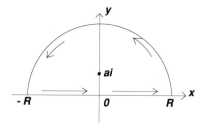

Figure 5.4

On $C^+(0, R)$, we have $z = Re^{i\theta}$ $(0 \le \theta \le \pi)$, and

$$\therefore \left| \int_{C^+(0,R)} \frac{e^{iz}}{z^2 + a^2} dz \right| = \left| \int_0^\pi \frac{e^{R(-\sin\theta + i\cos\theta)}}{R^2 e^{2i\theta} + a^2} \cdot iRe^{i\theta} d\theta \right|$$

$$\le \frac{R}{R^2 - a^2} \int_0^\pi e^{-R\sin\theta} d\theta$$

$$\le \frac{\pi R}{R^2 - a^2} \longrightarrow 0 \quad (R \to \infty).$$

As for the first integral on the left, we have

$$\int_{-R}^R \frac{e^{ix}}{x^2 + a^2} dx = 2 \int_0^R \frac{\cos x}{x^2 + a^2} dx \longrightarrow 2 \int_0^\infty \frac{\cos x}{x^2 + a^2} dx \quad (R \to \infty).$$

$$\therefore \int_0^\infty \frac{\cos x}{x^2 + a^2} dx = \frac{\pi e^{-a}}{2a}.$$

EXAMPLE.

$$\int_0^\infty \frac{\sin x}{x} dx = \frac{\pi}{2}.$$

In this case, if we use the most obvious complex function $\frac{\sin z}{z}$, which is an entire function (the origin is a removable singularity), we will not obtain the desired result. The trouble occurs because of the unboundedness of this function when $\Im z$ becomes large. However, note that $\frac{\cos x}{x}$ is an odd function; therefore, the Cauchy principal value of its integral, $p.v. \int_{-\infty}^\infty \frac{\cos x}{x} dx$, is 0. (Note that $\frac{\cos z}{z}$ has a simple pole at the origin.) Therefore, we have

$$\int_0^\infty \frac{\sin x}{x} dx = \frac{1}{2} p.v. \int_{-\infty}^\infty \frac{\sin x - i\cos x}{x} dx = \frac{1}{2i} p.v. \int_{-\infty}^\infty \frac{e^{ix}}{x} dx.$$

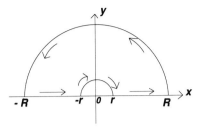

Figure 5.5

Hence, we proceed as follows: Around the path $\Gamma_{r,R}$ shown in Figure 5.5, we have

$$\int_{\Gamma_{r,R}} \frac{e^{iz}}{z} dz = 0.$$

This follows from the Cauchy-Goursat Theorem 4.3.1, for the integrand is analytic on $[\Gamma_{r,R}]$.

$$\therefore \int_r^R \frac{e^{ix}}{x} dx + \int_{C^+(0,R)} \frac{e^{iz}}{z} dz + \int_{-R}^{-r} \frac{e^{ix}}{x} dx - \int_{C^+(0,r)} \frac{e^{iz}}{z} dz = 0.$$

Because

$$\int_r^R \frac{e^{ix}}{x} dx + \int_{-R}^{-r} \frac{e^{ix}}{x} dx = \int_r^R \frac{e^{ix} - e^{-ix}}{x} dx = 2i \int_r^R \frac{\sin x}{x} dx,$$

we need to evaluate

$$\lim_{R \to \infty} \int_{C^+(0,R)} \frac{e^{iz}}{z} dz \quad \text{and} \quad \lim_{r \to 0} \int_{C^+(0,r)} \frac{e^{iz}}{z} dz.$$

Now, on $C^+(0, R)$ we have $z = R(\cos\theta + i\sin\theta)$ $(0 \le \theta \le \pi)$, and

$$\therefore \int_{C^+(0,R)} \frac{e^{iz}}{z} dz = i \int_0^\pi e^{-R(\sin\theta - i\cos\theta)} d\theta.$$

Because $\frac{\sin\theta}{\theta}$ is continuous and strictly positive for θ in the compact interval $[0, \frac{\pi}{2}]$, the minimum m of $\frac{\sin\theta}{\theta}$ is a strictly positive number.[1]

---

[1] Looking at the graph of $\sin\theta$ and $\frac{2\theta}{\pi}$, we easily see (and prove) that $m = \frac{2}{\pi}$; that is, we have the *Jordan inequality*:

$$\frac{\sin\theta}{\theta} \ge \frac{2}{\pi} \quad (0 \le \theta \le \frac{\pi}{2});$$

but the actual value of m is inessential—all we need is $m > 0$.

Therefore, using the formula $\sin(\pi - \theta) = \sin\theta$, we have

$$\left| \int_{C+(0,R)} \frac{e^{iz}}{z} dz \right| \leq \int_0^\pi e^{-R\sin\theta} d\theta = 2 \int_0^{\frac{\pi}{2}} e^{-R\sin\theta} d\theta$$

$$< 2 \int_0^{\frac{\pi}{2}} e^{-mR\theta} d\theta = \frac{2}{mR} \left(1 - e^{-\frac{mR\pi}{2}} \right)$$

$$\longrightarrow 0 \quad (R \to \infty).$$

(We could have used the Lebesgue bounded convergence theorem here.) Next, for z near the origin we have

$$\frac{e^{iz}}{z} = \frac{1}{z} \cdot \sum_{n=0}^\infty \frac{(iz)^n}{n!} = \frac{1}{z} + P(z),$$

where $P(z)$ is a power series with radius of convergence equal to ∞. Hence, *a fortiori*, $P(z)$ is a continuous function near the origin. Thus, $P(z)$ is bounded there. Let $|P(z)| < M$ for $|z| \leq 1$. Then

$$\left| \int_{C+(0,r)} P(z) \, dz \right| \leq M \cdot \pi r \longrightarrow 0 \quad (r \to 0),$$

and

$$-\int_{C+(0,r)} \frac{dz}{z} = -i \int_0^\pi d\theta = -\pi i.$$

Therefore,

$$\lim_{r \to 0} \left(-\int_{C+(0,r)} \frac{e^{iz}}{z} dz \right) = -\pi i.$$

Thus, we obtain

$$2i \int_0^\infty \frac{\sin x}{x} dx - \pi i = 0,$$

and finally

$$\int_0^\infty \frac{\sin x}{x} dx = \frac{\pi}{2}.$$

Remark.

$$\int_0^\infty \frac{\sin cx}{x} dx = \begin{cases} \dfrac{\pi}{2} & (c > 0); \\ 0 & (c = 0); \\ -\dfrac{\pi}{2} & (c < 0), \end{cases}$$

so that

$$\frac{2}{\pi} \int_0^\infty \frac{\sin cx}{x} dx = \operatorname{sgn} c \quad (c \in \mathbb{R}).$$

This integral is known as the *Dirichlet discontinuous factor*. It plays an important role in the theory of Fourier analysis.

EXAMPLE.

$$\int_{-\frac{\pi}{2}}^{\frac{\pi}{2}} \frac{d\theta}{a + b\sin\theta} = \frac{\pi}{\sqrt{a^2 - b^2}} \quad (a > b > 0).$$

When the integrand is a rational function of $\sin\theta$ and $\cos\theta$, it is a standard technique to convert the integrand into a rational function of z by using the substitutions

$$\sin\theta = \frac{1}{2i}\left(z - \frac{1}{z}\right), \quad \cos\theta = \frac{1}{2}\left(z + \frac{1}{z}\right) \quad \left(z = e^{i\theta}, \quad d\theta = \frac{dz}{iz}\right).$$

Making this substitution, we obtain

$$\begin{aligned}
\int_{-\frac{\pi}{2}}^{\frac{\pi}{2}} \frac{d\theta}{a + b\sin\theta} &= \frac{1}{2}\int_{-\pi}^{\pi} \frac{d\theta}{a + b\sin\theta} \\
&= \frac{1}{2}\int_{C(0,1)} \frac{1}{a + \dfrac{b}{2i}\left(z - \dfrac{1}{z}\right)} \cdot \frac{dz}{iz} \\
&= \int_{C(0,1)} \frac{dz}{bz^2 + 2aiz - b} \\
&= \int_{C(0,1)} \frac{dz}{b(z - \alpha)(z - \beta)},
\end{aligned}$$

where

$$\alpha = i\frac{-a + \sqrt{a^2 - b^2}}{b}, \quad \beta = i\frac{-a - \sqrt{a^2 - b^2}}{b}.$$

Because $\alpha \cdot \beta = -1$ and $|\alpha| < |\beta|$, we have $|\alpha| < 1$ and $|\beta| > 1$ (Figure 5.6). The only singularity of the integrand $f(z) = \frac{1}{b(z-\alpha)(z-\beta)}$ inside the unit circle is $z = \alpha$, which is a simple pole. Finally,

$$\begin{aligned}
\int_{-\frac{\pi}{2}}^{\frac{\pi}{2}} \frac{d\theta}{a + b\sin\theta} &= 2\pi i \cdot \operatorname{Res}(\alpha; f) \\
&= \frac{2\pi i}{b(\alpha - \beta)} = \frac{\pi}{\sqrt{a^2 - b^2}}.
\end{aligned}$$

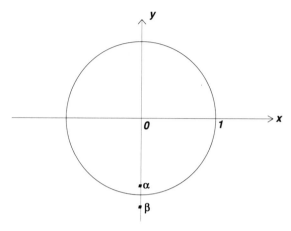

Figure 5.6

Remark. Differentiating both sides of the result with respect to a under the integral sign (why is this differentiation legitimate?), we obtain

$$\int_{-\frac{\pi}{2}}^{\frac{\pi}{2}} \frac{d\theta}{(a + b\sin\theta)^2} = \frac{\pi}{(a^2 - b^2)^{\frac{3}{2}}} \quad (a > b > 0).$$

EXAMPLE.

$$\int_0^{\pi} \frac{\cos n\theta}{1 - 2a\cos\theta + a^2}\,d\theta = \frac{\pi a^n}{1 - a^2} \quad (-1 < a < 1, \ n = 0, 1, 2, \ldots).$$

Because $(1 - 2a\cos\theta + a^2) = (e^{i\theta} - a)(e^{-i\theta} - a)$, the integrand is the real part of the coefficient of t^n in

$$\frac{\sum_{k=0}^{\infty}(tz)^k}{(z - a)\left(\dfrac{1}{z} - a\right)} \quad (z = e^{i\theta}).$$

If $|t| < 1$, then the function

$$\frac{1}{(z - a)(1 - az)(1 - tz)} \quad (z \in D)$$

has only one pole inside the unit circle, a simple one at a. Hence, by the Cauchy Residue Theorem 5.4.1,

$$\frac{1}{2\pi i} \int_{C(0,1)} \frac{dz}{(z-a)(1-az)(1-tz)} = \frac{1}{(1-a^2)(1-ta)}.$$

Using $z = e^{i\theta}$ to change the variable of integration, we obtain

$$\frac{1}{2\pi i} \int_{C(0,1)} \frac{dz}{(z-a)(1-az)(1-tz)}$$

$$= \frac{1}{2\pi} \int_0^{2\pi} \frac{d\theta}{(1 - 2a\cos\theta + a^2)(1 - te^{i\theta})}.$$

$$\therefore \quad \frac{1}{2\pi} \int_0^{2\pi} \frac{\sum_{n=0}^{\infty} t^n e^{in\theta}}{1 - 2a\cos\theta + a^2} d\theta = \frac{\sum_{n=0}^{\infty} t^n a^n}{1 - a^2}.$$

Comparing the coefficients of t^n on both sides, we obtain

$$\int_0^{2\pi} \frac{e^{in\theta}}{1 - 2a\cos\theta + a^2} d\theta = \frac{2\pi a^n}{1 - a^2}.$$

Finally, taking the real part of both sides, we obtain the desired result. *Question:* What if $a > 1$?

EXAMPLE.

$$\int_0^{\infty} \cos\left(x^2\right) dx = \int_0^{\infty} \sin\left(x^2\right) dx = \frac{\sqrt{\pi}}{2\sqrt{2}} \quad (Fresnel).$$

The function e^{iz^2} is entire; let us integrate this function around a sector of opening $\frac{\pi}{4}$ as shown in Figure 5.7. By the Cauchy-Goursat Theorem 4.3.1,

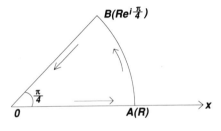

Figure 5.7

$$\int_0^R e^{ix^2}\, dx + \int_{\widehat{AB}} e^{iz^2}\, dz - \int_{[0,B]} e^{iz^2}\, dz = 0.$$

On $\widehat{AB}$, we have $z = R(\cos\theta + i\sin\theta)$ $(0 \le \theta \le \frac{\pi}{4})$,

$$\therefore \ \left| \int_{\widehat{AB}} e^{iz^2}\, dz \right| = \left| \int_0^{\frac{\pi}{4}} e^{iR^2(\cos 2\theta + i\sin 2\theta)} \cdot iRe^{i\theta}\, d\theta \right|$$

$$\le R \int_0^{\frac{\pi}{4}} e^{-R^2 \sin 2\theta}\, d\theta = \frac{R}{2} \int_0^{\frac{\pi}{2}} e^{-R^2 \sin t}\, dt$$

$$\le \frac{R}{2} \int_0^{\frac{\pi}{2}} e^{-mR^2 t}\, dt = \frac{R}{2} \cdot \frac{1}{mR^2} \left(1 - e^{-\frac{1}{2}m\pi R^2} \right)$$

$$\longrightarrow 0 \quad (R \to \infty).$$

On $[O, B]$, we have $z = re^{i\frac{\pi}{4}}$, $z^2 = ir^2$ $(0 \le r \le R)$, and so, as $R \to \infty$, we obtain

$$\int_0^\infty e^{ix^2}\, dx = \lim_{R\to\infty} \int_{[0,B]} e^{iz^2}\, dz = \int_0^\infty e^{-r^2} \cdot e^{i\frac{\pi}{4}}\, dr = e^{i\frac{\pi}{4}} \cdot \frac{\sqrt{\pi}}{2}.$$

Taking the real and imaginary parts of both sides, we obtain the desired results.

EXAMPLE.

$$\int_{-\infty}^\infty e^{-ax^2} \cos 2bx\, dx = \sqrt{\frac{\pi}{a}} e^{-\frac{b^2}{a}} \quad (a > 0,\ b > 0).$$

Because

$$e^{-ax^2} \cos 2bx = \Re\left\{ e^{-ax^2} \cdot e^{-i2bx} \right\} = \Re\left\{ e^{-a(x+i\frac{b}{a})^2} \right\} \cdot e^{-\frac{b^2}{a}},$$

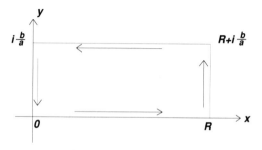

Figure 5.8

we integrate the entire function e^{-az^2} around the rectangle shown in Figure 5.8. We obtain

$$\int_0^R e^{-ax^2}\,dx + i\int_0^{\frac{b}{a}} e^{-a(R+iy)^2}\,dy + \int_R^0 e^{-a(x+i\frac{b}{a})^2}\,dx + i\int_{\frac{b}{a}}^0 e^{ay^2}\,dy = 0.$$

Taking the real part, we obtain

$$\int_0^R e^{-ax^2}\,dx + e^{-aR^2}\int_0^{\frac{b}{a}} e^{ay^2}\sin(2aRy)\,dy - e^{\frac{b^2}{a}}\int_0^R e^{-ax^2}\cos 2bx\,dx = 0.$$

The second integral is bounded, and so the second term tends to zero as $R \to \infty$; hence

$$e^{\frac{b^2}{a}}\int_0^\infty e^{-ax^2}\cos 2bx\,dx = \int_0^\infty e^{-ax^2}\,dx = \frac{1}{2}\sqrt{\frac{\pi}{a}},$$

which gives the desired result.

Note that no residue was used in the last two examples.

5.6   The Argument Principle

Suppose $f(z)$ is analytic and has a root of order m at a point a; then we can write $f(z) = (z-a)^m \cdot \varphi(z)$, where $\varphi(z)$ is analytic at a and $\varphi(a) \neq 0$. It follows that

$$\frac{f'(z)}{f(z)} = \frac{m}{z-a} + \frac{\varphi'(z)}{\varphi(z)},$$

and because $\frac{\varphi'(z)}{\varphi(z)}$ is analytic at a, the logarithmic derivative, $\frac{f'(z)}{f(z)}$, of $f(z)$ has a simple pole with residue m at a.

Therefore, if $f(z)$ is analytic on $[\Gamma]$, where Γ is a curve and $f(z)$ has no root on Γ, then, by the Residue Theorem 5.4.1,

$$\frac{1}{2\pi i}\int_\Gamma \frac{f'(z)}{f(z)}\,dz = N_0,$$

where N_0 is the total number of roots of $f(z)$ inside the curve Γ, with multiplicity counted. We obtain this result because the only singularities of $\frac{f'(z)}{f(z)}$ arise from the roots of $f(z)$ and the corresponding residue is just its multiplicity.

Now if $f(z)$ has a pole of order m, then we simply replace m by $-m$ and the roots by the poles in the paragraph above; we obtain

THEOREM 5.6.1 (Argument Principle) *If $f(z)$ is analytic except possibly for some poles in $[\Gamma]$, where Γ is a curve on which $f(z)$ has neither root nor pole, then*

$$\frac{1}{2\pi i}\int_\Gamma \frac{f'(z)}{f(z)}dz = N_0 - N_\infty,$$

where N_0 and N_∞ are, respectively, the number of roots and poles of $f(z)$ in (Γ), counting their multiplicities. In particular, if $f(z)$ is analytic, then

$$\frac{1}{2\pi i}\int_\Gamma \frac{f'(z)}{f(z)}dz = N_0.$$

But why is this known as the argument principle? Suppose $f(z)$ is expressed in polar form:

$$f(z) = Re^{i\theta}, \quad R = |f(z)|, \quad \theta = \arg f(z).$$

Then

$$f'(z)dz = df(z) = d\left(Re^{i\theta}\right) = e^{i\theta}(dR + iR\,d\theta);$$

hence

$$\frac{1}{2\pi i}\int_\Gamma \frac{f'(z)}{f(z)}dz = \frac{1}{2\pi i}\int_\Gamma \frac{dR}{R} + \frac{1}{2\pi}\int_\Gamma d\theta.$$

Now

$$\frac{1}{2\pi i}\int_\Gamma \frac{dR}{R} = [\text{Log } R]_\Gamma = [\text{Log } |f(z)|]_\Gamma = 0,$$

because $\text{Log } |f(z)|$ returns to its initial value when the point z traverses the entire curve Γ (Figure 5.9). Note that $w = f(z)$ will describe a closed arc in the w-plane when z traverses Γ in the z-plane (because $f(z)$ is single-valued).

On the other hand,

$$\int_\Gamma d\theta = [\theta]_\Gamma = [\arg f(z)]_\Gamma$$

and this time we cannot conclude that the integral is zero, for $\arg f(z)$ does not necessarily return to its initial value as z traverses the entire curve Γ. If z_0 is a point on Γ, then

$$\int_\Gamma d\theta = \theta^*(z_0) - \theta(z_0) = [\arg f(z_0)]^* - [\arg f(z_0)],$$

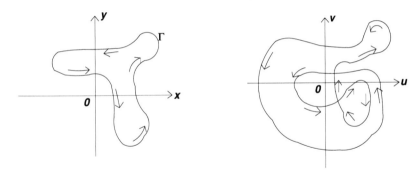

Figure 5.9

where * indicates the value after the curve Γ has been traversed. Thus

$$N_0 - N_\infty = \frac{1}{2\pi i} \int_\Gamma \frac{f'(z)}{f(z)} dz = \frac{1}{2\pi} [\arg f(z)]_\Gamma$$

$$= \frac{1}{2\pi} \cdot (\text{the increment of } \arg f(z) \text{ as } z \text{ traverses } \Gamma).$$

This also explains why the integral $\frac{1}{2\pi i} \int_\Gamma \frac{f'(z)}{f(z)} dz$ is an integer. Indeed, the various possible values of the argument of a complex number can differ from each other only by an integer multiple of 2π.

EXAMPLE. Let $f(z) = \prod_{j=1}^n (z - z_j)$; then

$$\frac{f'(z)}{f(z)} = \sum_{j=1}^n \frac{1}{z - z_j}.$$

Hence, if $r \neq |z_j|$ $(j = 1, 2, \ldots, n)$, then we have

$$\frac{1}{2\pi i} \int_{C(0,r)} \frac{f'(z)}{f(z)} dz = \frac{1}{2\pi i} \sum_{j=1}^n \int_{C(0,r)} \frac{dz}{z - z_j} = \sum_{z_j \in D(0,r)} 1$$

$$= \text{ the number of roots inside the circle } C(0, r).$$

Note that

$$\arg f(z) = \sum_{j=1}^n \arg(z - z_j),$$

and as z makes one full circle $C(0, r)$, $\arg(z - z_j)$ increases by 2π if z_j is inside the circle, but returns to its original value if z_j is outside the circle.

EXAMPLE. (Titchmarsh) Consider the problem of finding the general location of the roots of

$$f(z) = z^4 + z^3 + 4z^2 + 2z + 3.$$

For z on the positive real axis, $f(z) > 0$, hence there is no root on the positive real axis. To study the case that z is on the negative real axis, replace z by $-x$ and let x range over $[0, \infty)$. We obtain

$$f(-x) = x^4 - x^3 + 4x^2 - 2x + 3 = x^2(x^2 - x + 4) + (3 - 2x)$$
$$= x^2 \left\{ \left(x - \frac{1}{2} \right)^2 + \left(4 - \frac{1}{4} \right) \right\} + (3 - 2x),$$

and this expression is obviously positive for $0 \le x \le \frac{3}{2}$. Now, rewrite $f(-x)$ to obtain

$$f(-x) = (x^4 - x^3) + (4x^2 - 2x + 3)$$
$$= x^3(x - 1) + 4 \left(x - \frac{1}{4} \right)^2 + \left(3 - \frac{1}{4} \right) > 0 \quad (x > 1).$$

Therefore, $f(z) > 0$ for any real value of z.

On the imaginary axis, let $z = iy$; then

$$f(z) = y^4 - iy^3 - 4y^2 + 2iy + 3$$
$$= (y^4 - 4y^2 + 3) - i(y^3 - 2y)$$
$$= (y^2 - 3)(y^2 - 1) - iy(y^2 - 2).$$

If $f(z) = 0$ on the imaginary axis, then both the real and imaginary parts must be zero. But the real part is zero only at $y = \pm 1, \pm\sqrt{3}$, and the imaginary part is zero only at $y = 0$ and $\pm\sqrt{2}$, so $f(z) \ne 0$ on the imaginary axis.

Now the roots of a polynomial with real coefficients must either be real or occur in conjugate pairs. Because $f(z)$ has neither a real nor a purely imaginary root, we have the following three possibilities:

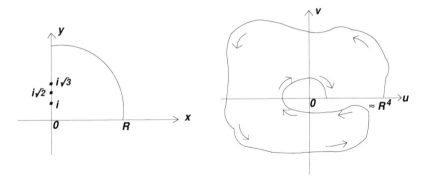

Figure 5.10

(a) Two roots in the first quadrant and two roots (conjugates of the first pair) in the fourth quadrant;

(b) Two roots in the second quadrant and two conjugate roots in the third quadrant;

(c) One root in each quadrant;

Thus it follows that if we can determine the number of roots in any given quadrant, then we know where all the roots are. We choose to examine the first quadrant. To this end we evaluate $f(z)$ along the real axis from $z = 0$ to $z = R$, then along the circle $C(0, R)$ to the "positive" imaginary axis, and back to the origin to see how many times $w = f(z)$ winds around the origin in the w-plane.

At $z = 0$, $f(z) = 3$; between $0 \le x < \infty$, $f(x)$ is never negative and if we choose R sufficiently large, then $f(R) \approx R^4$.

On $z = Re^{i\theta}$ $(0 \le \theta \le \frac{\pi}{2})$, z^4 transverses the circle $C(0, R^4)$; however, $\arg f(z)$ does not quite increase by 2π. In fact, $\Im f(iR) < 0$, and writing $f(z) = \rho e^{i\varphi}$ and letting $z = Re^{i\frac{\pi}{2}}$, we find that

$$\varphi = \arctan \frac{-R(R^2 - 2)}{(R^2 - 3)(R^2 - 1)} \approx -\arctan\left(\frac{1}{R}\right),$$

and so $f(iR)$ is in the fourth quadrant at some small angle below the positive real axis.

On the positive imaginary axis above $y = \sqrt{3}$, $\Re f(z) > 0$ and $\Im f(z) < 0$, and so $f(z)$ is in the fourth quadrant and crosses into the third quadrant when $y = \sqrt{3}$, as $f(i\sqrt{3}) = -i\sqrt{3}$.

For $\sqrt{2} < y < \sqrt{3}$, $\Re f(z) < 0$ and $\Im f(z) < 0$, and so $f(z)$ remains in the third quadrant and crosses into the second quadrant when $y = \sqrt{2}$, as $f(i\sqrt{2}) = -1$.

For $1 < y < \sqrt{2}$, $\Re f(z) < 0$ and $\Im f(z) > 0$, hence $f(z)$ remains in the second quadrant and crosses into the first quadrant when $y = 1$ as $f(i) = i$.

For $0 < y < 1$, $f(z)$ is in the first quadrant and $w = f(z)$ proceeds to the starting point $f(0) = 3$.

The results of all the above calculations are indicated in Figure 5.10. It is now evident that $w = f(z)$ has not wound around the origin as z has made the complete circuit and so, by the argument principle, there exists no root in the first quadrant. This result implies no root exists in the fourth quadrant, either. Therefore, we conclude that there are two roots in both the second and third quadrants.

DEFINITION 5.6.2 *A function $f(z)$ is said to be* **meromorphic** *in a region Ω if it has no singularities other than poles in Ω.*

THEOREM 5.6.3 *Let $f(z)$ be meromorphic and $g(z)$ analytic on $[\Gamma]$, where Γ is a curve. Then*

$$\frac{1}{2\pi i} \int_\Gamma g(z) \cdot \frac{f'(z)}{f(z)} dz = \sum_{j=1}^n g(a_j) - \sum_{k=1}^m g(b_k),$$

where $a_1, a_2, \ldots, a_n$ and $b_1, b_2, \ldots, b_m$ are the roots and poles, respectively, of $f(z)$ inside (Γ) (but no root or pole on Γ), and they occur as often as their multiplicity.

Proof. The proof is a slight modification of that of the argument principle, and hence is left to the reader. $\square$

Note that the argument principle may be used to count the number of times, N_α, a meromorphic function $f(z)$ takes the value α inside the curve. Indeed, because $f(z) - \alpha$ is zero if $f(z) = \alpha$, it follows that

$$N_\alpha - N_\infty = \frac{1}{2\pi i} \int_\Gamma \frac{f'(z)}{f(z) - \alpha} dz.$$

In particular, if $f(z)$ is analytic on $[\Gamma]$ (and, of course, never equal to α on Γ), then

$$N_\alpha = \frac{1}{2\pi i} \int_\Gamma \frac{f'(z)}{f(z) - \alpha} dz.$$

THOEREM 5.6.4 (Hurwitz) *Suppose a sequence $\{f_n(z)\}_{n=1}^{\infty}$ of analytic functions converges locally uniformly in a region Ω to a function $f(z)$ not identically zero, and $f(z_0) = 0$ ($z_0 \in \Omega$). Then z_0 is an accumulation point of roots of $f_n(z)$ ($n \in \mathbb{N}$).*

Proof. Note first that $f(z) \in H(\Omega)$, by Corollary 4.5.5. Because $f(z)$ is not identically 0, and the roots of the analytic function $f(z)$ are isolated, we may choose r so small that the closed disc $\bar{D}(z_0, r)$ is entirely inside the region Ω and $f(z)$ does not vanish on the closed disc other than at the point z_0. Inasmuch as

$$f_n(z) \longrightarrow f(z) \quad \text{uniformly on the circle } C(z_0, r),$$

the functions $f_n(z)$ will not vanish on the circle $C(z_0, r)$ for sufficiently large n, and by Corollary 4.5.5 we also have

$$f_n'(z) \longrightarrow f'(z) \quad \text{uniformly on the circle } C(z_0, r).$$

Therefore, $\frac{f_n'(z)}{f(z)} \longrightarrow \frac{f'(z)}{f(z)}$ uniformly on the circle $C(z_0, r)$ and

$$\lim_{n \to \infty} \frac{1}{2\pi i} \int_{C(z_0,r)} \frac{f_n'(z)}{f_n(z)} dz = \frac{1}{2\pi i} \int_{C(z_0,r)} \frac{f'(z)}{f(z)} dz = m,$$

where m is the multiplicity of the root of $f(z)$ at $z = z_0$. Because

$$\frac{1}{2\pi i} \int_{C(z_0,r)} \frac{f_n'(z)}{f_n(z)} dz$$

is an integer for each n, we obtain

$$\frac{1}{2\pi i} \int_{C(z_0,r)} \frac{f_n'(z)}{f_n(z)} dz = m \quad \text{for sufficiently large } n.$$

Thus $f_n(z)$ has m roots inside the circle $C(z_0, r)$ if n is sufficiently large. $\square$

Note carefully that we have proved much more than the statement of the theorem. Also, the condition that $f(z)$ is not identically zero is necessary. For, let $f_n(z) = \frac{e^z}{n}$ ($n \in \mathbb{N}$). Then $\{f_n(z)\}_{n=1}^{\infty}$ converges locally uniformly to $f(z) \equiv 0$ on the complex plane $\mathbb{C}$, while $f_n(z) \neq 0$

for any $z \in \mathbb{C}$ ($n \in \mathbb{N}$). Actually, we could take $f_n(z) = \frac{1}{n}$. (Then the convergence is uniform throughout the complex plane $\mathbb{C}$.)

DEFINITION 5.6.5 *A function* $f(z)$ *is said to be* **univalent** *on* Ω *if* $f(z_1) = f(z_2)$ ($z_1, z_2 \in \Omega$) *implies* $z_1 = z_2$. *(The German word* **schlicht** *is often used.)*

COROLLARY 5.6.6 *Suppose* $\{f_n(z)\}_{n=1}^{\infty}$ *is a sequence of functions analytic and univalent on a region* Ω, *converging locally uniformly to a nonconstant function* $f(z)$ *on* Ω. *Then* $f(z)$ *is also univalent on* Ω.

COROLLARY 5.6.7 *Suppose* z_0 *is a root of* $f(z) = \sum_{k=0}^{\infty} a_k z^k$ *inside the circle of convergence. Then* z_0 *is an accumulation point of the roots of the partial sums*

$$f_n(z) = \sum_{k=0}^{n} a_k z^k \quad (n \in \mathbb{N}).$$

Remark. Jentzsch showed that every point on the circle of convergence of a power series is an accumulation point of roots of the polynomials that are the partial sums of the power series.

THEOREM 5.6.8 (Rouché) *Suppose* $f(z)$ *and* $g(z)$ *are analytic on* $[\Gamma]$, *where* Γ *is a curve, and*

$$|f(z)| > |g(z)| \quad \text{for } z \text{ on } \Gamma.$$

Then the functions

$$f(z) + g(z) \quad \text{and} \quad f(z)$$

have the same number of roots inside the curve Γ.

    Proof. Let

$$N(t) = \frac{1}{2\pi i} \int_{\Gamma} \frac{f'(z) + tg'(z)}{f(z) + tg(z)} dz \quad (0 \le t \le 1).$$

Note that $f(z) + tg(z) \ne 0$ on Γ, because

$$|f(z)| > |g(z)| \ge |tg(z)| \quad \text{on } \Gamma.$$

We want to show that $N(0) = N(1)$. Because $N(t)$ is integer-valued, it will be sufficient to show that $N(t)$ depends continuously on t.

$$N(t_1) - N(t_2) = \frac{1}{2\pi i} \int_{\Gamma} \left\{ \frac{f'(z) + t_1 g'(z)}{f(z) + t_1 g(z)} - \frac{f'(z) + t_2 g'(z)}{f(z) + t_2 g(z)} \right\} dz$$

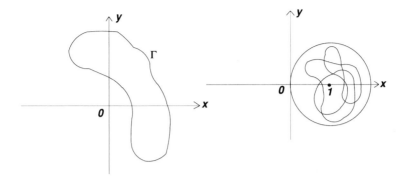

Figure 5.11

$$= \frac{1}{2\pi i} \int_\Gamma \frac{(t_1 - t_2)\{f(z)g'(z) - f'(z)g(z)\}}{\{f(z) + t_1 g(z)\}\{f(z) + t_2 g(z)\}} dz.$$

Therefore,

$$|N(t_1) - N(t_2)| \leq \frac{|t_1 - t_2|}{2\pi} \cdot \frac{ML}{m^2},$$

where

$$M = \max\{|f(z)g'(z) - f'(z)g(z)|\,;\, z \in \Gamma\},$$
$$m = \min\{|f(z)| - |g(z)|\,;\, z \in \Gamma\} > 0,$$

and L is the length of the curve Γ. This inequality shows that $N(t)$ is a (uniformly) continuous function of t.   □

Remark. Let us write

$$f(z) + g(z) = f(z) \cdot \left\{1 + \frac{g(z)}{f(z)}\right\},$$

$$\arg\{f(z) + g(z)\} = \arg f(z) + \arg\left\{1 + \frac{g(z)}{f(z)}\right\}.$$

Because $\left|\frac{g(z)}{f(z)}\right| < 1$ for z on the curve Γ, $1 + \frac{g(z)}{f(z)}$ remains inside the circle $C(1, 1)$ as z traverses the curve Γ (Figure 5.11). Therefore, the increment of $\arg\left\{1 + \frac{g(z)}{f(z)}\right\}$ is zero as z traverses Γ.

We can now give another proof of the Fundamental Theorem of Algebra 2.10.3. Suppose

$$p(z) = a_0 + a_1 z + \cdots + a_{n-1} z^{n-1} + a_n z^n \quad (a_n \neq 0,\ n \geq 1).$$

Let

$$f(z) = a_n z^n, \quad g(z) = a_0 + a_1 z + \cdots + a_{n-1} z^{n-1},$$

and take r so large that the absolute value of $f(z)$ dominates that of $g(z)$ on $C(0, r)$. (See the proof of Theorem 2.10.3.) By the Rouché Theorem 5.6.8, $p(z) = f(z) + g(z)$ and $f(z)$ have exactly the same number of roots inside the circle $C(0, r)$, and $f(z)$ has n. This shows that $p(z)$ not only has a root, it has n roots.

EXAMPLE. Let $f(z) = \left(z - \frac{1}{2}\right) \cdot \left(z + \frac{1}{2}\right)$. Then for z on the unit circle,

$$\left|z - \frac{1}{2}\right| \geq \frac{1}{2}, \quad \text{and} \quad \left|z + \frac{1}{2}\right| \geq \frac{1}{2}.$$

Thus we have, very crudely,

$$|f(z)| \geq \frac{1}{4} \quad \text{for } z \in C(0, 1).$$

Hence, if $|c| < \frac{1}{4}$, then, by the Rouché Theorem 5.6.8, the equations $f(z) - c = 0$ and $f(z) = 0$ have the same number of solutions; that is, $f(z)$ will take on the value c twice in the unit disc.

EXAMPLE. If $f(z) = (z + 2)(z + 3)$, then for z on the unit circle,

$$|z + 2| \geq 1, \quad |z + 3| \geq 2.$$

Hence

$$|f(z)| \geq 2 \quad \text{for } z \in C(0, 1).$$

Therefore, if $|c| < 2$, then $f(z) \neq c$ in the unit disc.

As an application of the Rouché Theorem 5.6.8, we prove the following useful result, known as the *open mapping theorem*.

THEOREM 5.6.9 *Suppose $f(z) \in H(\Omega)$ is not a constant. Then the image*

$$f(G) = \{w \in \mathbb{C}; \ w = f(z) \ \text{for some} \ z \in G\}$$

of any open set $G \subset \Omega$ is an open set.

Proof. It is sufficient to show that for an arbitrary point $a \in G$ there exists a positive number $\delta > 0$ such that $D(f(a), \delta) \subset f(G)$. Because

$f(z)$ is not constant, there exists a first nonzero coefficient of the Taylor-series expansion of $f(z)$ about the point a, say the nth.

$$f(z) = f(a) + \sum_{k=n}^{\infty} c_k(z-a)^k \quad (c_n \neq 0)$$

$$= f(a) + (z-a)^n \cdot \sum_{k=n}^{\infty} c_k(z-a)^{k-n}.$$

The function $g(z) = \sum_{k=n}^{\infty} c_k(z-a)^{k-n}$ is analytic in the closed disc $\bar{D}(a, r), 0 < r < \text{dist}(a, \partial\Omega)$, and

$$f(z) - f(a) = (z-a)^n \cdot g(z), \quad g(a) \neq 0.$$

(Actually, $g(z)$ can be "continued analytically" to Ω.) See Chapter 8. Because $g(z)$ is continuous, we may assume, by replacing r by a smaller value if necessary, that

$$g(z) \neq 0 \quad \text{for any } z \in \bar{D}(a, r).$$

Set $\delta = \min\{|(z-a)^n \cdot g(z)| \,;\, z \in C(a, r)\}$. Then, clearly, $\delta > 0$, for $C(a, r)$ is compact. Now, for arbitrary $\beta \in D(f(a), \delta)$, we have

$$|f(a) - \beta| < \delta \leq |(z-a)^n \cdot g(z)| \quad z \in C(a, r).$$

Therefore, by the Rouché Theorem 5.6.8,

$$f(z) - \beta = \{(z-a)^n \cdot g(z)\} + \{f(a) - \beta\}$$

has exactly n roots in $D(a, r)$. (Note that $n \geq 1$.) It follows that

$$D(f(a), \delta) \subset f(D(a, r)) \subset f(G). \qquad \square$$

Because the continuous image of a connected set is connected, we obtain

COROLLARY 5.6.10 *Suppose $f(z)$ is a nonconstant analytic function in a region Ω. Then the image*

$$f(\Omega) = \{w \in \mathbb{C} \,;\, w = f(z) \quad \text{for some } z \in \Omega\}$$

is a region.

Exercises

1. Expand the following functions in Laurent series centered at the origin.
   Indicate the annulus of convergence for each expansion.

   (a) $\dfrac{1}{z(z-1)(2-z)}$;

   (b) $\dfrac{7z-2}{z(z+1)(z-2)}$;

   (c) $\dfrac{1}{(z^2+1)(z-2)}$.

2. (a) Expand the function $e^{\frac{1}{z}}$ in Laurent series around the origin and
   show that

   $$\frac{1}{\pi} \int_0^\pi e^{\cos\theta} \cos(\sin\theta - n\theta)\, d\theta = \frac{1}{n!} \quad (n \geq 0).$$

   (b) Expand the function $e^{z+\frac{1}{z}}$ in Laurent series around the origin, and
   show that

   $$\frac{1}{2\pi} \int_0^{2\pi} e^{2\cos\theta} \cos(n\theta)\, d\theta = \sum_{k=0}^\infty \frac{1}{k!(n+k)!} \quad (n \geq 0).$$

3. Show that

   $$\frac{1}{2\pi i} \int_{C(0,1)} \frac{z^n e^{z\zeta}}{n!\,\zeta^{n+1}}\, d\zeta = \left(\frac{z^n}{n!}\right)^2,$$

   and

   $$\frac{1}{2\pi} \int_0^{2\pi} e^{2z\cos\theta}\, d\theta = \sum_{n=0}^\infty \left(\frac{z^n}{n!}\right)^2.$$

   Justify your steps.

4. Classify the singularities of the following functions, and find the singular part at each of the isolated singular points in $\hat{\mathbb{C}}$:

   (a) $\dfrac{\sin z}{z}$;

   (b) $\dfrac{1-\cos z}{z}$;

   (c) $\dfrac{1}{\sin z}$;

   (d) $\dfrac{z}{e^z-1}$;

(e) $\dfrac{(z^2 - 1)(z - 2)^3}{\sin \pi z}$;

(f) $\dfrac{1 - z^n}{1 - z^2}$ $(n \geq 1)$.

5. Discuss the behavior of the following functions at ∞.

(a) $\dfrac{1}{1 + z}$;

(b) $\dfrac{z^2}{z^3 + z + 1}$;

(c) $\dfrac{z^2}{e^z - 1}$.

6. Evaluate

(a) $\displaystyle\int_{C(0,e)} \dfrac{1 - \cos z}{(e^z - 1)\sin z}\,dz.$

(b) $\displaystyle\int_{C(0,\pi)} \dfrac{e^z - 1}{1 - \cos z}\,dz.$

(c) $\displaystyle\int_{C(0,e\pi)} \dfrac{\sin z}{1 - \cos z}\,dz.$

7. (a) Let $f(z)$ and $g(z)$ have poles of order m and n, respectively, at a point a. Discuss the behavior of the following functions at a:

$$f(z) \pm g(z), \quad f(z) \cdot g(z), \quad \dfrac{f(z)}{g(z)}.$$

(b) If the function $f(z)$ has a pole (of order m) at a and $p(z)$ is a polynomial (of degree n), show that the composite function $p(f(z))$ has a pole at a. What is its order?

8. Suppose $f(z)$ and $g(z)$ are entire functions such that $(f \circ g)(z)$ is a nonconstant polynomial. Show that both $f(z)$ and $g(z)$ are polynomials.

Hint: The Casorati-Weierstrass Theorem 5.2.4.

9. Find the singular part of the function $f(z) = \dfrac{1}{(1+z^3)^2}$ at $z = -1$.

10. Suppose $f(z)$ has an isolated singularity at a point a and, for some positive integer m, $(z - a)^m f(z)$ is bounded in a punctured disc $D'(a, r)$ $(r > 0)$. Show that a is either a removable singularity or a pole. If it is a pole, what can be said about its order?

11. Suppose $f(z)$ is analytic in the punctured unit disc $D'(0, 1)$ and that

$$\int \int_{D'(0,1)} |f'(z)|^2 \, dx \, dy < \infty.$$

Show that the origin is a removable singularity.

12. (a) Suppose $f(z)$ is bounded and analytic except at a finite number of points in the complex plane $\mathbb{C}$; show that $f(z)$ must be a constant.

    (b) Let $f(z)$ be analytic except at a finite number of poles and suppose that, for some positive numbers M, R and a positive integer n,

$$|f(z)| \leq M|z|^n \quad \text{whenever} \ |z| > R.$$

    Show that $f(z)$ must be a rational function.

    (c) Find all the entire functions $f(z)$ for which there exist positive constants M, R and a positive integer n such that

$$|f(z)| \geq M \cdot |z|^n \quad \text{whenever} \ |z| > R.$$

    (d) What if the word *entire* is replaced by *meromorphic* in (c)?

13. (a) Find all entire functions $f(z)$ for which there exists a positive constant M such that

$$|f(z)| \leq M \cdot |\sin z| \quad \text{for all} \ z \in \mathbb{C}.$$

    (b) Find all entire functions $f(z)$ for which there exists a positive constant M such that

$$|f(z)| \geq M \cdot |\sin z| \quad \text{for all} \ z \in \mathbb{C}.$$

14. Find the partial fraction expansion of $\frac{z+3}{z^3-z^2-2z}$.

15. (a) Find a rational function $f(z)$ that has simple roots at $z = 1$ and $z = 3$, simple poles at $z = \pm i$, and satisfies $\lim_{z \to \infty} f(z) = 1$.

    (b) Find a rational function $g(z)$ that has simple poles at $z = \pm(1+i)$ and double roots at $z = \pm(1 - i)$. Is such a rational function unique?

16. Show that the sum of the orders of all poles (including $z = \infty$) of a nonconstant rational function is the same as the sum of the orders of all roots.

17. A rational function has four roots of order 6, poles of order 3, 4, 7, and 8 (one each), but no other singularity in the plane $\mathbb{C}$. Discuss its behavior at ∞.

18. Given $2n$ complex numbers $a_1, a_2, \ldots, a_n, b_1, b_2, \ldots, b_n$, where the numbers a_j are all distinct, show that there exists a unique polynomial $p(z)$ of degree $\leq n - 1$ such that

$$p(a_j) = b_j \quad \text{for all} \ \ j = 1, 2, \ldots, n.$$

Hint. Consider a linear combination of the polynomials

$$p_k(z) = \prod_{j \neq k} (z - a_j) \quad (k = 1, 2, \ldots, n). \quad \text{(Lagrange)}$$

19. Find the fallacies:
    (a) Because

$$\sum_{k=0}^{\infty} z^k = \frac{1}{1 - z}$$

and

$$\sum_{k=-\infty}^{-1} z^k = \frac{1}{z} + \frac{1}{z^2} + \cdots + \frac{1}{z^k} + \cdots$$

$$= \frac{1}{z} \cdot \frac{1}{\left(1 - \dfrac{1}{z}\right)} = \frac{1}{z - 1},$$

we have

$$\sum_{-\infty}^{\infty} z^k = \sum_{-\infty}^{-1} z^k + \sum_{0}^{\infty} z^k$$

$$= \frac{1}{z - 1} + \frac{1}{1 - z} = 0 \quad \text{for all} \ z \in \mathbb{C}.$$

(b) Let $f(z) = \frac{1}{(z-1)(2z-1)}$. Then

$$f(z) = \frac{1}{z - 1} - \frac{2}{2z - 1} = -\sum_{k=0}^{\infty} z^k - \frac{1}{z\left(1 - \dfrac{1}{2z}\right)}$$

$$= -(1 + z + z^2 + z^3 + \cdots + z^k + \cdots)$$

$$- \frac{1}{z} \cdot \left\{ 1 + \frac{1}{2z} + \frac{1}{(2z)^2} + \cdots + \frac{1}{(2z)^k} + \cdots \right\}.$$

Therefore, the origin is an essential singularity of $f(z)$ with residue -1.

20. Show that, for any given $k \in \mathbb{R}$, there exist a pair of positive real numbers p and q such that

$$\lim_{R \to \infty} \int_{-pR}^{qR} \frac{x}{x^2 + 1} dx = k.$$

21. Evaluate the following definite integrals:

(a) $\displaystyle\int_{-\infty}^{\infty} \frac{x^2 - 3x + 2}{x^4 + 10x^2 + 9} dx$ $\left[\dfrac{5\pi}{12}\right]$

(b) $\displaystyle\int_{0}^{\infty} \frac{dx}{(x^2 + a^2)(x^2 + b^2)}$ $(a,\, b > 0)$. $\left[\dfrac{\pi}{2ab(a+b)}\right]$

(c) $\displaystyle\int_{0}^{\infty} \frac{x^{m-1}}{x^n + a^n} dx$ $(m < n,$ positive integers, $a > 0)$.

$\left[\dfrac{\pi}{na^{n-m} \sin \frac{m\pi}{n}}\right]$

(d) $\displaystyle\int_{0}^{\infty} \frac{x \sin mx}{x^2 + a^2} dx$ $(m,\, a > 0)$. $\left[\dfrac{\pi}{2} e^{-ma}\right]$

(e) $\displaystyle\int_{0}^{\infty} \frac{\sin mx}{x(x^2 + a^2)} dx$ $(m,\, a > 0)$. $\left[\dfrac{\pi}{2a^2}(1 - e^{-ma})\right]$

(f) $\displaystyle\int_{-\infty}^{\infty} \frac{\cos mx}{(x^2 + a^2)(x^2 + b^2)} dx$ $(a,\, b,\, m > 0,\ a \neq b)$.

$\left[\dfrac{\pi(ae^{mb} - be^{-ma})}{ab(a^2 - b^2)}\right]$

What if $a = b$?

(g) $\displaystyle\int_{0}^{\pi} \frac{d\theta}{(a - b\cos\theta)^2}$ $(a > b > 0)$. $\left[\dfrac{\pi a}{(a^2 - b^2)^{\frac{3}{2}}}\right]$

(h) $\displaystyle\int_{0}^{\pi} \frac{\sin^2 \theta d\theta}{a + b\cos\theta}$ $(a > b > 0)$. $\left[\dfrac{\pi}{b^2}(a - \sqrt{a^2 - b^2})\right]$

(i) $\displaystyle\int_{0}^{\pi} \frac{d\theta}{a + \sin^2 \theta}$ $(a > 0)$. $\left[\dfrac{\pi}{\sqrt{a(a+1)}}\right]$

(j) $\displaystyle\int_{0}^{2\pi} \frac{d\theta}{a^2 \cos^2 \theta + b^2 \sin^2 \theta}$ $(ab > 0)$. $\left[\dfrac{2\pi}{ab}\right]$

(k) $\displaystyle\int_{0}^{\pi} \frac{\sin n\theta}{\sin \theta} d\theta$ $(n \in \mathbb{N})$.   [0 if n is even; π if n is odd.]

(1) $\displaystyle\int_0^\pi \left(\frac{\sin n\theta}{\sin\theta}\right)^2 d\theta \quad (n \in \mathbb{N}). \quad [n\pi]$

22. Find the *Fourier transform* of the function $f(x) = e^{-\frac{x^2}{2}}$; that is, find the function $\hat{f}(t)$ defined for all $t \in \mathbb{R}$ by

$$\hat{f}(t) = \frac{1}{\sqrt{2\pi}} \int_{-\infty}^{\infty} e^{-\frac{x^2}{2}} \cdot e^{-itx} \, dx.$$

23. Show that

    (a) A function analytic on the Riemann sphere $\hat{\mathbb{C}}$ must be a constant.

    (b) A function meromorphic on the Riemann sphere $\hat{\mathbb{C}}$ must be a rational function.

    (c) A function meromorphic and univalent on the Riemann sphere $\hat{\mathbb{C}}$ must be a Möbius transformation.

24. Suppose $f(z)$ and $g(z)$ are analytic on $[\Gamma]$, where Γ is a curve, and

$$|f(z) + g(z)| < |f(z)| + |g(z)| \quad \text{for } z \text{ on } \Gamma.$$

Show that $f(z)$ and $g(z)$ have the same number of roots in (Γ).

25. Give an alternate proof of the Fundamental Theorem of Algebra 2.10.3 by computing the residue of the rational function $\frac{p'(z)}{p(z)}$ at $z = \infty$.

26. Let the residue of the function $f(z) = \frac{z^n e^{\frac{1}{z}}}{1+z}$ at the points $z = 0$ and $z = \infty$ be A and B, respectively. Show that if n is a nonnegative integer, then

$$A = (-1)^{n+1}\frac{1}{e} + \frac{1}{n!} - \frac{1}{(n-1)!} + \cdots + (-1)^n \frac{1}{1!};$$

$$A + B + (-1)^n \frac{1}{e} = 0.$$

What if n is a negative integer?

27. Suppose $f(z)$ is analytic in the closed unit disc and

$$|f(z)| < 1 \quad \text{for } |z| = 1.$$

Show that $f(z)$ has one and only one fixed point; that is, there exists a unique point z_0 in the unit disc such that $f(z_0) = z_0$.

28. Prove the Hurwitz Theorem 5.6.4 using the Rouché Theorem 5.6.8.

29. Show by an example that the Hurwitz Theorem 5.6.4 is false for real functions.

30. Is the converse of the Hurwitz Theorem 5.6.4 true?

31. Show that for every positive real number R there exists an integer N such that, for all $n > N$, all roots of the polynomial

$$1 + \frac{z}{1!} + \frac{z^2}{2!} + \cdots + \frac{z^n}{n!}$$

are outside the circle $C(0, R)$.

32. Suppose $|a| > e$ and n is a positive integer. Show that $e^z - az^n$ has n roots inside the unit circle.

33. Show that the equation $z + e^{-z} = a$ $(a > 1)$ has one solution in the right half-plane, and this solution is real.

34. (a) Find how many solutions of the equation

$$z^6 + 6z + 10 = 0$$

are in each quadrant.

(b) Show that $z^5 - 15z + 1 = 0$ has one root in the disc $D(0, \frac{1}{8})$ and four roots in the annulus $\{z \in \mathbb{C} ; \frac{3}{2} < |z| < 2\}$.

35. Prove the Open Mapping Theorem 5.6.9 using the Argument Principle 5.6.1.
Hint: $\frac{1}{2\pi i} \int_{C(z_0,r)} \frac{f'(z)}{f(z)-w} dz$ is an integer-valued continuous function of w in some neighborhood of $w_0 = f(z_0)$.

36. Suppose $f(z)$ is analytic on $[\Gamma]$, where Γ is a curve, and the image C of Γ under $f(z)$ is also a curve. Show that $f(z)$ maps (Γ) one-to-one onto (C).

Chapter 6

The Maximum Modulus Principle

6.1 The Maximum and Minimum Modulus Principles, Revisited

In the Cauchy Integral Formula 4.4.3,

$$f(a) = \frac{1}{2\pi i} \int_{C(a,r)} \frac{f(z)}{z - a} dz,$$

if we substitute $z - a = re^{i\theta}$, $dz = ire^{i\theta} d\theta$, we obtain the *Gauss mean-value theorem*

$$f(a) = \frac{1}{2\pi} \int_0^{2\pi} f(a + re^{i\theta}) d\theta.$$

This equality shows that the value of an analytic function at the center of a circle is equal to the average of its values on the circle, provided that the closed disc $\bar{D}(a, r)$ is in the region of analyticity. We then obtain the obvious corollary

$$|f(a)| \leq \frac{1}{2\pi} \int_0^{2\pi} |f(a + re^{i\theta})| d\theta.$$

Utilizing this observation, we can give an alternate proof of the Maximum and Minimum Modulus Principles 2.10.1. (See Remark after Theorem 4.4.4.)

THEOREM 6.1.1 (Maximum and Minimum Modulus Principles) *Assume that $f(z)$ is a nonconstant analytic function in a region Ω; then*

(a) its absolute value $|f(z)|$ has no local maximum in Ω; and

(b) $|f(z)|$ has no local minimum in Ω, either, provided $f(z)$ does not vanish in Ω.

Proof. Suppose that the modulus $|f(z)|$ attains a local maximum at $z = a$. Let Ω_0 be a subregion of Ω containing the point a where $|f(a)|$ is a *global* maximum. Set

$$S = \{z \in \Omega_0 \,; |f(z)| = |f(a)|\}.$$

Note that S is not empty, because $a \in S$. Because $|f(z)|$ is a continuous function of z, S is clearly closed in Ω_0. We claim that S is also open. Suppose $z_0 \in S$. Let δ be a positive number such that $D(z_0, \delta) \subset \Omega_0$. From the Gauss mean-value theorem we obtain

$$|f(a)| = |f(z_0)| \leq \frac{1}{2\pi} \int_0^{2\pi} \left|f(z_0 + re^{i\theta})\right| d\theta$$
$$\leq \frac{1}{2\pi} \int_0^{2\pi} |f(a)| \, d\theta = |f(a)|.$$

Because the two ends are equal, we must have equality everywhere. This implies that

$$|f(z_0 + re^{i\theta})| = |f(a)| \quad \text{for all } \theta \in [0, 2\pi],$$

because the modulus is a continuous function. But this equality is true for all $r \in [0, \delta)$, and so we have

$$|f(z)| \equiv |f(a)| \quad \text{for all } z \in D(z_0, \delta).$$

We have shown that $D(z_0, \delta) \subset S$, and so S is open. But Ω_0 is connected, hence we must have $S = \Omega_0$; that is,

$$|f(z)| \equiv |f(a)| \, (= \text{constant}) \quad \text{for all } z \in \Omega_0.$$

Hence (by Exercise 3.5),

$$f(z) \equiv \text{constant} \, (= k, \text{ say}) \quad \text{in } \Omega_0.$$

Let Ω_1 be a maximal subregion of Ω where $f(z) = k$. We claim that $\Omega_1 = \Omega$. Suppose $\Omega_1 \neq \Omega$. Then there must be a point $z_1 \in (\partial \Omega_1) \cap \Omega$. Let $D(z_1, \delta_1) \subset \Omega \, (\delta_1 > 0)$. Then $D(z_1, \delta_1) \cap \Omega_1$ is a nonempty open set where $f(z) = k$. But then, by Theorem 2.7.3 (see Remark after Theorem 4.4.4), we must have

$$f(z) = k \quad \text{for all } z \in D(z_1, \delta_1).$$

This result implies that $\Omega_1 \cup D(z_1, \delta_1)$ is a subregion of Ω greater than Ω_1 where $f(z) = k$, contradicting the maximality of Ω_1. This contradiction establishes our assertion.

To prove the minimum modulus principle, observe that if $f(z) \neq 0$ in Ω, then $\frac{1}{f(z)}$ is analytic there, and because the local minimum of $|f(z)|$ is attained at the same point as the local maximum of $\left|\frac{1}{f(z)}\right|$, it follows from the maximum modulus principle that $|f(z)|$ cannot have a local minimum unless $f(z) = 0$ somewhere in the region Ω. □

COROLLARY 6.1.2 *Let $f(z) \in H(\Omega)$, where Ω is a bounded region, and suppose $f(z)$ is continuous on $\bar{\Omega}$. Then $f(z)$ attains its maximum modulus on the boundary $\partial\Omega$ of Ω. Moreover, if $f(z) \neq 0$ in Ω, then $f(z)$ also attains its minimum modulus on the boundary.*

Remark. This corollary is *not* true if Ω is not bounded. For example, let $f(z) = e^z$ and $\Omega = \{z \in \mathbb{C}; \Re z > 0\}$. Then $\partial\Omega$ is the imaginary axis, where

$$|f(iy)| = \left|e^{iy}\right| = 1.$$

Yet $|f(z)| \longrightarrow \infty$ as $z \to \infty$ along the positive real axis. A counterexample for the case of the minimum modulus can be obtained by considering the exponential function on the left half-plane.

6.2 The Schwarz Lemma

As an application of the Maximum Modulus Principle 6.1.1, we shall prove the following very important theorem, known as the Schwarz lemma. (See Exercise 2.56.)

LEMMA 6.2.1 (**Schwarz**) *Let $f(z)$ be a bounded analytic function in the disc $D(0, R)$ that satisfies the condition $f(0) = 0$. Then*

$$|f'(0)| \leq \frac{M}{R} \quad \text{and} \quad |f(z)| \leq \frac{M}{R}|z| \quad z \in D(0, R),$$

where $M = \sup\{|f(z)|; z \in D(0, R)\}$. Furthermore, if $|f'(0)| = \frac{M}{R}$, or $|f(z)| = \frac{M}{R}|z|$ holds at even one point in addition to $z = 0$, then the function is a dilation:

$$f(z) = e^{i\alpha}\frac{M}{R}z \quad \text{for some } \alpha \in \mathbb{R}.$$

Proof. Expanding the function $f(z)$ in Taylor series around the origin, we have, for $z \in D(0, R)$,

$$f(z) = a_1 z + a_2 z^2 + \cdots + a_n z^n + \cdots \quad (a_0 = f(0) = 0).$$

Thus

$$g(z) = \frac{f(z)}{z} = a_1 + a_2 z + \cdots + a_n z^{n-1} + \cdots$$

is analytic in $D(0, R)$ (after removing the removable singularity at the origin). For an arbitrary $z_0 \in D'(0, R)$, choose $0 < r < R$ so that $z_0 \in D(0, r)$; applying the Maximum Modulus Principle 6.1.1 to $g(z)$ on $\bar{D}(0, r)$, we obtain

$$\left| \frac{f(z_0)}{z_0} \right| = |g(z_0)| \leq \max\{|g(z)| ; z \in \bar{D}(0, r)\}$$

$$= \max\{|g(z)| ; z \in C(0, r)\}$$

$$= \max\left\{ \left| \frac{f(z)}{z} \right| ; z \in C(0, r) \right\} \leq \frac{M}{r}.$$

Letting $r \to R$, we obtain the desired inequality.

If the equality holds at some point $z \in D'(0, R)$, then $\left| \frac{f(z)}{z} \right|$ attains its maximum in the interior; therefore $g(z) = \frac{f(z)}{z}$ must reduce to a constant, by the Maximum Modulus Principle 6.1.1.

The case $z_0 = 0$ is implicit in the proof above if we observe that $f'(0) = a_1 = g(0)$.   □

Expressed graphically, the Schwarz lemma says that if the surface $w = |f(z)|$ is inside the cylinder $\sqrt{x^2 + y^2} \leq R$, below the plane $w = M$ and passes through the origin, then the surface $w = |f(z)|$ can never go above the conical surface $w = \frac{M}{R}\sqrt{x^2 + y^2}$, and if the surface $w = |f(z)|$ meets the conical surface at even one point in addition to the origin, then it lies *entirely* on the conical surface (Figure 6.1). By the contrapositive, if $w = |f(z)|$ is below the conical surface at even one point, then $w = |f(z)|$ is below the conical surface everywhere (except at the origin).

Although simple, the Schwarz Lemma 6.2.1 is very useful and has many far-reaching applications. We now present one of the most important and striking of these applications. Before stating it as Theorem 6.2.2, we recall that an analytic function $f(z)$ maps a region Ω_1 *conformally* onto a region

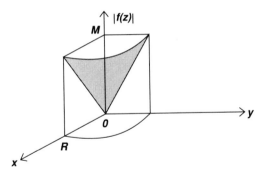

Figure 6.1

Ω_2 if $f(z)$ is a bijection between the regions Ω_1 and Ω_2 (and so $f'(z) \neq 0$ for any $z \in \Omega_1$).

THEOREM 6.2.2 *A conformal mapping* $w = f(z)$ *of the unit disc onto itself must be a Möbius transformation.*

Proof. Let $w_0 = f(0)$. Then $|w_0| < 1$, and the Möbius transformation

$$\zeta = Tw = \frac{w - w_0}{1 - \bar{w}_0 w},$$

maps the unit disc onto itself, carrying w_0 to the origin $\zeta = 0$. Thus the function

$$\zeta = g(z) = (T \circ f)(z)$$

maps the unit disc conformally onto itself and $g(0) = 0$. Thus the Schwarz Lemma 6.2.1 is applicable, and we obtain

$$|\zeta| = |g(z)| \leq |z| \quad \text{for all } z \in D.$$

On the other hand,

$$z = g^{-1}(\zeta) = (T \circ f)^{-1}(\zeta),$$

is also a conformal mapping of the unit disc onto itself, carrying the origin $\zeta = 0$ to the origin $z = 0$. Therefore, the Schwarz Lemma 6.2.1 is again applicable, and we obtain

$$|z| = |g^{-1}(\zeta)| \leq |\zeta| \quad (\zeta \in D).$$

It follows that

$$|g(z)| = |\zeta| = |z| \quad \text{for all } z \in D,$$

which means that the function $\frac{g(z)}{z}$ (is analytic and) attains its maximum modulus at all interior points (of the unit disc), and so $\frac{g(z)}{z}$ must be a constant (of modulus 1). It follows that

$$(T \circ f)(z) = g(z) = e^{i\alpha} z \quad (\alpha \in \mathbb{R}),$$

and hence

$$f(z) = T^{-1}(e^{i\alpha} z) = \frac{e^{i\alpha} z + w_0}{e^{i\alpha} \bar{w}_0 z + 1}. \qquad \square$$

Now, let us drop the condition $f(0) = 0$ in the Schwarz Lemma 6.2.1. Suppose $f(z)$ is a nonconstant bounded analytic function in $D(0, R)$ such that $w_0 = f(z_0)$, $z_0 \in D(0, R)$. Then $|w_0| < M = \sup\{|f(z)|; z \in D(0, R)\}$, and

$$\zeta = Tz = \frac{R(z - z_0)}{R^2 - \bar{z}_0 z}$$

is a Möbius transformation that maps the disc $D_z(0, R)$ onto $D_\zeta(0, 1)$ with z_0 mapping to the origin. (The index z in $D_z(0, R)$ indicates that the disc is in the z-plane, and so on.) Similarly,

$$w_1 = Sw = \frac{M(w - w_0)}{M^2 - \bar{w}_0 w}$$

is a Möbius transformation that maps the disc $D_w(0, M)$ onto $D_{w_1}(0, 1)$ with w_0 mapping to the origin. Then it is clear that the function $(S \circ f \circ T^{-1})(\zeta)$ maps the unit disc in the ζ-plane to that of the w_1-plane and satisfies the hypothesis of the Schwarz Lemma 6.2.1. Thus, we have

$$\left| (S \circ f \circ T^{-1})(\zeta) \right| \le |\zeta|, \quad \zeta \in D_\zeta(0, 1),$$

from which it follows that

$$|(S \circ f)(z)| \le \left| \frac{R(z - z_0)}{R^2 - \bar{z}_0 z} \right|, \quad z \in D_z(0, R).$$

Rewriting, we obtain

$$\left| \frac{M\{f(z) - f(z_0)\}}{M^2 - \overline{f(z_0)} f(z)} \right| \le \left| \frac{R(z - z_0)}{R^2 - \bar{z}_0 z} \right|, \quad z \in D(0, R).$$

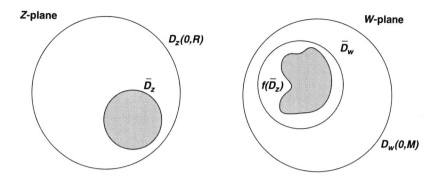

Figure 6.2

The equality holds if and only if

$$\frac{M\{f(z) - f(z_0)\}}{M^2 - \overline{f(z_0)}f(z)} = e^{i\alpha}\frac{R(z - z_0)}{R^2 - \bar{z}_0 z} \quad (\alpha \in \mathbb{R});$$

that is, $f(z)$ must be a Möbius transformation mapping the disc $D_z(0, R)$ onto $D_w(0, M)$.

Let us discuss the geometric significance of this result. The set of all points z satisfying the inequality

$$\left|\frac{R(z - z_0)}{R^2 - \bar{z}_0 z}\right| \leq r \quad (0 < r < R)$$

is an Apollonius disc $\bar{D}_z$ (Figure 6.2). The inequality

$$\left|\frac{M\{f(z) - f(z_0)\}}{M^2 - \overline{f(z_0)}f(z)}\right| \leq \left|\frac{R(z - z_0)}{R^2 - \bar{z}_0 z}\right| \quad z \in D(0, R)$$

says that if $z \in \bar{D}_z$, then $w = f(z)$ is in the Apollonius disc $\bar{D}_w$ given by

$$\left|\frac{M\{w - f(z_0)\}}{M^2 - \overline{f(z_0)}w}\right| \leq r.$$

Moreover, unless

$$\frac{M\{f(z) - f(z_0)\}}{M^2 - \overline{f(z_0)}f(z)} = e^{i\alpha}\frac{R(z - z_0)}{R^2 - \bar{z}_0 z} \quad (\alpha \in \mathbb{R}),$$

the image $w = f(z)$ will never touch the boundary of $\bar{D}_w$. If

$$\frac{M\{f(z) - f(z_0)\}}{M^2 - \overline{f(z_0)}f(z)} = e^{i\alpha}\frac{R(z - z_0)}{R^2 - \bar{z}_0 z} \quad (\alpha \in \mathbb{R}),$$

then $\bar{D}_z$ is mapped conformally onto $\bar{D}_w$, and $f(z)$ is a Möbius transformation.

COROLLARY 6.2.3 *If $f(z)$ is a bounded analytic function in the disc $D(0, R)$ and $f(z_0) = 0$, then*

$$|f(z)| \leq MR\left|\frac{z - z_0}{R^2 - \bar{z}_0 z}\right| \quad z \in D(0, R),$$

where $M = \sup\{|f(z)|\,;\, z \in D(0, R)\}$. The equality holds if and only if

$$f(z) = e^{i\alpha}MR\frac{z - z_0}{R^2 - \bar{z}_0 z} \quad (\alpha \in \mathbb{R}).$$

For simplicity in notation, let us consider the normalized case

$$|f(z)| \leq 1 \quad \text{on } D;$$

that is, $M = 1$, $R = 1$ in the Schwarz Lemma 6.2.1. Suppose $z_1, z_2, \ldots,$ z_n are roots of $f(z)$ in the unit disc. These are not necessarily all the roots of $f(z)$ in the unit disc, and if z_j is a root of multiplicity k, then we may write z_j up to k times (i.e., *k times or fewer*):

$$f(z_1) = f(z_2) = \cdots = f(z_n) = 0.$$

Then the (finite) *Blaschke product*

$$B(z) = \prod_{j=1}^{n} \frac{z - z_j}{1 - \bar{z}_j z}$$

is analytic in the unit disc, and

$$|B(z)| = 1 \quad \text{for } z \in C(0, 1),$$

because $\left|\frac{z - z_j}{1 - \bar{z}_j z}\right| = 1$ when $|z| = 1$. Therefore, the Maximum Modulus Principle 6.1.1 implies

$$|B(z)| < 1 \quad \text{for } z \in D.$$

Clearly, $B(z) \neq 0$ for $z \neq z_i$ ($j = 1, 2, \ldots, n$). Thus $\frac{f(z)}{B(z)}$ is analytic when $z \neq z_j$. At $z = z_j$, we have $f(z_j) = B(z_j) = 0$, but the multiplicity of the root z_j for $B(z)$ is not greater than that for $f(z)$, and so the singularity is removable; hence $\frac{f(z)}{B(z)}$ is also analytic at $z = z_j$ ($j = 1, 2, \ldots, n$).

Suppose $z_j \in D(0, r)$ ($j = 1, 2, \ldots, n; 0 < r < 1$); then, recalling that $M = 1$,

$$\sup\left\{\left|\frac{f(z)}{B(z)}\right|; z \in D(0, r)\right\} = \sup\left\{\left|\frac{f(z)}{B(z)}\right|; z \in C(0, r)\right\}$$

$$\leq \sup\left\{\left|\frac{1}{B(z)}\right|; z \in C(0, r)\right\}.$$

Because $|B(z)| \longrightarrow 1$ (as $|z| \to 1$), we obtain, by letting $r \to 1$ in the above inequality,

$$\left|\frac{f(z)}{B(z)}\right| \leq 1 \quad z \in D.$$

As before, the equality holds if and only if

$$f(z) = e^{i\alpha} B(z) \quad (\alpha \in \mathbb{R}).$$

THEOREM 6.2.4 *Suppose $f(z)$ is analytic in the unit disc D with*

$$|f(z)| < 1 \quad (z \in D),$$

and $z_1, z_2, \ldots, z_n$ are roots of $f(z)$ in the unit disc D. Then

$$|f(z)| \leq \left|\frac{z - z_1}{1 - \bar{z}_1 z} \cdot \frac{z - z_2}{1 - \bar{z}_2 z} \cdots \frac{z - z_n}{1 - \bar{z}_n z}\right| \quad z \in D.$$

Moreover, the equality holds if and only if

$$f(z) = e^{i\alpha} \prod_{j=1}^{n} \frac{z - z_j}{1 - \bar{z}_j z} \quad (\alpha \in \mathbb{R}).$$

In Section 7.5, we shall discuss (infinite) Blaschke products.

6.3   The Three-Circle Theorem

As another application of the Maximum Modulus Principle 6.1.1, we prove the following theorem, known as the *three-circle theorem*:

THEOREM 6.3.1 (Hadamard) *Suppose* $f(z) \in H(\Omega) \cap C(\bar{\Omega})$, *where* $\Omega = \{z \in \mathbb{C}; \, a < |z| < b\}$, *and let*

$$M(r) = \max\{|f(z)| \, ; \, |z| = r\} \quad (a \le r \le b).$$

Then

$$\begin{vmatrix} \log a & \log M(a) & 1 \\ \log r & \log M(r) & 1 \\ \log b & \log M(b) & 1 \end{vmatrix} \ge 0.$$

Equivalently,

$$\log M(r) \le \frac{\log b - \log r}{\log b - \log a} \log M(a) + \frac{\log r - \log a}{\log b - \log a} \log M(b);$$

that is,

$$M(r) \le M(a)^\theta \cdot M(b)^{1-\theta},$$

where $\theta = \frac{\log b - \log r}{\log b - \log a}$ *depends on* a, b, *and* r, *but is independent of* $f(z)$, *and is between 0 and 1.*

In other words, $\log M(r)$ *is a convex function of* $\log r$; *that is, the graph of* $\log M(r)$ *(versus* $\log r$) *has the property that the part of the graph between* $(\log c, \, \log M(c))$ *and* $(\log d, \, \log M(d))$ $(a \le c < d \le b)$ *lies below (or coincides with) the line segment joining these two endpoints (Figure 6.3).*

Proof. Set $g(z) = z^\lambda f(z)$, where λ is a real number to be determined later. As z traverses any full circle $C(0, \, r)$, $g(z)$ becomes $e^{2\pi i \lambda}$ times its initial value, and so $g(z)$ is not single-valued in the annulus Ω (unless λ is an integer). However, recalling that the Maximum Modulus Principle 6.1.1 asserts that a nonconstant analytic function has no *local* maximum modulus, and because $|g(z)| = |z|^\lambda \cdot |f(z)|$ is single-valued, we obtain

$$|z|^\lambda \cdot |f(z)| \le \max\left\{a^\lambda M(a), \, b^\lambda M(b)\right\} \quad (a \le |z| \le b).$$

Hence

$$r^\lambda M(r) \le \max\left\{a^\lambda M(a), \, b^\lambda M(b)\right\} \quad (a \le r \le b);$$

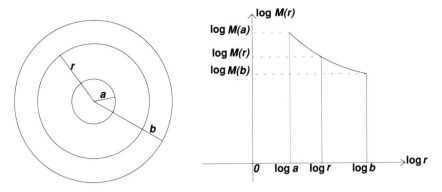

Figure 6.3

that is,

$$M(r) \le \max\left\{ \left(\frac{a}{r}\right)^\lambda M(a), \; \left(\frac{b}{r}\right)^\lambda M(b) \right\} \quad (a \le r \le b).$$

Setting aside the trivial case $f(z) \equiv 0$, $M(a)$, $M(r)$, and $M(b)$ are all positive, and we may choose (uniquely) the real number λ such that

$$a^\lambda M(a) = b^\lambda M(b); \quad \text{that is,} \quad \lambda = -\frac{\log M(b) - \log M(a)}{\log b - \log a}.$$

We then have

$$
\begin{aligned}
M(r) &\le \left(\frac{a}{r}\right)^\lambda \cdot M(a) \\
&= \exp\left\{ (\log r - \log a)\frac{\log M(b) - \log M(a)}{\log b - \log a} \right\} \cdot M(a) \\
&= \left\{ \frac{M(b)}{M(a)} \right\}^{\frac{\log r - \log a}{\log b - \log a}} \cdot M(a) \\
&= M(a)^{\frac{\log b - \log r}{\log b - \log a}} \cdot M(b)^{\frac{\log r - \log a}{\log b - \log a}}.
\end{aligned}
$$

Note that we obtain the same result if we begin with $M(r) \le \left(\frac{b}{r}\right)^\lambda M(b)$.

□

6.4   A Maximum Theorem for an Unbounded Region

One of the weak formulations of the maximum modulus principle is the following: A function $f(z)$ analytic on a bounded region Ω and continuous on $\bar{\Omega}$ assumes its maximum modulus on $\partial\Omega$; unless $f(z)$ is constant the maximum is assumed *only* on $\partial\Omega$. In short, if $f(z) \in H(\Omega) \cap C(\bar{\Omega})$ where Ω is a bounded region, then

$$\max_{z\in\bar{\Omega}} |f(z)| = \max_{z\in\partial\Omega} |f(z)|.$$

It is natural to ask whether some "reasonable" extension of this theorem to unbounded domains exists.

Consider first the following example. Let $f(z) = \exp\{\exp(-iz)\}$ on the (closed) strip $\{z \in \mathbb{C}; |\Re z| \le \frac{\pi}{2}\}$. On the edges $f(z)$ becomes $\exp\{\mp i\exp(y)\}$. Because the exponent $\mp i\exp(y)$ is purely imaginary, $|f(z)| \equiv 1$, but on the imaginary axis $f(z)$ becomes $\exp\{\exp(y)\}$, and so $|f(z)|$ is unbounded; indeed, this behavior occurs on *every* vertical line in the *open* strip. Thus, in this case we may say that the maximum modulus principle fails "*spectacularly.*" (In fact, $|f(z)|$ *exceeds* 1 everywhere in the open strip, exactly contrary to the maximum modulus principle.)

This example suggests the following question: Suppose we add the hypothesis that $|f(z)|$ is *bounded* in Ω; can we then conclude that $\sup_{z\in\bar{\Omega}} |f(z)|$ *equals* $\sup_{z\in\partial\Omega} |f(z)|$? (Note that we speak of $\sup |f(z)|$ rather than $\max |f(z)|$, because $\bar{\Omega}$ and $\partial\Omega$, although closed, are not compact.) The answer is, in fact, affirmative in the case of a parallel strip; this is shown in the present section. In Section 6.6, two somewhat deeper theorems are presented.

An extensive literature has grown up around problems of this type; they are often said to be "of Phragmén-Lindelöf type," after two major investigators, Phragmén and Lindelöf. The theorem that is now presented may be considered a suitable introduction to this branch of complex analysis.

THEOREM 6.4.1 *Let Ω be an open strip $\{z \in \mathbb{C}; a < \Re z < b\}$ and let $f(z)$ be analytic and bounded on Ω and continuous on $\bar{\Omega}$. Then*

$$\sup_{z\in\partial\Omega} |f(z)| = \sup_{z\in\bar{\Omega}} |f(z)|.$$

Proof. Let $\sup_{z\in\partial\Omega} |f(z)| = M_1$ and $\sup_{z\in\bar{\Omega}} |f(z)| = M_2 < \infty$. It is obvious that $M_1 \le M_2$, and so it suffices to eliminate the inequality $M_1 < M_2$. By means of a clever artifice the problem is reduced to an appeal to the maximum modulus principle for a *bounded* region.

Let z_0 be any real number in Ω; that is, $z_0 = c$, $a < c < b$. (It is then a trivial matter to generalize to *any* point in Ω.) Choose any positive number ϵ (however small) and any positive number A; later an appropriate choice of A will be made. Consider the rectangular region $\{z \in \Omega \, ; \, -A < \Im z < A\}$ and let

$$g(z) = \frac{f(z)}{1 + \epsilon(z - a)}$$

in the closure of this region. Then $\Re\{1 + \epsilon(z - a)\} \geq 1$, and so $|1 + \epsilon(z - a)| \geq 1$ in this region; on the vertical edges $|g(z)| \leq |f(z)| \leq M_1$, and on the horizontal edges $|1 + \epsilon(z - a)| \geq |1 \pm i\epsilon A| = \sqrt{1 + \epsilon^2 A^2}$, and therefore $|g(z)| \leq \frac{M_2}{\sqrt{1 + \epsilon^2 A^2}}$. Now choose A so large that $\frac{M_2}{\sqrt{1 + \epsilon^2 A^2}} < M_1$. (This is possible, for M_2 is assumed finite.) Then everywhere on the boundary of the rectangular region the inequality $|g(z)| \leq M_1$ is satisfied. Then $|g(z)| \leq M_1$ holds everywhere in the interior also, and at z_0 in particular. Therefore $|f(z_0)| \leq M_1|1 + i\epsilon(z_0 - a)| \leq M_1|1 + i\epsilon(b - a)|$. Because this inequality makes no reference to A, we may let ϵ approach zero, and so we obtain $|f(z_0)| \leq M_1$. $\qquad\square$

Of course, the restriction to a *vertical* strip is entirely a matter of convenience.

6.5 The Three-Line Theorem

In the situation described in Theorem 6.4.1, it appears plausible that if x is close to a (or b) then $L(x)$ is close to $L(a)$ (or $L(b)$). (Here and below, $L(x) = \sup_{\Re z = x} |f(z)|$.) The following theorem asserts that $L(x)$ is a logarithmically convex function of x in the interval $a \leq x \leq b$.

THOEREM 6.5.1 (Doetsch) *Suppose* $f(z) \in H(\Omega) \cap C(\bar{\Omega})$, *where* $\Omega = \{z \in \mathbb{C} \, ; \, a < \Re z < b\}$, *and is bounded in* $\bar{\Omega}$. *Let*

$$L(x) = \sup\{|f(z)| \, ; \, \Re z = x\} \quad (a \leq x \leq b).$$

Then $\log L(x)$ *is a convex function of* x; *that is,*

$$\log L(x) \leq \theta \log L(a) + (1 - \theta) \log L(b),$$

or equivalently,

$$L(x) \leq L(a)^\theta \cdot L(b)^{1-\theta} \quad (a \leq x \leq b),$$

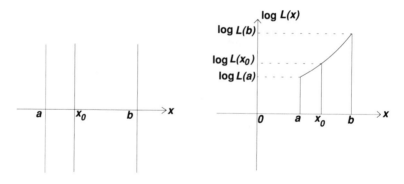

Figure 6.4

where $\theta = \frac{b-x}{b-a}$ depends on a, b, and x, but is independent of the function $f(z)$ (Figure 6.4).

Proof. Let us introduce the auxiliary function $g(z) = e^{\lambda z} \cdot f(z)$, where λ is a real number to be specified later. Then $g(z)$ satisfies the hypothesis of Theorem 6.4.1, and

$$\sup_{z \in \bar{\Omega}} |g(z)| = \sup_{z \in \partial\Omega} |g(z)|$$

$$= \max \left\{ \sup_{x=a} |g(z)|, \sup_{x=b} |g(z)| \right\}$$

$$= \max\{e^{\lambda a} L(a),\ e^{\lambda b} L(b)\}$$

(because $|e^{\lambda z}| = e^{\lambda a}$ on the left edge and $e^{\lambda b}$ on the right edge). Now let us choose λ so that the two numbers that appear above become equal; that is,

$$e^{\lambda a} L(a) = e^{\lambda b} L(b); \quad \text{namely,} \quad \lambda = -\frac{\log L(b) - \log L(a)}{b - a}.$$

Then, by Theorem 6.4.1, $e^{\lambda x} L(x)$ cannot exceed the common value of $e^{\lambda a} L(a)$ and $e^{\lambda b} L(b)$; this relation works out precisely to

$$L(x) \le L(a)^{\theta} \cdot L(b)^{1-\theta}. \qquad \square$$

This theorem bears a strong resemblance, both in content and in method of proof, to the Three-Circle Theorem 6.3.1. However, in the present case the hypothesis of boundedness of the function $f(z)$ is essential, to compensate for the noncompactness of the (closed) strip. We might add that the Three-Line Theorem 6.5.1 is needed to prove the important *Riesz-Thorin interpolation theorem* in functional analysis.

Remark. A function convex on an *open* interval $a < x < b$ must be continuous; however, a function convex on a *closed* interval $a \le x \le b$ may be discontinuous at either or both endpoints, but the reader is advised to prove that, under the hypotheses of Theorem 6.5.1, $L(x)$ remains continuous at the endpoints, a and b.

6.6   The Phragmén-Lindelöf Theorems

We now return to the type of problem that was introduced in Section 6.4; as stated there, we shall present two theorems that are nontrivial extensions of Theorem 6.4.1.

First let us introduce a notation that will be used throughout the rest of this book. Suppose a function $f(z)$ is bounded on a set $E \subset \mathbb{C}$. We define the *supremum norm*, $\|f\|_E$, of the function $f(z)$ on the set E by

$$\|f\|_E = \sup\{|f(z)|; \ z \in E\}.$$

It is simple to verify that this is actually a norm:

(a) $\|f\|_E \ge 0$, and $\|f\|_E = 0$ if and only if $f(z) \equiv 0$ on E.
(b) $\|kf\|_E = |k| \cdot \|f\|_E$ $(k \in \mathbb{C})$.
(c) $\|f + g\|_E \le \|f\|_E + \|g\|_E$.   (*Triangle inequality.*)

THEOREM 6.6.1 (Phragmén-Lindelöf) *Let* $\Omega = \{z \in \mathbb{C}; |\Re z| < \frac{\pi}{2}\}$ *and* $f(z) \in H(\Omega) \cap C(\bar{\Omega})$. *Suppose there exist positive constants* M, a *and* $\lambda < 1$ *such that*

$$\|f\|_{\partial\Omega} \le M \quad and \quad |f(z)| \le \exp\left\{ae^{\lambda|\Im z|}\right\} \quad (z \in \Omega).$$

Then
$$\|f\|_\Omega \le M.$$

Remark. The theorem can easily be generalized to a strip of the form $\{z \in \mathbb{C}; \ a < \Re z < b\}$.

Proof. Choose $\mu > 0$ such that $\lambda < \mu < 1$. Given $\epsilon > 0$, define an entire function

$$g(z) = \exp\{-\epsilon \cos(\mu z)\}.$$

For $z = x + iy \in \bar{\Omega}$, we have

$$\Re\{\cos(\mu z)\} = \cos(\mu x) \cdot \cosh(\mu y) \ge \cos\left(\frac{\mu\pi}{2}\right) \cdot \cosh(\mu y)$$
$$= \delta \cdot \cosh(\mu y) \quad (\because \ 0 < \mu < 1),$$

where $\delta = \cos \frac{\mu\pi}{2} > 0$. Therefore

$$|g(z)| \leq \exp\{-\epsilon\delta \cosh(\mu y)\} \leq 1 \qquad (z \in \bar{\Omega}).$$

It follows that $\|f \cdot g\|_{\partial\Omega} \leq M$, and

$$|f(z) \cdot g(z)| \leq \exp\left\{ae^{\lambda|y|} - \epsilon\delta \cosh(\mu y)\right\} \qquad (z = x + iy \in \Omega).$$

(Note that the right-hand side is independent of $\Re z = x$.) Because $\lambda < \mu$ and $\epsilon\delta > 0$, we have

$$\lim_{y \to \pm\infty} |f(z) \cdot g(z)| = 0.$$

Hence, for any $z_0 \in \Omega$, we may choose R so large that $R > |\Im z_0|$ *and*

$$|f(z) \cdot g(z)| \leq M \qquad (\text{whenever } |\Im z| > R).$$

We have shown that
$$|f(z) \cdot g(z)| \leq M$$
on the boundary of the rectangle with vertices at $\pm\frac{\pi}{2} \pm iR$. By the maximum modulus principle, the last inequality is valid throughout the rectangle. In particular,
$$|f(z_0) \cdot g(z_0)| \leq M.$$
Because $\lim_{\epsilon \to 0+} g(z_0) = 1$, we obtain

$$|f(z_0)| \leq M.$$

But z_0 is an arbitrary point in Ω, and so we conclude that

$$\|f\|_\Omega \leq M. \qquad\qquad \square$$

THEOREM 6.6.2 (Phragmén-Lindelöf) *Suppose $f(z) \in H(\Omega) \cap C(\bar{\Omega})$, where $\Omega = \{z \in \mathbb{C}; \, |\arg z| < \alpha\}$, and let $f(z)$ satisfy the following conditions:*

(a) The function $f(z)$ is bounded on the boundary of Ω:

$$\|f\|_{\partial\Omega} \leq M \quad \text{for some positive constant } M;$$

(b) For any $\epsilon > 0$, there exist positive constants K_ϵ and r_ϵ such that

$$|f(z)| \leq K_\epsilon \cdot \exp\{\epsilon|z|^\rho\} \quad \text{for all } z \in \Omega \text{ satisfying } |z| \geq r_\epsilon,$$

where $\rho = \frac{\pi}{2\alpha}$.

Then

$$\|f\|_\Omega \leq M.$$

Remark. The theorem can easily be generalized to a region of the form

$$\{z \in \mathbb{C};\ \theta - \alpha < \arg z < \theta + \alpha\}.$$

Proof. Given $\epsilon > 0$, choose $\delta > \epsilon$ and let

$$F(z) = f(z) \cdot \exp\{-\delta z^\rho\},$$

where we choose the branch of z^ρ that maps the positive real axis onto itself. Then, for $z = re^{i\theta} \in \Omega$, we have

$$\begin{aligned}
|F(z)| &= |f(z)| \cdot \exp\{-\delta r^\rho \cos(\rho\theta)\} \\
&\leq |f(z)| \cdot \exp\{-\delta r^\rho \cos(\rho\alpha)\} = |f(z)|.
\end{aligned}$$

Moreover, for $z \in \partial\Omega$,

$$|F(z)| \leq |f(z)| \leq M.$$

Now, for $z = r > r_\epsilon$ (note that z is on the positive real axis), we have

$$|F(r)| \leq K_\epsilon \cdot \exp\{(\epsilon - \delta)r^\rho\} \longrightarrow 0 \quad (r \to \infty),$$

because $\delta > \epsilon$. It follows that

$$M' = \max\{|F(r)|\,;\, 0 \leq r < \infty\}$$

is attained at some point $r_0 < \infty$. Set

$$\begin{aligned}
M_0 &= \max\{M,\ M'\}; \\
\Omega_1 &= \{z \in \mathbb{C};\ 0 < \arg z < \alpha\}; \\
\Omega_2 &= \{z \in \mathbb{C};\ -\alpha < \arg z < 0\}.
\end{aligned}$$

Choose γ such that $\rho = \frac{\pi}{2\alpha} < \gamma < \frac{\pi}{\alpha}$, and for $\eta > 0$, let

$$G(z) = F(z) \cdot \exp\left\{ -\eta \left(z \cdot e^{-\frac{\alpha}{2}i} \right)^{\gamma} \right\} \quad (z \in \Omega_1).$$

If $z = re^{i\theta} \in \Omega_1$, then

$$\cos \gamma \left(\theta - \frac{\alpha}{2} \right) \geq \cos \frac{\gamma\alpha}{2} > 0.$$

Because $\gamma > \rho$, we have

$$|G(z)| \leq |F(z)| \cdot \exp\left\{ -\eta r^{\gamma} \cos \gamma \left(\theta - \frac{\alpha}{2} \right) \right\}$$

$$\leq K_{\epsilon} \cdot \exp\left\{ \epsilon r^{\rho} - \eta r^{\gamma} \cos \frac{\gamma\alpha}{2} \right\} \longrightarrow 0 \quad (r \to \infty).$$

Moreover,

$$|G(z)| \leq |F(z)| \quad \text{for } z \in \bar{\Omega}_1,$$

and hence

$$|G(z)| \leq M_0 \quad \text{for } z \in \bar{\Omega}_1.$$

Letting $\eta \to 0$, we obtain

$$|F(z)| \leq M_0 \quad \text{for } z \in \bar{\Omega}_1.$$

Similarly,

$$|F(z)| \leq M_0 \quad \text{for } z \in \bar{\Omega}_2.$$

$$\therefore |F(z)| \leq M_0 \quad \text{for all } z \in \bar{\Omega}.$$

Suppose $M_0 = M' \ (\geq M)$; then because $|F(r_0)| = M_0$, by the Maximum Modulus Principle 6.1.1, $f(z) = M_0 e^{i\beta} e^{\delta z^{\rho}}$ (for some $\beta \in \mathbb{R}$), which contradicts the assumption that, for any $\epsilon > 0$,

$$|f(z)| \leq K_{\epsilon} \cdot \exp\left\{ \epsilon |z|^{\rho} \right\}.$$

Hence $M_0 = M$, and so

$$|f(z)| \cdot |\exp\left\{ -\delta z^{\rho} \right\}| = |F(z)| \leq M \quad \text{for all } z \in \Omega.$$

Letting $\delta \to 0$, we obtain

$$|f(z)| \leq M \quad \text{for all } z \in \Omega. \qquad \square$$

Exercises

1. (a) Let $P_1, P_2, \ldots, P_n$ be n arbitrary points of the plane. If a variable point P is confined to a closed bounded set E, show that the product

$$\prod_{k=1}^{n} \overline{PP_k}$$

attains its maximum if P is some point of the boundary of E, where $\overline{PP_k}$ denotes the distance between points P_k and P ($k = 1, 2, \ldots, n$).

   (b) Suppose $\triangle ABC$ is an equilateral triangle inscribed in the unit circle $C(0, 1)$. Find the maximum value of

$$\overline{PA} \cdot \overline{PB} \cdot \overline{PC}$$

   where P is a variable point in $\bar{D}(0, 2)$.

2. Suppose $f_j(z) \in H(\Omega)$ ($j = 1, 2, \ldots, n$). Show that the function

$$\varphi(z) = \sum_{j=1}^{n} |f_j(z)|^2$$

   has no local maximum in the region Ω unless all the functions $f_j(z)$ ($j = 1, 2, \ldots, n$) reduce to constant functions.

3. (a) Let $f(z) \in H(\Omega) \cap C(\bar{\Omega})$, where Ω is a bounded region. If $|f(z)|$ is constant on $\partial\Omega$, show that, unless $f(z)$ reduces to a constant, it must have at least one root in Ω.

   (b) Show that a function $f(z)$ meromorphic in the closed unit disc and satisfying
$$|f(z)| \equiv \text{constant} \quad \text{on} \quad C(0, 1),$$
   must be a rational function.

   (c) Suppose $f(z)$ is an entire function such that $|f(z)| = 1$ on $C(0, 1)$. Show that $f(z) = kz^n$ for some constant k, $|k| = 1$, and some nonnegative integer n.

4. Prove the Maximum Modulus Principle 6.1.1

   (a) using the Rouché Theorem 5.6.8;

   (b) using the Open Mapping Theorem 5.6.9.

5. (Fundamental Theorem of Algebra) Let $p(z) = a_0 + a_1 z + \cdots + a_{n-1} z^{n-1} + z^n$, and $c = \max\{|a_0|, |a_1|, \ldots, |a_{n-1}|\}$. Show that all the roots of $p(z)$ are in $\bar{D}(0, c+1)$.

Hint: For $|z| > 1$ (and $c \neq 0$),

$$|p(z)| \geq |z|^n - c \frac{|z|^n - 1}{|z| - 1} > |z|^n \left(1 - \frac{c}{|z| - 1}\right).$$

Thus

$$|p(z)| > 0 \quad \text{for } |z| > c + 1.$$

On the other hand, if $z \in C(0, 2c + 1)$, then

$$|p(z)| > (2c + 1)^n \cdot \frac{1}{2} > c \geq |p(0)|.$$

Therefore, $p(z)$ has a root in the disc $D(0, 2c + 1)$.

6. Prove, by using the part of the Schwarz Lemma 6.2.1 that involves the derivative, that a conformal mapping of the unit disc onto itself must be a Möbius transformation.

7. (a) If $f(z)$ is analytic and bounded in the disc $D(0, R)$ and satisfies the conditions $f(0) = f'(0) = \cdots = f^{(k)}(0) = 0$, show that

$$|f(z)| \leq \frac{M}{R^{k+1}} |z|^{k+1} \quad z \in D(0, R),$$

(where $M = \sup\{|f(z)|; z \in D(0, R)\}$).

(b) If the equality holds at some point in the punctured disc $D'(0, R)$, show that

$$f(z) = e^{i\alpha} \frac{M}{R^{k+1}} z^{k+1} \quad (\alpha \in \mathbb{R}).$$

(c) What can be said about $|f^{(k+1)}(0)|$?

8. If $f(z)$ is analytic in the unit disc and $f(0) = 0$, show that

$$f(z) + f(z^2) + f(z^3) + \cdots + f(z^n) + \cdots$$

converges locally uniformly to an analytic function in the unit disc.

9. Suppose $f_j(z) \in H(D)$ maps the unit disc conformally to Ω_j ($j = 1, 2$), where $\Omega_1 \subset \Omega_2$ and $f_1(0) = f_2(0)$. Show that

$$|f_1'(0)| \leq |f_2'(0)|,$$

and equality holds if and only if $\Omega_1 = \Omega_2$.

10. (Carathéodory) If $f(z)$ is analytic and $|f(z)| \leq 1$ in the unit disc, show that

$$\left| \frac{f(z) - f(z_0)}{z - z_0} \right| \leq \frac{2}{1 - |z_0| \cdot |z|},$$

and

$$|f'(z_0)| \leq \frac{1 - |f(z_0)|^2}{1 - |z_0|^2} \leq \frac{1}{1 - |z_0|^2}.$$

11. Let $f(z)$ be analytic in the unit disc, $\Re f(z) \geq 0$, and $f(0) > 0$. Show that

$$f(0) \cdot \frac{1 - |z|}{1 + |z|} \leq |f(z)| \leq f(0) \cdot \frac{1 + |z|}{1 - |z|} \qquad z \in D.$$

Hint: First assume $f(0) = 1$, and apply the Schwarz lemma to $g(z) = \frac{f(z)-1}{f(z)+1}$.

12. Let $f(z)$ be an analytic function mapping the unit disc into itself. Suppose $f(z)$ has two or more distinct fixed points in the unit disc, show that $f(z)$ must be the identity function $f(z) = z$, $z \in D$.

13. How do you interpret the limiting case $a = 0$ and $f(0) = 0$ in the three-circle theorem?

14. Consider the function $\tan z$ in the strip $-\frac{\pi}{2} < \Re z < \frac{\pi}{2}$. Let

$$L(x) = \sup\{|\tan(x + iy)|; \ y \in \mathbb{R}\} \qquad \left(-\frac{\pi}{2} < x < \frac{\pi}{2}\right),$$

and confirm that $\log L(x)$ is a convex function of x for $-\frac{\pi}{2} < x < \frac{\pi}{2}$.

Chapter 7

Entire and Meromorphic Functions

7.1 The Mittag-Leffler Theorem

Let us recall the definition of meromorphic functions.

DEFINITION 7.1.1 *A function is* **meromorphic** *in a region Ω if the function is analytic in Ω except for poles. More specifically, a function $f(z)$ is meromorphic in Ω if, to every point $a \in \Omega$, there exists a disc $D(a, \delta) \subset \Omega$ such that either $f(z)$ is analytic in the disc or else $f(z)$ is analytic in the punctured disc $D'(a, \delta)$ and the isolated singularity is a pole.*

We call a function *meromorphic* (without specifying the region) if it is meromorphic in the whole finite complex plane $\mathbb{C}$. Examples of meromorphic functions are given by $\tan z$, $\cot z$, and $\frac{z}{e^z - 1}$; in each of these cases the point at infinity is an accumulation point of poles, hence is a nonisolated singularity.

Suppose $f(z)$ and $g(z)$ are two meromorphic functions that happen to have the same poles and singular parts. Then their difference is an entire function. For example, if $f(z) = \frac{1}{\sin z}$, $g(z) = \frac{1}{\sin z} + e^z$, then $f(z) - g(z) = -e^z$ is entire. This observation leads to the following: *The most general meromorphic function with prescribed singular parts is a particular meromorphic function having these prescribed singular parts plus an arbitrary entire function.*

We now ask the question: How much "*bad*" behavior can we prescribe and still get a meromorphic function? For example, can we find a function having simple poles with residue 1 at every integer point? This particular case can be answered "Yes," for $f(z) = \pi \cot \pi z$ is such a function. Therefore, the general solution to the problem is $\pi \cot \pi z + g(z)$, where $g(z)$ is an arbitrary entire function.

Is it possible to find a meromorphic function with poles at $\left\{\frac{1}{n} \,;\, n \in \mathbb{N}\right\}$? Here the answer is "No," for the origin, being an accumulation point of poles, must be a nonisolated singular point. This observation implies that if there is an accumulation point of the poles of $f(z)$, then this accumulation point must be the point at infinity.

Now, suppose we wish to construct a meromorphic function having poles at a finite number of points, say $b_1, b_2, \ldots, b_n$, with singular parts $P(z; b_j)$ $(j = 1, 2, \ldots, n)$. Because $P(z; b_j)$ is analytic everywhere except at the point b_j, the sum $\sum_{j=1}^{n} P(z; b_j)$ is a specific meromorphic function having the prescribed behavior at each of the points b_j $(j = 1, 2, \ldots, n)$. Hence we need only add an arbitrary entire function to obtain the general solution.

What if the number of poles is infinite? Let us try a similar procedure to construct a meromorphic function having a simple pole with residue 1 at each integer point. We know that such meromorphic functions exist in abundance. From the prescription, the singular part at $z = n$ is $P(z; n) = \frac{1}{z-n}$, but the sum of the singular parts,

$$\sum_{n=-\infty}^{\infty} P(z; n) = \sum_{n=-\infty}^{\infty} \frac{1}{z-n},$$

converges nowhere! Thus it is apparent that although we must modify this series so that it becomes convergent, this modification must be made in such a manner as not to introduce new singularities or affect the given singular parts. This objective is achieved in the proof of the following

THEOREM 7.1.2 (Mittag-Leffler) *Let $\{b_k\}_{k=1}^{\infty}$ be a sequence of distinct points having no accumulation point in the finite complex plane $\mathbb{C}$, and $\{P_k(z)\}_{k=1}^{\infty}$ be a sequence of polynomials without constant terms. Then there exists a meromorphic function $f(z)$ having the singular part*

$$P_k\left(\frac{1}{z - b_k}\right) \quad at \ b_k \ (k \in \mathbb{N})$$

and no other singularity in the finite complex plane $\mathbb{C}$.

Proof. Because $\{b_k\}_{k=1}^{\infty}$ has no finite accumulation point, any circle with center at the origin has at most a finite number of b_k on it. We can thus rearrange the numbering of b_k so that

$$|b_1| \leq |b_2| \leq \cdots \leq |b_k| \leq \cdots.$$

We also temporarily assume that $b_1 \neq 0$. Choose a convergent series of positive numbers $\sum_{k=1}^{\infty} c_k$. Because each singular part $P_k \left(\frac{1}{z-b_k} \right)$ is an analytic function in the disc $D(0, |b_k|)$, it can be expanded in a Taylor series about the origin:

$$P_k \left(\frac{1}{z - b_k} \right) = a_0^{(k)} + a_1^{(k)} z + \cdots + a_n^{(k)} z^n + \cdots, \quad z \in D(0, |b_k|).$$

This series is then absolutely and uniformly convergent on $D \left(0, \frac{|b_k|}{2} \right)$. Let $Q_k(z)$ be a partial sum

$$Q_k(z) = \sum_{j=0}^{n_k} a_j^{(k)} z^j,$$

where n_k is chosen so large that

$$\left\| P_k \left(\frac{1}{z - b_k} \right) - Q_k(z) \right\|_{\bar{D}(0, \frac{|b_k|}{2})} < c_k.$$

Now consider the series

$$\sum_{k=1}^{\infty} \left\{ P_k \left(\frac{1}{z - b_k} \right) - Q_k(z) \right\}.$$

Inside any given circle of radius R, say, there are only a finite number of points at which the sum has poles; only those $P_k \left(\frac{1}{z-b_k} \right)$ with $b_k \in D(0, R)$ contribute poles to the sum. Thus, for any value of R, we can break up the series into two parts,

$$\sum_{|b_k| \leq 2R} \left\{ P_k \left(\frac{1}{z - b_k} \right) - Q_k(z) \right\} + \sum_{|b_k| > 2R} \left\{ P_k \left(\frac{1}{z - b_k} \right) - Q_k(z) \right\},$$

such that the second part has no singularity in the disc $D(0, R)$, and

$$\left\| P_k \left(\frac{1}{z - b_k} \right) - Q_k(z) \right\|_{\bar{D}(0, R)} < c_k \quad \text{for } |b_k| > 2R.$$

Hence the "tail" part of the series is absolutely and uniformly convergent for $z \in \bar{D}(0, R)$, and so it is an analytic function in $D(0, R)$, and the finite sum $\sum_{|b_k|<2R}$ is a rational function with the prescribed behavior at the points b_k inside $D(0, R)$.

Because R is arbitrary, we can let it grow as large as we please, and we see that the Mittag-Leffler series

$$\sum_{k=1}^{\infty} \left\{ P_k \left(\frac{1}{z - b_k} \right) - Q_k(z) \right\}$$

converges locally uniformly in the region $\mathbb{C} \backslash \{b_k\}_{k=1}^{\infty}$. It has the prescribed behavior everywhere in the finite complex plane $\mathbb{C}$ except perhaps at the origin. If we now affix to the sum one more term, viz., the singular part at $z = 0$ if the origin is a prescribed pole, we obtain a function with all the desired properties.   □

Remark. If $P_k(z) = z$ for all k and if, for some fixed integer n,

$$\sum_{k=1}^{\infty} \frac{1}{|b_k|^{n+1}} < \infty \qquad \left(\sum_{k=1}^{\infty} \frac{1}{|b_k|^n} = \infty \right),$$

then it is sufficient to take $Q_k(z)$ to be the first n terms in the Taylor-series expansion of $P_k \left(\frac{1}{z-b_k} \right) = \frac{1}{z-b_k}$ around the origin; that is,

$$Q_k(z) = -\frac{1}{b_k} - \frac{z}{b_k^2} - \frac{z^2}{b_k^3} - \cdots - \frac{z^{n-1}}{b_k^n} \qquad (k \in \mathbb{N}).$$

That is, in this case n_k can be chosen independent of k. The proof of this statement is an immediate consequence of the equality

$$P_k \left(\frac{1}{z - b_k} \right) - Q_k(z) = \frac{z^n}{b_k^n (z - b_k)}.$$

This gives the estimate

$$\left\| P_k \left(\frac{1}{z - b_k} \right) - Q_k(z) \right\|_{\bar{D}(0,R)} \leq \frac{2 R^n}{|b_k|^{n+1}} \qquad (z \in \bar{D}(0, R), \; |b_k| > 2R).$$

We now return to a meromorphic function having simple poles with residue 1 at each integer point. The singular parts are

$$P(z; n) = \frac{1}{z - n} \qquad (n \in \mathbb{Z})$$

$$= -\frac{1}{n} - \frac{z}{n^2} - \frac{z^2}{n^3} - \cdots \qquad (|z| < n, \; n \neq 0).$$

Because $\sum_{n \neq 0} \frac{1}{n^2} < \infty$, we follow the remark above and form the series

$$g(z) = \frac{1}{z} + \sum_{n \neq 0} \left(\frac{1}{z - n} + \frac{1}{n} \right)$$

$$= \frac{1}{z} + \sum_{n \neq 0} \frac{z}{n(z - n)}.$$

Let us carry out in detail for this example the procedure sketched above. Take any $R > 0$ and consider the sum

$$\sum_{|n| > 2R} \frac{z}{n(z - n)}, \quad z \in \bar{D}(0, R).$$

This sum is dominated by the series

$$\sum_{|n| > 2R} \frac{R}{|n| \cdot |z - n|} \le \sum_{|n| > 2R} \frac{2R}{n^2}, \quad z \in \bar{D}(0, R),$$

and the latter clearly converges. This implies, by the Weierstrass M-test 2.4.5, that $\sum_{|n| > 2R} \frac{z}{n(z-n)}$ is uniformly convergent in $\bar{D}(0, R)$, and hence the sum is analytic. Because we may have omitted at most a finite number of terms, we let $R \to \infty$ and conclude that

$$g(z) = \frac{1}{z} + \sum_{n \neq 0} \left(\frac{1}{z - n} + \frac{1}{n} \right)$$

is analytic in $\mathbb{C}$ except at the integer points, where the singularities are simple poles with residue 1.

   Remark. If we agree to sum symmetrically; that is, to regard the series as the limit of the symmetric partial sums

$$\frac{1}{z} + \sum_{0 < |n| < N} \left(\frac{1}{z - n} + \frac{1}{n} \right),$$

then the terms $\frac{1}{n}$ and $-\frac{1}{n}$ will cancel in pairs, and hence we can write the sum without these terms. However, the series $\sum_{-\infty}^{\infty} \frac{1}{z-n}$ does not converge if we do *not* sum symmetrically. Recall, as in the case of improper integrals, that we have, *by definition*,

$$\sum_{k=-\infty}^{\infty} f_k(z) = \lim_{\substack{M \to \infty \\ N \to \infty}} \sum_{k=-M}^{N} f_k(z),$$

where M and N tend to ∞ *independently*. In other words, $\sum_{k=-\infty}^{\infty} f_k(z)$ converges if and only if both $\sum_{k=0}^{\infty} f_k(z)$ and $\sum_{k=1}^{\infty} f_{-k}(z)$ converge.

Now, set

$$h(z) = \pi \cot \pi z - g(z)$$

$$= \pi \cot \pi z - \left\{ \frac{1}{z} + {\sum_{-\infty}^{\infty}}' \left(\frac{1}{z-n} + \frac{1}{n} \right) \right\},$$

where $\sum'$ stands for $\sum_{n \neq 0}$, as customary. Then $h(z)$ is an entire function. Let us determine this function. Because the convergence is locally uniform in $\mathbb{C} \setminus \mathbb{Z}$, we may differentiate termwise (by Corollary 4.5.5), and so we obtain

$$h'(z) = -\frac{\pi^2}{\sin^2 \pi z} + \sum_{n=-\infty}^{\infty} \frac{1}{(z-n)^2}.$$

We claim that this entire function is bounded. Because the right-hand side is a periodic function of period 1, it is sufficient to show its boundedness in a strip of width 1, say

$$\left\{ z \in \mathbb{C}; \ -\frac{1}{2} \leq \Re z \leq \frac{1}{2} \right\}.$$

Because of the symmetry, $h'(\bar{z}) = \overline{h'(z)}$, it is sufficient to show that $h'(z)$ is bounded in the upper half-strip

$$\left\{ z \in \mathbb{C}; \ -\frac{1}{2} \leq \Re z \leq \frac{1}{2}, \ \Im z \geq 0 \right\}.$$

Furthermore, it is sufficient to show that $h'(z)$ is bounded in the half-strip

$$\left\{ z \in \mathbb{C}; \ -\frac{1}{2} \leq \Re z \leq \frac{1}{2}, \ \Im z \geq 1 \right\},$$

for, being an entire function, $h'(z)$ is bounded on any compact set; in particular, on the square

$$\left\{ z \in \mathbb{C}; \ -\frac{1}{2} \leq \Re z \leq \frac{1}{2}, \ 0 \leq \Im z \leq 1 \right\}.$$

We remark that we cannot show boundedness by proceeding in steps of compact sets, for the sequence of bounds could approach infinity. Instead, we treat each of the terms separately.

$$\left| -\frac{\pi^2}{\sin^2 \pi z} \right| = \pi^2 \left| \frac{2i}{e^{i\pi z} - e^{-i\pi z}} \right|^2 \le \frac{4\pi^2}{||e^{i\pi z}| - |e^{-i\pi z}||^2}$$

$$= \frac{4\pi^2}{(e^{\pi y} - e^{-\pi y})^2} \le \frac{4\pi^2}{(e^{\pi} - e^{-\pi})^2} \quad (y \ge 1).$$

Actually, the next-to-last term tends to zero uniformly in $-\frac{1}{2} \le x \le \frac{1}{2}$ as $y \to \infty$. Now, for z in the half-strip under consideration, we have

$$\left| \frac{1}{(z-n)^2} \right| = \frac{1}{(x-n)^2 + y^2} \le \begin{cases} \dfrac{1}{(n-\frac{1}{2})^2 + 1} & (n > 0); \\[2ex] \dfrac{1}{(n+\frac{1}{2})^2 + 1} & (n < 0); \\[2ex] 1 & (n = 0). \end{cases}$$

Because the series

$$\sum_{n<0} \frac{1}{(n+\frac{1}{2})^2 + 1} + 1 + \sum_{n>0} \frac{1}{(n-\frac{1}{2})^2 + 1} = 1 + 2\sum_{n>0} \frac{1}{(n-\frac{1}{2})^2 + 1}$$

converges, the entire function $h'(z)$ is bounded, and now, by the Liouville Theorem 4.4.9, $h'(z) \equiv$ constant. Given $\epsilon > 0$, there exists a positive integer N such that, for z in the strip,

$$\left| \sum_{|n|>N} \frac{1}{(z-n)^2} \right| \le \sum_{|n|>N} \left| \frac{1}{(z-n)^2} \right| \le 2 \sum_{|n|>N} \frac{1}{(n-\frac{1}{2})^2 + 1} < \frac{\epsilon}{2}.$$

Clearly,

$$\left| \sum_{|n| \le N} \frac{1}{(z-n)^2} \right| < \frac{\epsilon}{2} \quad \text{for } \Im z = y \quad \text{sufficiently large,}$$

and we obtain $h'(z) \equiv 0$. That is,

$$\frac{\pi^2}{\sin^2 \pi z} = \sum_{n=-\infty}^{\infty} \frac{1}{(z-n)^2}.$$

It follows that

$$h(z) = \pi \cot \pi z - \left\{ \frac{1}{z} + \sum_{n=1}^{\infty} \left(\frac{1}{z-n} + \frac{1}{z+n} \right) \right\} \equiv c \quad \text{(constant)}.$$

We evaluate the constant c by noting that $h(z)$ is an odd function:

$$c = h(-z) = -h(z) = -c. \quad \therefore \ c = 0.$$

We obtain

$$\pi \cot \pi z = \frac{1}{z} + \sideset{}{'}\sum_{-\infty}^{\infty} \left(\frac{1}{z-n} + \frac{1}{n} \right)$$

$$= \frac{1}{z} + \sum_{n=1}^{\infty} \left(\frac{1}{z-n} + \frac{1}{z+n} \right)$$

$$= \frac{1}{z} + 2z \sum_{n=1}^{\infty} \frac{1}{z^2 - n^2}.$$

Using these results, we can deduce some remarkable identities. For example, setting $z = \frac{1}{4}$ in the partial fraction expansion of $\pi \cot \pi z$, we obtain

$$\pi \cot \frac{\pi}{4} = 4 + \sum_{n=1}^{\infty} \left(\frac{1}{\frac{1}{4} - n} + \frac{1}{\frac{1}{4} + n} \right)$$

$$= 4 + \left(-\frac{1}{3/4} + \frac{1}{5/4} \right) + \left(-\frac{1}{7/4} + \frac{1}{9/4} \right)$$

$$+ \left(-\frac{1}{11/4} + \frac{1}{13/4} \right) + \cdots$$

$$= 4 \left\{ 1 + \left(-\frac{1}{3} + \frac{1}{5} \right) + \left(-\frac{1}{7} + \frac{1}{9} \right) + \left(-\frac{1}{11} + \frac{1}{13} \right) + \cdots \right\}$$

$$= 4 \left\{ 1 - \frac{1}{3} + \frac{1}{5} - \frac{1}{7} + \frac{1}{9} - + \cdots \right\}.$$

$$\therefore \ \frac{\pi}{4} = 1 - \frac{1}{3} + \frac{1}{5} - \frac{1}{7} + \frac{1}{9} - \frac{1}{11} + \frac{1}{13} - + \cdots.$$

Setting $z = \frac{1}{2}$ in the partial fraction expansion of $\frac{\pi^2}{\sin^2 \pi z}$, we obtain

$$\pi^2 = \sum_{-\infty}^{\infty} \frac{1}{(\frac{1}{2} - n)^2} = \sum_{-\infty}^{\infty} \frac{4}{(2n-1)^2} = 4 \sum_{k:\text{odd}} \frac{1}{k^2} = 8 \sum_{\substack{k:\text{odd} \\ k>0}} \frac{1}{k^2},$$

and hence

$$\frac{\pi^2}{8} = \frac{1}{1^2} + \frac{1}{3^2} + \frac{1}{5^2} + \frac{1}{7^2} + \cdots.$$

Rewriting the partial fraction expansion of $\frac{\pi^2}{\sin^2 \pi z}$, we obtain

$$\sum_{-\infty}^{\infty}{}' \frac{1}{(z-n)^2}$$

$$= \frac{\pi^2}{\sin^2 \pi z} - \frac{1}{z^2} = \frac{\pi^2 z^2 - \sin^2 \pi z}{z^2 \cdot \sin^2 \pi z}$$

$$= \pi^2 \cdot \frac{t^2 - \sin^2 t}{t^2 \cdot \sin^2 t} \qquad (t = \pi z)$$

$$= \pi^2 \cdot \frac{(t - \sin t) \cdot (t + \sin t)}{(t \sin t)^2}$$

$$= \pi^2 \cdot \frac{\left\{ t - \left(t - \dfrac{t^3}{3!} + \dfrac{t^5}{5!} - + \cdots \right) \right\} \cdot \left\{ t + \left(t - \dfrac{t^3}{3!} + \dfrac{t^5}{5!} - + \cdots \right) \right\}}{\left\{ t \left(t - \dfrac{t^3}{3!} + \dfrac{t^5}{5!} - + \cdots \right) \right\}^2}$$

$$= \pi^2 \cdot \frac{\dfrac{2}{3!}t^4 + O(t^6)}{t^4 + O(t^6)} = \frac{\pi^2}{3} + O(t^2) \quad (t \text{ small}).$$

Letting $t = \pi z$ approach zero, we obtain

$$\sum_{n=1}^{\infty} \frac{1}{n^2} = \frac{\pi^2}{6}.$$

Similarly, differentiating both sides of the partial fraction expansion of $\frac{\pi^2}{\sin^2 \pi z}$ twice and letting z approach zero, we obtain

$$\sum_{n=1}^{\infty} \frac{1}{n^4} = \frac{\pi^4}{90}.$$

Remarks. (a) Because

$$\pi z \cdot \cot \pi z = 1 - 2 \sum_{n=1}^{\infty} \frac{z^2}{n^2 - z^2}$$

and

$$\frac{z^2}{n^2 - z^2} = \left(\frac{z}{n}\right)^2 + \left(\frac{z}{n}\right)^4 + \cdots + \left(\frac{z}{n}\right)^{2k} + \cdots \quad (z \in D,\ n \in \mathbb{N}),$$

we obtain, by the Weierstrass Double Series Theorem 4.5.6,

$$\pi z \cdot \cot \pi z = 1 - 2 \sum_{k=1}^{\infty} S_{2k} z^{2k} \quad (z \in D),$$

where

$$S_m = \sum_{n=1}^{\infty} \frac{1}{n^m}.$$

(b) Heuristically, integrating

$$\pi \cot \pi z - \frac{1}{z} = \sum_{-\infty}^{\infty}{}' \left(\frac{1}{z - n} + \frac{1}{n}\right)$$

from 0 to z, we obtain

$$\log \frac{\sin \pi \zeta}{\zeta} \bigg|_0^z = \sum_{-\infty}^{\infty}{}' \left[\log(\zeta - n) + \frac{\zeta}{n}\right]_0^z ;$$

that is,

$$\log \frac{\sin \pi z}{\pi z} = \sum_{-\infty}^{\infty}{}' \left\{\log\left(1 - \frac{z}{n}\right) + \frac{z}{n}\right\}.$$

Exponentiating, we obtain

$$\sin \pi z = \pi z \cdot \prod_{-\infty}^{\infty}{}' \left\{\left(1 - \frac{z}{n}\right) \cdot e^{\frac{z}{n}}\right\},$$

and then

$$\sin \pi z = \pi z \cdot \prod_{n=1}^{\infty} \left(1 - \frac{z^2}{n^2}\right),$$

by grouping the factors in pairs. This factorization of $\sin \pi z$ turns out to be correct, as we shall see in Section 7.6.

7.2   A Theorem of Weierstrass

Suppose we are given a sequence of distinct points $\{a_j\}_{j=1}^{\infty}$. Is it possible to find an entire function with roots of prescribed multiplicity $\{k_j\}_{j=1}^{\infty}$ at

these and only these points? If the sequence is finite, there is no problem.
Given the finite sequence $\{a_j\}_{j=1}^n$ we simply form the polynomial

$$f(z) = \prod_{j=1}^n (z - a_j)^{k_j} .$$

If the sequence is infinite, then we have the following

THEOREM 7.2.1 (Weierstrass) *Let $\{a_j\}_{j=1}^\infty$ be a sequence of distinct
points having no finite accumulation point, and let a sequence of positive
integers $\{k_j\}_{j=1}^\infty$ be given. Then there exists an entire function having roots
of multiplicity k_j at a_j for all $j \in \mathbb{N}$, and nowhere else.*

This theorem is a corollary of the Mittag-Leffler Theorem 7.1.2, although
it was first proved independently by Weierstrass. The first proof we give is
Mittag-Leffler's own.

Proof. Suppose there is such a function, say $f(z)$; then its logarithmic
derivative $\frac{f'(z)}{f(z)}$ is meromorphic, and at a_j its singular part is $\frac{k_j}{z-a_j}$ (Cf.
the Argument Principle 5.6.1). Now let $g(z)$ be a meromorphic function
having the singular part $\frac{k_j}{z-a_j}$ at each a_j. Such a function exists, by the
Mittag-Leffler Theorem 7.1.2.

To see what we must do with $g(z)$, consider the following[1]: Suppose
$f(z) = z^4$; then $\frac{f'(z)}{f(z)} = \frac{4}{z}$.

$$\int \frac{4}{z} dz = 4 \log z = \log z^4.$$

$$f(z) = z^4 = \exp \left\{ \int \frac{4}{z} dz \right\} = \exp \left\{ \int \frac{f'(z)}{f(z)} dz \right\} .$$

This simple illustration suggests that we should look for $f(z)$ by considering $\exp \left\{ \int g(z) \, dz \right\}$.

Therefore, let ζ and z be any points other than $a_1, a_2, \ldots, a_n, \ldots$, and
form the function

$$G(z) = \int_\zeta^z g(t) \, dt,$$

where the path of integration does not pass through any of the points a_j
(Figure 7.1). The function $G(z)$ is not single-valued, for its value depends

---

[1] Remember: *Just because mathematics does not stink like some other subjects, that
does not mean you don't have to do experiments!*

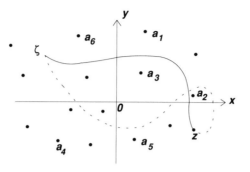

Figure 7.1

on the path of integration, but any two branches of $G(z)$ differ by an integer multiple of $2\pi i$. Thus the function

$$h(z) = \exp\{G(z)\}$$

is single-valued. Because $G(z)$ is analytic at every point at which it is defined, $h(z)$ is analytic and does not vanish for any z except possibly for $z = a_j\ (j \in \mathbb{N})$.

We wish to show that $h(z)$ is a function of the kind we desire, and so we must show that the singularities at a_j can be removed in such a way that $h(z)$ has a root of multiplicity k_j at $a_j\ (j \in \mathbb{N})$. For this purpose, it is sufficient to show that the function

$$H(z) = (z - a_j)^{-k_j} \cdot h(z)$$

has neither root nor pole at a_j. It is analytic except possibly at $a_1, a_2, \ldots$, and

$$H'(z) = (z - a_j)^{-k_j} \cdot h'(z) - k_j(z - a_j)^{-k_j - 1} \cdot h(z).$$

Therefore,

$$\frac{H'(z)}{H(z)} = \frac{h'(z)}{h(z)} - \frac{k_j}{z - a_j}.$$

But

$$h'(z) = G'(z) \cdot \exp\{G(z)\} = G'(z) \cdot h(z)$$

and

$$G'(z) = \frac{d}{dz} \int_\zeta^z g(t)\, dt = g(z).$$

From this result it follows that

$$h'(z) = g(z) \cdot h(z).$$

Hence

$$\frac{H'(z)}{H(z)} = g(z) - \frac{k_j}{z - a_j}.$$

However, $g(z)$ has a simple pole at a_j with residue k_j; thus $\frac{H'(z)}{H(z)}$ is analytic at a_j, and so $H(z)$ has neither root nor pole at a_j. Because $h(z)$ has no other root, it is an entire function satisfying all the prescribed conditions. $\square$

What happens if there are two functions $h_1(z)$ and $h_2(z)$ that satisfy the conditions in the Weierstrass Theorem 7.2.1? Then the quotient $\frac{h_1(z)}{h_2(z)}$ is an entire function without roots. Therefore, the most general function that satisfies the condition in the Weierstrass Theorem 7.2.1 is any particular solution times an entire function without root. Of course, this statement is the obvious analog of the earlier statement concerning the general solution of the Mittag-Leffler problem. But what is the most general entire function having no root? This question is answered in the following

COROLLARY 7.2.2 *If $f(z)$ is an entire function without root, then there exists an entire function $G(z)$ such that*

$$f(z) = e^{G(z)}.$$

Proof. The proof is implicit in the proof of the Weierstrass Theorem 7.2.1. Without loss of generality, we may assume $f(0) = 1$. The function $\frac{f'(z)}{f(z)}$, being a quotient of two entire functions, can have no singularity other than poles in the finite complex plane, but poles occur only at points where $f(z)$ vanishes and, by assumption, $f(z)$ has no root. Thus $\frac{f'(z)}{f(z)}$ is an entire function. Let

$$G(z) = \int_0^z \frac{f'(t)}{f(t)} dt, \quad h(z) = e^{G(z)}.$$

Note here that $G(z)$ is single-valued and defined everywhere as an entire function. We have

$$h'(z) = G'(z) \cdot e^{G(z)} = \frac{f'(z)}{f(z)} \cdot h(z) \quad \text{for all } z \in \mathbb{C},$$

from which it follows that

$$f(z)h'(z) - f'(z)h(z) = 0 \quad \text{for all } z \in \mathbb{C}.$$

Thus

$$\left\{\frac{h(z)}{f(z)}\right\}' = 0 \quad \text{for all } z \in \mathbb{C}; \quad \text{hence} \quad \frac{h(z)}{f(z)} \equiv c \quad \text{(constant)}.$$

But

$$f(0) = 1 = h(0) \quad \text{and so} \quad f(z) = h(z) = e^{G(z)}. \qquad \square$$

We give two examples of the use of the Weierstrass Theorem 7.2.1 and the Mittag-Leffler Theorem 7.1.2. The first one is an important example of an "interpolation theorem."

COROLLARY 7.2.3 *Let* $\{a_j\}_{j=1}^{\infty}$ *be a sequence of distinct points having no finite accumulation point and* $\{\zeta_j\}_{j=1}^{\infty}$ *a completely arbitrary sequence of complex numbers (with repetition permitted). Then there exists an entire function* $f(z)$ *such that*

$$f(a_j) = \zeta_j \quad (j \in \mathbb{N}).$$

Proof. Form an entire function $g(z)$ having simple roots at each a_j. Such a function exists, by the Weierstrass Theorem 7.2.1. Then $g'(a_j) \neq 0$ ($j \in \mathbb{N}$), because the roots are simple. Also, form a meromorphic function $h(z)$ with simple poles having $\frac{\zeta_j}{g'(a_j)}$ as the residue at a_j ($j \in \mathbb{N}$). Existence of such a function is guaranteed by the Mittag-Leffler Theorem 7.1.2. Their product $f(z) = g(z) \cdot h(z)$ is analytic except perhaps at a_j ($j \in \mathbb{N}$), for $g(z)$ is entire and $h(z)$ is analytic except at a_j ($j \in \mathbb{N}$). We assert that $f(z)$ is in fact an entire function which satisfies the prescribed conditions. For, near a_j, we have

$$g(z) = (z - a_j) \cdot g_j(z), \quad h(z) = \frac{h_j(z)}{z - a_j},$$

where both $g_j(z)$ and $h_j(z)$ are analytic at a_j; moreover,

$$g_j(a_j) = g'(a_j), \quad h_j(a_j) = \frac{\zeta_j}{g'(a_j)}.$$

Therefore, $f(z) = g_j(z) \cdot h_j(z)$ is analytic at a_j, and

$$f(a_j) = g_j(a_j) \cdot h_j(a_j) = g'(a_j) \cdot \frac{\zeta_j}{g'(a_j)} = \zeta_j. \qquad \square$$

COROLLARY 7.2.4 *Every meromorphic function can be expressed as the quotient of two entire functions.*

Proof. Given a meromorphic function $f(z)$, form an entire function $g(z)$ with the poles of $f(z)$ for roots with the same multiplicity. Such a function exists by the Weierstrass Theorem 7.2.1. Arguing as in the proof of Corollary 7.2.3, we can show that the product $h(z) = f(z) \cdot g(z)$ is an entire function. We now have the desired representation, $f(z) = \frac{h(z)}{g(z)}$. $\square$

7.3   Extensions of Theorems of Mittag-Leffler and Weierstrass

THEOREM 7.3.1 (Mittag-Leffler) *Let $\{b_k\}_{k=1}^{\infty}$ be a sequence of distinct points in a region $\Omega \subset \hat{\mathbb{C}}$ having no accumulation point in Ω. Given a sequence $\{P_k(z)\}_{k=1}^{\infty}$ of polynomials without constant term, there exists a meromorphic function on Ω having a pole with the singular part*

$$P_k \left(\frac{1}{z - b_k} \right) \quad \text{at } b_k \ (\text{for each } k \in \mathbb{N}),$$

and no other singularity in Ω.

Proof. We may assume that Ω is not $\hat{\mathbb{C}}$ or $\mathbb{C}$, for these cases have been discussed already. We also temporarily assume that none of b_k are 0 or ∞. Let

$$E_1 = \{b_k; \ |b_k| \cdot \delta_k \geq 1\}, \quad E_2 = \{b_k; \ |b_k| \cdot \delta_k < 1\},$$

where δ_k is the distance from b_k to the boundary $\partial \Omega$ of Ω (because of our assumption that $\Omega \neq \hat{\mathbb{C}}$ or $\mathbb{C}$, we have $\delta_k < \infty$, and because Ω is open we have $\delta_k > 0$ for all $k \in \mathbb{N}$).

We claim that the set E_1 has no finite accumulation point. For, if c $(\neq \infty)$ were an accumulation point of E_1, then

$$|b_k| < |c| + 1 \quad \text{for infinitely many } k,$$

and because $|b_k| \cdot \delta_k \geq 1$, we have

$$\delta_k \geq \frac{1}{|c| + 1} \quad \text{for infinitely many } k.$$

Therefore,

$$\text{dist}(c, \partial \Omega) \geq \frac{1}{|c| + 1} > 0.$$

It follows that the point c is an interior point of Ω, contradicting the assumption that $\{b_k\}_{k=1}^\infty$ has no accumulation point in Ω. Thus the set E_1 has no finite accumulation point, and hence, by Theorem 7.1.2, we can find a meromorphic function $f_1(z)$ with the prescribed singular parts at each of the points in E_1.

In the case of E_2, we claim that $\delta_k \longrightarrow 0$ $(k \to \infty)$ if there are infinitely many points in E_2. (In case E_2 consists of only finitely many points, the proof of the theorem is immediate.) For otherwise, there exists $\epsilon > 0$ such that $\delta_k > \epsilon > 0$ for infinitely many k, and therefore $|b_k| < \frac{1}{\epsilon}$ for infinitely many k. Therefore, as before, E_2 has a finite accumulation point in Ω.

Now, for each k, let $c_k \in \partial\Omega$ such that (see Exercise 2.10)

$$|b_k - c_k| = \delta_k = \mathrm{dist}(b_k, \partial\Omega)$$

(if there are two or more such points, take an arbitrary one), and expand $P_k\left(\frac{1}{z-b_k}\right)$ in Laurent series with center at c_k in the region $\{z \in \mathbb{C};\ |z - c_k| > \delta_k\}$:

$$P_k\left(\frac{1}{z - b_k}\right) = A_0^{(k)} + \frac{A_1^{(k)}}{z - c_k} + \frac{A_2^{(k)}}{(z - c_k)^2} + \cdots.$$

The right-hand side converges uniformly outside any disc $D(c_k, r)$ $(r > \delta_k)$, so choose a partial sum $Q_k(z)$ of the expansion above such that

$$\left| P_k\left(\frac{1}{z - b_k}\right) - Q_k(z) \right| < \frac{1}{2^k}, \quad z \notin D(c_k, 2\delta_k).$$

Then

$$f_2(z) = \sum_{\substack{k=1 \\ b_k \in E_2}}^{\infty} \left\{ P_k\left(\frac{1}{z - b_k}\right) - Q_k(z) \right\}$$

is a meromorphic function with the prescribed singular parts at each of the points in E_2. Because $\delta_k \longrightarrow 0$ $(k \to \infty)$, if K is an arbitrary compact subset of Ω we can choose N sufficiently large such that the discs $D(c_k, 2\delta_k)$ are disjoint from K for all $k \geq N$. It follows that

$$\sum_{\substack{k \geq N \\ b_k \in E_2}}^{\infty} \left\{ P_k\left(\frac{1}{z - b_k}\right) - Q_k(z) \right\}$$

converges uniformly on K. Thus $f_1(z) + f_2(z)$ has the prescribed behavior everywhere in Ω except possibly at 0 and ∞. Finally, we add to this sum the singular parts at 0 and ∞ if they are prescribed poles, and we obtain a meromorphic function with all the desired properties. □

The Weierstrass Theorem 7.2.1 can also be extended to a general region:

THEOREM 7.3.2 *Let $\{a_k\}_{k=1}^{\infty}$ be a sequence of distinct points in a region Ω having no accumulation point in Ω, and $\{m_k\}_{k=1}^{\infty}$ a sequence of positive integers. Then there exists a function analytic in Ω with a root of multiplicity m_k at a_k for all $k \in \mathbb{N}$, and nowhere else.*

   Proof. The proof is left as Exercise 7.33. □

Using this theorem, one can show that given an arbitrary region Ω there exists a function analytic in Ω that cannot be extended analytically to any region properly containing Ω; that is, the boundary of the region Ω is the "*natural boundary*" of this function. (Cf. Theorem 8.2.3.)

7.4   Infinite Products

A naïve definition of convergence of an infinite product

$$\prod_{k=1}^{\infty} a_k \quad (a_k \in \mathbb{C},\ k \in \mathbb{N})$$

would be: Set

$$p_n = \prod_{k=1}^{n} a_k \quad (k \in \mathbb{N}).$$

If the sequence $\{p_n\}_{n=1}^{\infty}$ of *partial products* converges to a (finite) limit, we shall say that the infinite product

$$\prod_{k=1}^{\infty} a_k$$

is convergent and its value is the limit above.

   This definition has some defects:

(i) Consider the product $\prod_{k=0}^{\infty} k$. We have $p_n = 0$ for all n, and so the product is convergent, but the product $\prod_{k=1}^{\infty} k$, obtained by deleting the first factor, is divergent. This is in contrast to the situation in infinite series where removal of a finite number of terms does not alter the convergence of the resulting series.

(ii) Next, consider the infinite product $\prod_{k=1}^{\infty} \frac{k}{k+1}$. Here

$$p_n = \prod_{k=1}^{n} \frac{k}{k+1} = \frac{1}{2} \cdot \frac{2}{3} \cdot \frac{3}{4} \cdots \frac{n}{n+1} = \frac{1}{n+1};$$

hence $\lim_{n \to \infty} p_n = 0$. Thus the product converges to zero although none of the factors is zero. This situation is rather awkward and inconvenient for our purpose, as we shall see in the sequel.

Therefore we shall adopt the following more restrictive

DEFINITION 7.4.1 *An infinite product $\prod_{k=1}^{\infty} a_k$ is said to be* **convergent** *if*

1^0 *At most a finite number of factors a_k are zero;*
2^0 *Partial products obtained after removing the zero factors (if any) approach a limit;*
3^0 *The limit mentioned in 2^0 is not zero.*

     Then, if the product is convergent, its value is defined to be the limit mentioned in 2^0 provided no factor is zero; otherwise its value is zero.

EXAMPLE.

(a) The product $0 \cdot 1 \cdot 0 \cdot 1 \cdot 0 \cdot 1 \cdots$ is divergent because it does not satisfy 1^0.
(b) The product $0 \cdot 0 \cdot 0 \cdot \frac{1}{2} \cdot \frac{2}{3} \cdot \frac{3}{4} \cdot \frac{4}{5} \cdots$ satisfies 1^0 and 2^0, but not 3^0, and is therefore divergent.
(c) The product $0 \cdot 0 \cdot 0 \cdot 0 \cdot 0 \cdot 1 \cdot 1 \cdot 1 \cdots$ is convergent to zero, because $1^0, 2^0$ and 3^0 are satisfied.
(d) The product $\prod_{k=2}^{\infty} \frac{k^2-1}{k^2}$ has partial products

$$
\begin{aligned}
p_n &= \frac{1 \cdot 3}{2^2} \cdot \frac{2 \cdot 4}{3^2} \cdots \frac{(n-1)(n+1)}{n^2} \\
&= \frac{1}{2} \left(\frac{3}{2} \cdot \frac{2}{3} \right) \left(\frac{4}{3} \cdot \frac{3}{4} \right) \cdots \left(\frac{n}{n-1} \cdot \frac{n-1}{n} \right) \frac{n+1}{n} \\
&= \frac{1}{2} \cdot \frac{n+1}{n}.
\end{aligned}
$$

$$\therefore \ \lim_{n \to \infty} p_n = \lim_{n \to \infty} \frac{n+1}{2n} = \frac{1}{2}.$$

Thus the product converges to $\frac{1}{2}$.

It follows from the definition that

(i) If a product is convergent, then a finite number of factors may be removed and the resulting product will still be convergent.

(ii) If a product converges to zero, then at least one of the factors is zero.

(iii) An infinite product $\prod_{k=1}^{\infty} a_k$ converges if and only if there exists an integer N such that the sequence $\{\prod_{k=N}^{m} a_k\}_{m=N}^{\infty}$ converges to a *nonzero* limit.

THEOREM 7.4.2 (Cauchy Criterion) *A necessary and sufficient condition for the convergence of the infinite product* $\prod_{k=1}^{\infty} a_k$ *is that, for every* $\epsilon > 0$, *there exists an integer* N *such that*

$$\left| \left(\prod_{k=n+1}^{m} a_k \right) - 1 \right| < \epsilon \quad \text{whenever } m > n \geq N.$$

Proof. Necessity. Suppose that the product $\prod_{k=1}^{\infty} a_k$ converges. Because a finite number of zero factors can be ignored, we assume that none of the factors is zero. Then

$$p_n = \prod_{k=1}^{n} a_k \longrightarrow p \neq 0 \quad (\text{as } n \to \infty).$$

Hence the sequence of partial products $\{p_n\}_{n=1}^{\infty}$ converges, and, by the Cauchy Criterion 2.1.2 for sequences, given $\epsilon > 0$, there exists N such that

$$|p_n| > \frac{|p|}{2}, \quad |p_m - p_n| < \epsilon \cdot \frac{|p|}{2} \quad \text{whenever } m > n \geq N.$$

Thus

$$\left| \left(\prod_{k=n+1}^{m} a_k \right) - 1 \right| = \left| \frac{p_m}{p_n} - 1 \right| = \frac{|p_m - p_n|}{|p_n|} < \frac{\epsilon \cdot \dfrac{|p|}{2}}{\dfrac{|p|}{2}} = \epsilon$$

$$(m > n \geq N).$$

Sufficiency. Assume that for every $\epsilon > 0$ there exists N such that

$$\left| \left(\prod_{k=n+1}^{m} a_k \right) - 1 \right| < \epsilon \quad \text{for all } m > n \geq N.$$

Then the condition 1^0 in the definition of convergence of an infinite product is clearly satisfied.

Next, set $\epsilon = \frac{1}{2}$; then, from the hypotheses of the theorem, there exists N such that

$$\left| \left(\prod_{k=N}^{m} a_k \right) - 1 \right| < \frac{1}{2} \quad \text{whenever } m \geq N;$$

that is,

$$\frac{1}{2} < \left| \prod_{k=N}^{m} a_k \right| < \frac{3}{2} \quad (m \geq N).$$

Hence, if $\{\prod_{k=N}^{m} a_k\}_{m=N}^{\infty}$ has a limit (as $m \to \infty$), then this limit is not zero.

Now let $\epsilon > 0$ be arbitrary. For this ϵ, there exists N' such that

$$\left| \left(\prod_{k=n+1}^{m} a_k \right) - 1 \right| < \epsilon \quad \text{whenever } m > n \geq N'.$$

$$\therefore \left| \prod_{k=N}^{m} a_k - \prod_{k=N}^{n} a_k \right| = \left| \prod_{k=N}^{n} a_k \right| \cdot \left| \left(\prod_{k=n+1}^{m} a_k \right) - 1 \right| < \frac{3}{2}\epsilon,$$

provided $m > n \geq \max\{N, N'\}$. Thus, by the sequence version of the Cauchy Criterion 2.1.2, the sequence $\{\prod_{k=N}^{m} a_k\}_{m=N}^{\infty}$ is convergent. Because it has a limit different from zero, the original product is also convergent. □

Taking $m = n + 1$, we obtain

COROLLARY 7.4.3 *A necessary condition for convergence of the product* $\prod_{k=1}^{\infty} a_k$ *is that*

$$\lim_{k \to \infty} a_k = 1.$$

The divergence of

$$\prod_{k=1}^{\infty} \left(1 + \frac{1}{k} \right) \quad \text{or} \quad \prod_{k=1}^{\infty} \left(1 - \frac{1}{k} \right)$$

shows that the condition in the corollary is not sufficient.

On account of this corollary, it is convenient to write

$$a_k = 1 + c_k, \quad c_k \longrightarrow 0 \quad (k \to \infty),$$

and

$$\prod_{k=1}^{\infty} a_k = \prod_{k=1}^{\infty} (1 + c_k).$$

DEFINITION 7.4.4 *If the infinite product*

$$\prod_{k=1}^{\infty} (1 + |c_k|)$$

converges, then we say that the product $\prod_{k=1}^{\infty} (1 + c_k)$ **converges absolutely.**

EXAMPLE. The product $\prod_{n=2}^{\infty} \left(1 - \frac{(-1)^n}{n}\right) = \frac{1}{2}$ does not converge absolutely.

THEOREM 7.4.5 *An absolutely convergent infinite product is convergent.*
    Proof. This follows from the Cauchy Criterion 7.4.2 and the inequality:

$$\left| \left\{ \prod_{k=n+1}^{m} (1 + c_k) \right\} - 1 \right| \leq \left\{ \prod_{k=n+1}^{m} (1 + |c_k|) \right\} - 1. \qquad \square$$

THEOREM 7.4.6 *The infinite product* $\prod_{k=1}^{\infty} (1 + c_k)$ *converges absolutely if and only if the infinite series* $\sum_{k=1}^{\infty} c_k$ *converges absolutely.*
    Proof. Assume that $\sum_{k=1}^{\infty} |c_k| < \infty$. Set

$$p_n = \prod_{k=1}^{n} (1 + |c_k|) \quad (n \in \mathbb{N}).$$

Then, clearly, $p_n \geq 1$ for all $n \in \mathbb{N}$, and the sequence $\{p_n\}_{n=1}^{\infty}$ is monotone nondecreasing, so it is sufficient to show that the sequence is bounded above. We have

$$p_n = \prod_{k=1}^{n} (1 + |c_k|) \le \prod_{k=1}^{n} \exp\{|c_k|\}$$

$$= \exp\left\{\sum_{k=1}^{n} |c_k|\right\} \le \exp\left\{\sum_{k=1}^{\infty} |c_k|\right\} \quad (n \in \mathbb{N}),$$

where we have used the inequality $1 + x \le e^x$ $(x \in \mathbb{R})$.

The converse follows from the obvious inequality

$$\sum_{k=n+1}^{m} |c_k| \le \prod_{k=n+1}^{m} (1 + |c_k|) - 1$$

and the Cauchy criteria for series and products. $\qquad\square$

Remark. The theorem is no longer true if the word "*absolutely*" is deleted, for the convergence of $\prod_{k=1}^{\infty}(1 + c_k)$ neither implies that of $\sum_{k=1}^{\infty} c_k$ nor conversely. (See Exercise 7.18.)

COROLLARY 7.4.7 *The convergence and the value of an absolutely convergent infinite product are independent of the order of its factors.*

Remark. This is an analog of the theorem obtained by replacing "product" with "series" and "factors" with "summands."

Proof. Suppose $p = \prod_{k=1}^{\infty} (1 + c_k)$ is absolutely convergent. Then we have

$$\sum_{k=1}^{\infty} |c_k| < \infty.$$

Let $p' = \prod_{k=1}^{\infty} (1 + c_k')$ be an infinite product obtained by changing the order of factors. Then, as sets, we have

$$\{c_k ; \, k \in \mathbb{N}\} = \{c_k' ; \, k \in \mathbb{N}\},$$

and by the remark above about absolutely convergent series,

$$\sum_{k=1}^{\infty} |c_k'| = \sum_{k=1}^{\infty} |c_k| < \infty.$$

It follows from Theorem 7.4.6 that the infinite product $p' = \prod_{k=1}^{\infty} (1 + c_k')$ converges absolutely. Set

$$p_n = \prod_{k=1}^{n} (1 + c_k), \quad p_n' = \prod_{k=1}^{n} (1 + c_k').$$

Because $p_n \longrightarrow p$ ($n \to \infty$) and $p'_m \longrightarrow p'$ ($m \to \infty$), for given $\epsilon > 0$ there exists N' such that

$$|p_n - p| < \epsilon \quad \text{and} \quad |p'_m - p'| < \epsilon \quad \text{(whenever } m, n \geq N').$$

Moreover, by the absolute convergence of $\prod_{k=1}^{\infty} (1 + c_k)$ there exists a natural number $N \geq N'$ satisfying

$$\left\{ \prod_{k=N}^{n} (1 + |c_k|) \right\} - 1 < \epsilon \quad \text{(whenever } n \geq N).$$

Letting $n \to \infty$, we obtain

$$\left\{ \prod_{k=N}^{\infty} (1 + |c_k|) \right\} - 1 \leq \epsilon.$$

Now, by choosing m ($\geq N$) sufficiently large, we can make every term $c_1, c_2, \ldots, c_N$ appear in $c'_1, c'_2, \ldots, c'_m$; hence

$$\left| \frac{p'_m}{p_N} - 1 \right| \leq \left\{ \prod_{k=N+1}^{\infty} (1 + |c_k|) \right\} - 1 \leq \epsilon.$$

(We have tacitly assumed that $p \neq 0$; a minor modification of our argument covers the case $p = 0$.) It follows that

$$|p'_m - p_N| \leq \epsilon \cdot |p_N| \leq \epsilon \cdot (|p| + \epsilon).$$

Therefore,

$$\begin{aligned}|p - p'| &\leq |p - p_N| + |p_N - p'_m| + |p'_m - p'| \\ &< \epsilon + \epsilon(|p| + \epsilon) + \epsilon = \epsilon(|p| + 2 + \epsilon).\end{aligned} \qquad \square$$

EXAMPLE. (Euler) We now turn to the *Euler identity*. First label the primes[2]

---

[2]   Euclid proved that *the sequence of primes never terminates.*
Euclid's Proof. Assume the number of primes is finite, say $p_1, p_2, \ldots, p_N$. Let

$$q = (p_1 p_2 \cdots p_N) + 1.$$

Then q is certainly larger than the largest prime, hence, by hypothesis, q cannot be a prime; that is, it is a composite number. But then it must be divisible by at least one prime. Choose

$$p_1 = 2, \quad p_2 = 3, \quad p_3 = 5, \quad p_4 = 7, \dots, \text{etc.}$$

Because the sequence of primes never terminates, we may consider the infinite product

$$\prod_{k=1}^{\infty} \frac{1}{1 - p_k^{-s}} \quad (s > 1).$$

Here

$$c_k = \frac{1}{1 - p_k^{-s}} - 1 = \frac{1}{p_k^s - 1}, \quad \text{so that } 0 < c_k < \frac{2}{p_k^s} \quad (k \geq 2).$$

The series $\sum_{k=1}^{\infty} \frac{1}{p_k^s}$ converges, for it is just a part of the (absolutely) convergent series $\sum_{k=1}^{\infty} \frac{1}{k^s}$, and we have shown that the product $\prod_{k=1}^{\infty} \frac{1}{1-p_k^{-s}}$ converges absolutely for $s > 1$.

We now consider a typical factor $\frac{1}{1-p_k^{-s}}$. Because $0 < p_k^{-s} < 1$ (for $s > 1$), we obtain

$$\frac{1}{1 - p_k^{-s}} = 1 + p_k^{-s} + p_k^{-2s} + \cdots,$$

$$\therefore \prod_{k=1}^{\infty} \frac{1}{1 - p_k^{-s}} = \prod_{k=1}^{\infty} \left(1 + p_k^{-s} + p_k^{-2s} + \cdots \right).$$

If there existed a *"distributive law for an infinite product of infinite series"* then the right-hand side would become

$$\sum (p_\alpha^a \cdot p_\beta^b \cdots p_\lambda^l)^{-s},$$

any prime p_k. Then

$$p_k | (p_1 p_2 \cdots p_N), \quad \text{but } p_k \nmid 1.$$

Therefore

$$p_k \nmid q \quad (k = 1, 2, \dots, N),$$

and so q is a composite number that is not divisible by any prime; but this result is absurd.

Remark. The expression q can be used to exhibit new primes given an initial set; for if q is not a prime, then its factorization must contain a new prime. For example, let $\{2, 3\}$ be the initial set. We have

$$
\begin{aligned}
2 \cdot 3 + 1 &= 7 \quad \text{(prime)}; \\
2 \cdot 3 \cdot 7 + 1 &= 43 \quad \text{(prime)}; \\
2 \cdot 3 \cdot 7 \cdot 43 + 1 &= 1807 = 13 \cdot 139; \\
2 \cdot 3 \cdot 7 \cdot 13 \cdot 43 \cdot 139 + 1 &= 3263443 \quad \text{(prime)};
\end{aligned}
$$

$$\cdots \quad \cdots$$

where α, β, ..., λ are distinct and a, b, ..., l are nonnegative integers. Thus, by the *unique factorization theorem* for natural numbers, we obtain, *formally*,

$$\prod_{k=1}^{\infty} \frac{1}{1 - p_k^{-s}} = \sum_{n=1}^{\infty} \frac{1}{n^s}.$$

This is the renowned *Euler identity*, which equates an infinite product involving all primes and a series involving all positive integers. We now proceed to justify the result which we have obtained heuristically. First, consider the product of the first m factors; that is,

$$\prod_{k=1}^{m} \frac{1}{1 - p_k^{-s}} = \prod_{k=1}^{m} \left(1 + p_k^{-s} + p_k^{-2s} + \cdots\right) = \sum{}' \frac{1}{n^s},$$

where the summation $\sum'$ runs through all the natural numbers n whose prime factors are less than or equal to p_m. Because the set of such numbers includes all the natural numbers less than or equal to p_m, we obtain

$$\sum_{n=1}^{p_m} \frac{1}{n^s} < \prod_{k=1}^{m} \frac{1}{1 - p_k^{-s}} = \sum_{n=1}^{p_m} \frac{1}{n^s} + \sum{}'' \frac{1}{n^s}.$$

Because the series $\sum'' \frac{1}{n^s}$ is a subseries of the (absolutely) convergent series $\sum_{n > p_m} \frac{1}{n^s}$, we can make it as small as we please by taking m sufficiently large, and we obtain the desired result. We repeat, for emphasis, that the Euler identity has been demonstrated only with the restriction $s > 1$.

If $s = 1$, then the inequality

$$\sum_{n=1}^{p_m} \frac{1}{n} < \prod_{k=1}^{m} \frac{1}{1 - p_k^{-1}}$$

still holds. It is now evident that, with suitable interpretation, we are justified in writing

$$\sum_{n=1}^{\infty} \frac{1}{n} = \prod_{k=1}^{\infty} \left(1 + \frac{1}{p_k - 1}\right)$$

in the sense that the series and the product both diverge to ∞. This fact, in turn, implies the divergence of the series $\sum_{k=1}^{\infty} \frac{1}{p_k - 1}$; but $\frac{1}{p_k - 1} \leq \frac{2}{p_k}$, and so we have the divergence of $\sum_{k=1}^{\infty} \frac{1}{p_k}$. (Again, this divergence shows that there are infinitely many primes. Since $\sum_{n=1}^{\infty} \frac{1}{n^2}$ does converge, we see that, in some sense, the squares are sparser than the primes.)

It should be remarked that the function

$$\zeta(s) = \sum_{n=1}^{\infty} \frac{1}{n^s} \quad (\Re s > 1)$$

is the restriction to $\Re s > 1$ of a meromorphic function known as the *Riemann zeta function* (it turns out that $\zeta(s) - \frac{1}{s-1}$ is entire). This function has inspired the most celebrated conjecture of modern mathematics—the *Riemann Hypothesis*—that all the nontrivial roots of $\zeta(s)$ (i.e., other than the simple roots situated at negative even integers) are on the line $\Re s = \frac{1}{2}$. Hardy proved that $\zeta(s)$ has infinitely many roots on the *critical line*. Hadamard and de la Vallée Poussin also proved, independently, that $\zeta(s)$ has no root on $\Re s = 1$, but the problem of whether or not $\zeta(s)$ has roots in the half-plane $\Re s > \frac{1}{2}$ still remains open.

In passing we mention another celebrated theorem closely related to the Riemann Hypothesis.

Prime Number Theorem (Hadamard-de la Vallée Poussin) *If $\pi(x)$ denotes the number of primes less than or equal to x, then*

$$\pi(x) \sim \frac{x}{\log x}; \quad \text{that is,} \quad \lim_{x \to \infty} \frac{\pi(x)}{x/\log x} = 1.$$

Gauss guessed this relation after looking at a table of primes in the first $10\,000$ integers. As an example, let $x = 100$; then

$$\frac{x}{\log x} \approx \frac{100}{4.6} \approx 22, \quad \text{but } \pi(100) = 25.$$

For $x = 10^9$,

$$\frac{x}{\log x} \approx \frac{10^9}{20.7} \approx 48 \times 10^6, \quad \pi(10^9) \approx 50.5 \times 10^6.$$

These two cases show that the convergence of $\frac{\pi(x)}{x/\log x}$, while assured, is very slow.

We now return to the general topic of infinite products. If $\{f_n(z)\}_{n=1}^{\infty}$ is a sequence of functions defined in some region Ω, then we can consider the problem of local uniform convergence of the infinite product

$$\prod_{n=1}^{\infty} \{1 + f_n(z)\}.$$

DEFINITION 7.4.8 *Let* $\{f_k(z)\}_{k=1}^{\infty}$ *be a sequence of functions defined in a region* Ω*. The infinite product* $\prod_{k=1}^{\infty}\{1 + f_k(z)\}$ *is said to be* **locally uniformly convergent** *on* Ω *if for every compact subset* $K \subset \Omega$ *there exist at most finitely many factors in the product that have roots in* K*, and if these factors (if any) are deleted, the partial products of the remaining infinite product converge uniformly on* K *to a function that does not vanish on* K*.*

THEOREM 7.4.9 (Cauchy Criterion) *A necessary and sufficient condition that an infinite product*

$$\prod_{n=1}^{\infty}\{1 + f_n(z)\}$$

converges locally uniformly on Ω *is that, given any* $\epsilon > 0$ *and an arbitrary compact subset* K *of* Ω*, there exists an integer* $N = N(K, \epsilon)$ *such that*

$$\left\|\left[\prod_{k=n+1}^{m}\{1 + f_k(z)\}\right] - 1\right\|_K < \epsilon \quad \text{for } m > n \geq N.$$

THEOREM 7.4.10 *Let* $\{f_n(z)\}_{n=1}^{\infty}$ *be a sequence of functions continuous on a region* Ω*. If the series*

$$|f_1(z)| + |f_2(z)| + \cdots + |f_n(z)| + \cdots$$

converges locally uniformly on Ω*, then*

(a) The infinite product

$$p(z) = \prod_{n=1}^{\infty}\{1 + f_n(z)\}$$

converges locally uniformly to a continuous function on Ω*.*

(b) In particular, if $f_n(z) \in H(\Omega)$ $(n \in \mathbb{N})$*, then the product is an analytic function that has a root in* Ω *where and only where at least one of the factors in the product has a root.*

(c) Moreover, if the product does not vanish, then the logarithmic derivative $\frac{p'(z)}{p(z)}$ *of the product* $p(z)$ *is the sum of the logarithmic derivatives of the individual factors; that is,*

$$\frac{p'(z)}{p(z)} = \sum_{n=1}^{\infty}\frac{f_n'(z)}{1 + f_n(z)},$$

and the convergence is uniform on every compact subset of Ω not containing roots of $p(z)$.

Proof. Let K be an arbitrary compact subset of Ω, and $\{p_n(z)\}_{n=1}^{\infty}$ the sequence of partial products. On account of the uniform convergence on K, $\sum_{n=1}^{\infty} |f_n(z)|$ defines a continuous function; let σ be its maximum value on K. Then, for every $n \in \mathbb{N}$ and $z \in K$,

$$|p_n(z)| = \left| \prod_{k=1}^{n} \{1 + f_k(z)\} \right| \leq \prod_{k=1}^{n} \{1 + |f_k(z)|\}$$

$$\leq \prod_{k=1}^{n} \exp\{|f_k(z)|\} = \exp\left\{ \sum_{k=1}^{n} |f_k(z)| \right\} \leq e^{\sigma}.$$

Therefore,

$$|p_n(z) - p_{n-1}(z)| = |p_{n-1}(z) \cdot f_n(z)| \leq e^{\sigma} |f_n(z)| \quad (z \in K).$$

Now, by the uniform convergence of the series on K, given $\epsilon > 0$, there exists an integer N such that

$$\sum_{n>N} |f_n(z)| < \epsilon \quad \text{for all } z \in K.$$

Then, for $m > n \geq N$, and $z \in K$, we have

$$|p_n(z) - p_m(z)| \leq |p_n(z) - p_{n+1}(z)| + |p_{n+1}(z) - p_{n+2}(z)|$$
$$+ \cdots + |p_{m-1}(z) - p_m(z)|$$
$$\leq e^{\sigma} \{|f_{n+1}(z)| + |f_{n+2}(z)| + \cdots + |f_m(z)|\} < e^{\sigma} \cdot \epsilon.$$

Thus, by the Cauchy Criterion 2.4.1, we have the desired local uniform convergence of the product, and hence $p(z)$ is continuous on Ω.

Part (b) concerning the analyticity is a consequence of Corollary 4.5.5, and the assertion about the roots follows from the definition of the convergence of infinite products.

As for Part (c), because $p_n(z) \longrightarrow p(z)$ locally uniformly on Ω, by Corollary 4.5.5 we also have $p'_n(z) \longrightarrow p'(z)$ locally uniformly there. Thus

$$\frac{p'_n(z)}{p_n(z)} \longrightarrow \frac{p'(z)}{p(z)} \quad \text{locally uniformly provided } p(z) \neq 0.$$

However, a simple computation gives

$$\frac{p'_n(z)}{p_n(z)} = \sum_{k=1}^{n} \frac{f'_k(z)}{1 + f_k(z)},$$

and so

$$\frac{p'(z)}{p(z)} = \sum_{k=1}^{\infty} \frac{f'_k(z)}{1 + f_k(z)} \quad \text{provided } p(z) \neq 0. \qquad \Box$$

EXAMPLE. Consider

$$f(z) = \prod_{n=1}^{\infty} \left(1 + \frac{z^2}{n^2}\right).$$

Fix an arbitrary $R > 0$. If $z \in D(0, R)$, then we have $\left|\frac{z^2}{n^2}\right| < \frac{R^2}{n^2}$; and because $\sum_{n=1}^{\infty} \frac{R^2}{n^2} < \infty$, the infinite product converges uniformly in the disc $D(0, R)$. Because $R > 0$ is arbitrary, we conclude that $f(z)$ is an entire function.

7.5   Blaschke Products

We know that the roots of an analytic function cannot accumulate inside the region of analyticity unless the function is identically zero. Furthermore, if the function is not identically zero, then the number of roots inside the region is (at most) countable. (Why?) More can be said if the function is bounded.

THEOREM 7.5.1 (Blaschke) *Let $f(z)$ be a bounded analytic function, not identically zero, in the unit disc D. If $\{a_n\}_{n=1}^{\infty}$ is the sequence of roots of $f(z)$ in the unit disc D, each repeated as often as its multiplicity, then the product*

$$\prod_{n=1}^{\infty} |a_n|$$

is convergent; equivalently,

$$\sum_{n=1}^{\infty} (1 - |a_n|) < \infty.$$

Proof. If $f(z)$ has only a finite number of zeros, then there is nothing to prove. Thus we assume that $f(z)$ has infinitely many zeros. Moreover, we may also assume, without loss of generality, that $\|f\|_D \leq 1$. Then, by Theorem 6.2.4, we have

$$|f(z)| \leq \prod_{k=1}^{n} \left| \frac{z - a_k}{1 - \bar{a}_k z} \right| \quad (z \in D).$$

Setting $z = 0$, we obtain

$$|f(0)| \leq \prod_{k=1}^{n} |a_k|.$$

Set $p_n = \prod_{k=1}^{n} |a_k|$ ($n \in \mathbb{N}$). Then, we have

$$p_1 \geq p_2 \geq \cdots \geq p_n \geq \cdots \geq |f(0)|.$$

Thus, $\lim_{n \to \infty} p_n$ exists, and $\lim_{n \to \infty} p_n \geq |f(0)|$. Therefore, if $f(0) \neq 0$, we are done. If $f(0) = 0$, we may write (since $f(z)$ is not identically zero)

$$f(z) = z^k \cdot g(z),$$

where k is the multiplicity of the root of $f(z)$ at the origin, and $g(z)$ is a bounded analytic function satisfying $g(0) \neq 0$ having the same roots as $f(z)$ except at the origin. We may now apply to $g(z)$ what we have already proved for $f(z)$ to obtain the desired conclusion. □

COROLLARY 7.5.2 *Let $\{a_n\}_{n=1}^{\infty}$ be a sequence of points in the unit disc D. If the series $\sum_{n=1}^{\infty} (1 - |a_n|)$ diverges, then the only bounded analytic function in D vanishing at $\{a_n\}_{n=1}^{\infty}$ is the zero function.*

Note that the condition that the function be bounded cannot be deleted. (See the Weierstrass Theorem 7.3.2.)

Now let us consider the converse. Given a sequence $\{a_n\}_{n=1}^{\infty}$ of points in the unit disc D such that $\sum_{n=1}^{\infty} (1 - |a_n|) < \infty$, can we find a *bounded* analytic function $f(z)$ whose roots are precisely $\{a_n\}_{n=1}^{\infty}$?

A natural candidate of such a function would be

$$\prod_{n=1}^{\infty} \frac{z - a_n}{1 - \bar{a}_n z}.$$

However, this infinite product does not converge, in general. Fortunately, if we simply rotate each factor in the product appropriately, then the new infinite product *does* converge.

THEOREM 7.5.3 (Blaschke) *Let* $\{a_n\}_{n=1}^{\infty}$ *be a sequence of nonzero complex numbers in the unit disc* D *(repetition allowed). Then a necessary and sufficient condition that the infinite product*

$$\prod_{n=1}^{\infty} \frac{|a_n|}{a_n} \cdot \frac{a_n - z}{1 - \bar{a}_n z}$$

should converge locally uniformly in the unit disc D *is that the product*

$$\prod_{n=1}^{\infty} |a_n|$$

should converge; equivalently,

$$\sum_{n=1}^{\infty} (1 - |a_n|) < \infty.$$

When this condition is satisfied, the product defines an analytic function $B(z)$ *whose roots are precisely* $\{a_n\}_{n=1}^{\infty}$. *Moreover,*

$$\|B(z)\|_D = 1.$$

Proof. Necessity is immediate; simply substitute $z = 0$. Conversely, assume that

$$\sum_{n=1}^{\infty} (1 - |a_n|) < \infty;$$

then for $z \in \bar{D}(0, r)$ $(0 \le r < 1)$ we have

$$\left| 1 - \frac{|a_n|}{a_n} \cdot \frac{a_n - z}{1 - \bar{a}_n z} \right| = \left| \frac{(a_n + |a_n|z)(1 - |a_n|)}{a_n(1 - \bar{a}_n z)} \right| \le \frac{2(1 - |a_n|)}{1 - r}.$$

Thus, the infinite product converges uniformly on each disc $\bar{D}(0, r)$ $(0 < r < 1)$, and hence, by Theorem 7.4.10, defines an analytic function $B(z)$ in the unit disc D. Furthermore, each a_n is a root of $B(z)$ and $B(z) \ne 0$ elsewhere. Clearly,

$$\|B(z)\|_D = 1,$$

because the corresponding equality for the partial products is true.   □

DEFINITION 7.5.4 *A function of the form*

$$B(z) = z^m \cdot \prod_{n=1}^{\infty} \frac{|a_n|}{a_n} \cdot \frac{a_n - z}{1 - \bar{a}_n z} \quad (z \in D)$$

is called a **Blaschke product**, *where m is a nonnegative integer,* $a_n \in D$, $a_n \neq 0$ $(n \in \mathbb{N})$, *and*

$$\sum_{n=1}^{\infty} (1 - |a_n|) < \infty.$$

7.6   The Factorization of Entire Functions

With our knowledge of infinite products, let us return to the Weierstrass Theorem 7.2.1. The problem is to find an entire function $f(z)$ with prescribed roots:

$$a_1, a_2, \ldots, a_k, \ldots.$$

If $a_k \neq 0$ $(k \in \mathbb{N})$, a natural candidate would be

$$f(z) = \prod_{k=1}^{\infty} \left(1 - \frac{z}{a_k}\right).$$

The product converges and defines an entire function if $|a_k|$ tends to ∞ so rapidly that the series $\sum_{k=1}^{\infty} \frac{1}{|a_k|}$ is convergent (by Theorem 7.4.10). However, if the series $\sum_{k=1}^{\infty} \frac{1}{|a_k|}$ is not convergent, then the product may be divergent and need not define *any* function. As in the case of Mittag-Leffler series, we must modify each factor so that the product converges without introducing new roots.

We start from the following imitation of the proof of Corollary 7.2.2. Because $E_0(z) = 1 - z$ is analytic and has no root in the unit disc D,

$$\int_0^z \frac{E_0'(t)}{E_0(t)} dt = \int_0^z \frac{-dt}{1-t}$$

$$= -\int_0^z (1 + t + t^2 + \cdots + t^n + \cdots) dt$$

$$= -\left(z + \frac{z^2}{2} + \frac{z^3}{3} + \cdots + \frac{z^{n+1}}{n+1} + \cdots\right)$$

is analytic and single-valued in the unit disc D. Note that the integration can be along any path in the unit disc D.

$$\therefore\ 1 - z\ =\ E_0(z) = \exp\left\{\int_0^z \frac{E_0'(t)}{E_0(t)}dt\right\}$$

$$= \exp\left\{-z - \frac{z^2}{2} - \frac{z^3}{3} - \cdots - \frac{z^n}{n} - \cdots\right\}\quad (z \in D).$$

Hence we expect that the sequence of *entire* functions (called the *Weierstrass primary factors*)

$$E_n(z) = (1 - z)\cdot\exp\left\{z + \frac{z^2}{2} + \cdots + \frac{z^n}{n}\right\}\quad (n \in \mathbb{N})$$

converges locally uniformly to the constant function 1 in the unit disc D. Indeed, we have the following

LEMMA 7.6.1

$$|1 - E_n(z)| \le |z|^{n+1}\quad z \in \bar{D}(0, 1).$$

Proof. This inequality is trivial when $n = 0$. Suppose $n \ge 1$. Then

$$E_n'(z)\ =\ -\exp\left\{z + \frac{z^2}{2} + \cdots + \frac{z^n}{n}\right\}$$

$$+ (1 - z)(1 + z + \cdots + z^{n-1})\cdot\exp\left\{z + \frac{z^2}{2} + \cdots + \frac{z^n}{n}\right\}$$

$$= \{-1 + (1 - z^n)\}\cdot\exp\left\{z + \frac{z^2}{2} + \cdots + \frac{z^n}{n}\right\}$$

$$= -z^n\cdot\exp\left\{z + \frac{z^2}{2} + \cdots + \frac{z^n}{n}\right\}.$$

This shows that the Taylor coefficients (around the origin) of the entire function

$$1 - E_n(z)\ =\ E_n(0) - E_n(z) = -\int_0^z E_n'(t)\, dt$$

$$= \int_0^z t^n\cdot\exp\left\{t + \frac{t^2}{2} + \cdots + \frac{t^n}{n}\right\} dt$$

are all nonnegative real numbers *and* that $1 - E_n(z)$ has a root of multiplicity $n + 1$ at the origin. Therefore,

$$f(z) = \frac{1 - E_n(z)}{z^{n+1}}$$

is an entire function whose Taylor coefficients (around the origin) are also all real and nonnegative. It follows from the triangle inequality that for $z \in \bar{D}$ we have

$$|f(z)| \le f(1); \quad \text{that is,} \quad \left| \frac{1 - E_n(z)}{z^{n+1}} \right| \le \frac{1 - E_n(1)}{1^{n+1}} = 1,$$

which establishes the lemma. $\square$

We are now ready for the second proof of the Weierstrass Theorem 7.2.1.

THEOREM 7.6.2 (Weierstrass) *Let $\{a_k\}_{k=1}^{\infty}$ be a sequence of points having no accumulation point in the finite complex plane $\mathbb{C}$ (repetition allowed). Then there exists an entire function having precisely $\{a_k\}_{k=1}^{\infty}$ as its roots with multiplicity equal to the number of times a_k appears in the sequence. If $a_k \ne 0$ ($k \in \mathbb{N}$), then it is given by*

$$\prod_{k=1}^{\infty} E_{n_k}\left(\frac{z}{a_k}\right),$$

where $\{n_k\}_{k=1}^{\infty}$ is a suitably chosen sequence of nonnegative integers.
    Proof. Without loss of generality, we may assume that

$$0 < |a_1| \le |a_2| \le \cdots \le |a_k| \le \cdots.$$

Consider the product

$$\prod_{k=1}^{\infty} E_{n_k}\left(\frac{z}{a_k}\right).$$

We want to show that the product converges for an appropriately chosen sequence $\{n_k\}_{k=1}^{\infty}$. From the lemma, we have

$$\left| 1 - E_{n_k}\left(\frac{z}{a_k}\right) \right| \le \left| \frac{z}{a_k} \right|^{n_k+1} \quad \text{for } z \in \bar{D}(0, |a_k|).$$

Therefore, for each k, we choose an exponent n_k such that the series

$$\sum_{k=1}^{\infty} \left| \frac{z}{a_k} \right|^{n_k+1}$$

converges locally uniformly in the complex plane $\mathbb{C}$. For example, $n_k + 1 = k$ will always suffice. Indeed, for any $R > 0$, if $z \in \bar{D}(0, R)$ then

$$\sum_{k=1}^{\infty} \left| \frac{z}{a_k} \right|^k = \sum_{|a_k| \leq 2R} \left| \frac{z}{a_k} \right|^k + \sum_{|a_k| > 2R} \left| \frac{z}{a_k} \right|^k$$

$$\leq \sum_{|a_k| \leq 2R} \left| \frac{z}{a_k} \right|^k + \sum_{|a_k| > 2R} \left(\frac{1}{2} \right)^k.$$

Because $a_k \to \infty$, the first summation has a finite number of terms and the second series is convergent. Having chosen $\{n_k\}_{k=1}^{\infty}$ with the required property, we have, for an arbitrary positive number R and for $|a_k| > 2R$,

$$\left| 1 - E_{n_k}\left(\frac{z}{a_k} \right) \right| \leq \left| \frac{z}{a_k} \right|^{n_k+1} \leq \left| \frac{R}{a_k} \right|^{n_k+1} \qquad z \in \bar{D}(0, R).$$

The convergence of the series $\sum_{|a_k| > 2R} \left| \frac{R}{a_k} \right|^{n_k+1}$ implies the uniform convergence of the series $\sum_{|a_k| > 2R} \left| \frac{z}{a_k} \right|^{n_k+1}$ in the closed disc $\bar{D}(0, R)$, and therefore, by Theorem 7.4.10, the product

$$\prod_{|a_k| > 2R} E_{n_k}\left(\frac{z}{a_k} \right)$$

converges uniformly and defines an analytic function in $D(0, R)$.

Because we have omitted only a finite number of factors,

$$\prod_{k=1}^{\infty} E_{n_k}\left(\frac{z}{a_k} \right)$$

is analytic on $D(0, R)$, and because R is arbitrary, the product defines an entire function. Clearly, it has the required roots with the required multiplicities.   □

Remark. For certain sequences $\{a_k\}_{k=1}^{\infty}$ of points, the integers $\{n_k\}_{k=1}^{\infty}$ can be chosen independent of k. (For example, if $a_k = k$ ($k \in \mathbb{N}$), then

$n_k = 1$ for all $k \in \mathbb{N}$ will suffice.) The resulting product then assumes the form

$$\prod_{k=1}^{\infty} E_n \left(\frac{z}{a_k} \right) \quad \text{for some nonnegative integer } n.$$

If the smallest acceptable value of n is used in this case, then the product is called the *Weierstrass canonical product* associated with the sequence $\{a_k\}_{k=1}^{\infty}$.

Combining this result with the discussion preceding Corollary 7.2.2, we obtain

THEOREM 7.6.3 (Weierstrass Factorization Theorem) *Suppose $f(z)$ is an entire function not identically zero, having $\{a_k\}_{k=1}^{\infty}$ as its nonzero roots, each repeated as often as its multiplicity. Then there exists an entire function $g(z)$ and a sequence $\{n_k\}_{k=1}^{\infty}$ of nonnegative integers such that*

$$f(z) = e^{g(z)} \cdot z^n \cdot \prod_{k=1}^{\infty} E_{n_k} \left(\frac{z}{a_k} \right),$$

where n is the multiplicity of the root of $f(z)$ at the origin (if $f(0) \neq 0$, then $n = 0$).

Remark. The factorization is *not* unique. However, if the roots of $f(z)$ satisfy the condition that $\sum_{k=1}^{\infty} \frac{1}{|a_k|^{n+1}} < \infty$ for some n, then a canonical product exists, and thus a uniquely determined factorization of $f(z)$ can be achieved.

EXAMPLE. Consider the entire function $\sin \pi z$. It has simple roots at the integers, and $\sum_{n \neq 0} \frac{1}{n^2} < \infty$. Thus $\sum_{n \neq 0} \left| \frac{z}{n} \right|^2$ converges locally uniformly in $\mathbb{C}$, and we can use $E_1 \left(\frac{z}{n} \right)$ as our factors:

$$\sin \pi z = e^{g(z)} \cdot z \cdot \prod_{-\infty}^{\infty}{}' \left\{ \left(1 - \frac{z}{n} \right) \cdot e^{\frac{z}{n}} \right\}.$$

Using

$$\left| 1 - \left(1 - \frac{z}{n} \right) \cdot e^{\frac{z}{n}} \right| \leq \left| \frac{z}{n} \right|^2 \quad z \in \bar{D}(0, |n|), \ n \neq 0,$$

and the local uniform convergence of $\sum_{n \neq 0} \left| \frac{z}{n} \right|^2$ in the complex plane $\mathbb{C}$, we may take the logarithmic derivative, by part (c) of Theorem 7.4.10, and we obtain

$$\pi \cot \pi z = g'(z) + \frac{1}{z} + \sum_{-\infty}^{\infty}{}' \left\{ \frac{1}{z-n} + \frac{1}{n} \right\}.$$

But this series (less $g'(z)$) is just the series obtained for $\pi \cot \pi z$ in the example after the Mittag-Leffler Theorem 7.1.2. Thus $g'(z) = 0$, and hence $g(z) = $ constant.

Now, divide the original product by z; then we have

$$\frac{\sin \pi z}{z} = C \cdot \prod_{-\infty}^{\infty}{}' \left\{ \left(1 - \frac{z}{n}\right) \cdot e^{\frac{z}{n}} \right\}.$$

Letting $z \to 0$ on the left-hand side, we obtain

$$\lim_{z \to 0} \frac{\sin \pi z}{z} = \pi.$$

The product on the right-hand side is an entire function, and so we may substitute $z = 0$, obtaining $C = \pi$; hence[3]

$$\sin \pi z = \pi z \cdot \prod_{-\infty}^{\infty}{}' \left\{ \left(1 - \frac{z}{n}\right) \cdot e^{\frac{z}{n}} \right\}.$$

Pairing factors, we obtain

$$\sin \pi z = \pi z \cdot \prod_{n=1}^{\infty} \left(1 - \frac{z^2}{n^2}\right).$$

Substituting $z = \frac{1}{2}$, we obtain the *Wallis formula*:

$$\frac{2}{\pi} = \prod_{n=1}^{\infty} \left\{ 1 - \frac{1}{(2n)^2} \right\}.$$

---

[3] The convergence of a two-sided infinite product $\prod_{n=-\infty}^{\infty} a_n$ means the existence of the limit $\lim_{M,N \to \infty} \prod_{n=-M}^{N} a_n$, where M and N tend to ∞ *independently*; that is, the two-sided infinite product converges if and only if both limits

$$\lim_{M \to \infty} \prod_{n=-M}^{-1} a_n \quad \text{and} \quad \lim_{N \to \infty} \prod_{n=0}^{N} a_n$$

exist. In particular, if the two-sided product converges, then the limit of the "*symmetric product*" $\lim_{M \to \infty} \prod_{n=-M}^{M} a_n$ also exists and is equal to $\prod_{n=-\infty}^{\infty} a_n$. But the convergence of the symmetric product does not imply the convergence of the two-sided product, in general.

Note: Suppose

$$1 - a_1x^2 + a_2x^4 - + \cdots + (-1)^n a_n x^{2n}$$
$$= \left(1 - \frac{x^2}{\alpha_1^2}\right)\left(1 - \frac{x^2}{\alpha_2^2}\right) \cdots \left(1 - \frac{x^2}{\alpha_n^2}\right);$$

then

$$a_1 = \frac{1}{\alpha_1^2} + \frac{1}{\alpha_2^2} + \cdots + \frac{1}{\alpha_n^2}.$$

By analogy, from

$$\prod_{n=1}^{\infty}\left(1 - \frac{z^2}{n^2}\right) = \frac{\sin \pi z}{\pi z} = 1 - \frac{(\pi z)^2}{3!} + \frac{(\pi z)^4}{5!} - + \cdots,$$

Euler concluded (correctly but not rigorously) that

$$\sum_{n=1}^{\infty} \frac{1}{n^2} = \frac{\pi^2}{6}.$$

7.7   The Jensen Formula

We have seen in Section 6.1 that we can use the Cauchy Integral Formula 4.4.3 of a function $g(z)$ that is analytic in $\bar{D}(0, R)$ $(R > 0)$ to obtain

$$g(0) = \frac{1}{2\pi i} \int_{C(0,R)} \frac{g(z)}{z} dz = \frac{1}{2\pi i} \int_0^{2\pi} \frac{g(Re^{i\theta})}{Re^{i\theta}} i Re^{i\theta} d\theta$$
$$= \frac{1}{2\pi} \int_0^{2\pi} g(Re^{i\theta}) d\theta.$$

Taking the real part of both sides in the equality above, we obtain

$$u(0) = \frac{1}{2\pi} \int_0^{2\pi} u(Re^{i\theta}) d\theta,$$

where $u(z)$ is the real part of $g(z)$. Thus the value of a harmonic function at the center of a circle is given by the average of its values on the circle (for a harmonic function can be considered locally as the real part of an analytic function).

We now derive the important *Jensen formula*.

THEOREM 7.7.1 (Jensen) *Let $f(z)$ be analytic, $f(0) \neq 0$, and let a_1, $a_2, \ldots, a_n$ be the roots of $f(z)$ in the disc $\bar{D}(0, R)$ (each a_k being repeated as often as its multiplicity). Then*

$$\frac{1}{2\pi} \int_0^{2\pi} \mathrm{Log}\,|f(Re^{i\theta})|\,d\theta = \mathrm{Log}\,|f(0)| + \sum_{k=1}^{n} \mathrm{Log}\,\frac{R}{|a_k|}.$$

Proof. For clarity in presentation, we consider three cases, each of the first two being a particular case of the one immediately following it.

(a) If $f(z)$ has no root in $\bar{D}(0, R)$, then there exists a function $g(z)$ analytic in the disc such that $e^{g(z)} = f(z)$; in particular, $g(z) = \log f(z)$. Its real part is $\mathrm{Log}\,|f(z)|$, and hence we obtain

$$\mathrm{Log}\,|f(0)| = \frac{1}{2\pi} \int_0^{2\pi} \mathrm{Log}\,|f(Re^{i\theta})|\,d\theta,$$

which is the desired result for this case.

(b) Suppose there are roots at $a_1, a_2, \ldots, a_n$ inside the disc $D(0, R)$, but none on the boundary circle $C(0, R)$. Then we can use a finite Blaschke product to define the function

$$F(z) = f(z) \cdot \left\{ \prod_{k=1}^{n} \frac{R(a_k - z)}{R^2 - \bar{a}_k z} \right\}^{-1},$$

which is analytic on the disc $\bar{D}(0, R)$ and has no root there. By what we have established in (a):

$$\frac{1}{2\pi} \int_0^{2\pi} \mathrm{Log}\,|F(Re^{i\theta})|\,d\theta = \mathrm{Log}\,|F(0)| = \mathrm{Log}\,|f(0)| - \mathrm{Log}\,\left| \prod_{k=1}^{n} \frac{a_k}{R} \right|.$$

Because the finite Blaschke product has modulus identically equal to 1 on the circle $C(0, R)$ (this is why we used the Blaschke product), we have

$$|F(z)| = |f(z)| \quad \text{for } z \in C(0, R),$$

and hence

$$\frac{1}{2\pi} \int_0^{2\pi} \mathrm{Log}\,|f(Re^{i\theta})|\,d\theta = \mathrm{Log}\,|f(0)| + \sum_{k=1}^{n} \mathrm{Log}\,\frac{R}{|a_k|}.$$

(c) The formula holds even if $f(z)$ *does* have roots on the circle $C(0, R)$. Let $a_1, a_2, \ldots, a_n$ be the roots of $f(z)$ in $D(0, R)$, and $a_{n+1}, a_{n+2}, \ldots,$ a_m be those on the circle $C(0, R)$. Set

$$F(z) = f(z) \cdot \left\{ \prod_{k=1}^{n} \frac{R(a_k - z)}{R^2 - \bar{a}_k z} \cdot \prod_{k=n+1}^{m} \left(1 - \frac{z}{a_k}\right) \right\}^{-1}.$$

Then the function $F(z)$ is analytic in the closed disc $\bar{D}(0, R)$ and has no root there. Thus

$$\frac{1}{2\pi} \int_0^{2\pi} \text{Log} |F(Re^{i\theta})| \, d\theta = \text{Log} |F(0)|,$$

and it follows that

$$\frac{1}{2\pi} \int_0^{2\pi} \text{Log} |f(Re^{i\theta})| \, d\theta - \sum_{k=n+1}^{m} \frac{1}{2\pi} \int_0^{2\pi} \text{Log} \left|1 - \frac{Re^{i\theta}}{a_k}\right| \, d\theta$$

$$= \text{Log} |f(0)| + \sum_{k=1}^{n} \text{Log} \frac{R}{|a_k|}.$$

If we set $a_k = Re^{i\theta_k}$ $(k = n + 1, n + 2, \ldots, m)$, then

$$\int_0^{2\pi} \text{Log} \left|1 - \frac{Re^{i\theta}}{a_k}\right| \, d\theta = \int_0^{2\pi} \text{Log} \left|1 - e^{i(\theta - \theta_k)}\right| \, d\theta,$$

and so it is sufficient to show that

$$\int_0^{2\pi} \text{Log} \left|1 - e^{i\theta}\right| \, d\theta = 0.$$

This is a problem in elementary calculus. Because (Figure 7.2)

$$\left|1 - e^{i\theta}\right| = \left|e^{i\frac{\theta}{2}}\right| \cdot \left|e^{-i\frac{\theta}{2}} - e^{i\frac{\theta}{2}}\right| = \left|-2i \sin \frac{\theta}{2}\right| = 2 \left|\sin \frac{\theta}{2}\right|,$$

we have

$$\int_0^{2\pi} \text{Log} |1 - e^{i\theta}| \, d\theta = \int_0^{2\pi} \text{Log} \, 2 \, d\theta + \int_0^{2\pi} \text{Log} \sin \frac{\theta}{2} \, d\theta$$

$$= 2\pi \text{Log} \, 2 + 2 \int_0^{\pi} \text{Log} \sin t \, dt \quad \left(\frac{\theta}{2} = t\right).$$

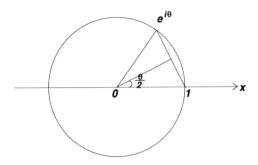

Figure 7.2

It remains to show that

$$\int_0^\pi \text{Log} \sin t \, dt = -\pi \, \text{Log} \, 2.$$

Now,

$$
\begin{aligned}
\int_0^\pi \text{Log} \sin t \, dt &= \int_0^\pi \text{Log} \left(2 \sin \frac{t}{2} \cdot \cos \frac{t}{2} \right) dt \\
&= \int_0^\pi \text{Log} \, 2 \, dt + \int_0^\pi \text{Log} \sin \frac{t}{2} \, dt + \int_0^\pi \text{Log} \cos \frac{t}{2} \, dt \\
&= \pi \text{Log} \, 2 + 2 \int_0^{\frac{\pi}{2}} \text{Log} \sin t \, dt + \int_0^\pi \text{Log} \cos \frac{\pi - \varphi}{2} \, d\varphi \\
&\qquad\qquad\qquad\qquad\qquad\qquad\qquad\qquad (t = \pi - \varphi) \\
&= \pi \text{Log} \, 2 + \int_0^\pi \text{Log} \sin t \, dt + \int_0^\pi \text{Log} \sin \frac{\varphi}{2} \, d\varphi \\
&= \pi \text{Log} \, 2 + 2 \int_0^\pi \text{Log} \sin t \, dt.
\end{aligned}
$$

Solving for $\int_0^\pi \text{Log} \sin t \, dt$, we obtain the desired result. Note that this integral is improper, but its convergence is obvious. $\square$

Remarks. (a) If $f(z)$ has a root of multiplicity m at the origin, by applying the above result to the function

$$g(z) = f(z) \cdot \left(\frac{z}{R} \right)^{-m},$$

we obtain

$$\frac{1}{2\pi} \int_0^{2\pi} \text{Log} \left| g(Re^{i\theta}) \right| d\theta = \text{Log} |g(0)| + \sum_{k=1}^n \text{Log} \frac{R}{|a_k|};$$

that is,

$$\frac{1}{2\pi}\int_0^{2\pi}\text{Log}\left|f(Re^{i\theta})\right|\,d\theta = \text{Log}\left|\frac{f(z)}{z^m}\right|_{z=0} + m\,\text{Log}\,R + \sum_{k=1}^n\text{Log}\frac{R}{|a_k|}.$$

(b) The Jensen formula can be put in another useful form. Let

$$0 < |a_1| \leq |a_2| \leq \cdots \leq |a_n| \leq \cdots,$$

and $n(r)$ denote the number of roots of $f(z)$ in the disc $\bar{D}(0, r)$. Then, for $|a_n| \leq R < |a_{n+1}|$, we have

$$\begin{aligned}
\sum_{k=1}^n\text{Log}\frac{R}{|a_k|} &= n\text{Log}R - \sum_{k=1}^n\text{Log}|a_k| \\
&= n(\text{Log}R - \text{Log}|a_n|) + \sum_{k=1}^{n-1}k(\text{Log}|a_{k+1}| - \text{Log}|a_k|) \\
&= n\int_{|a_n|}^R\frac{dr}{r} + \sum_{k=1}^{n-1}k\int_{|a_k|}^{|a_{k+1}|}\frac{dr}{r}.
\end{aligned}$$

Because

$$n(r) = \begin{cases} n & \text{for}\quad |a_n| \leq r \leq R < |a_{n+1}|, \\ k & \text{for}\qquad |a_k| \leq r < |a_{k+1}| \qquad (\text{if } |a_k| < |a_{k+1}|) \end{cases}$$

(the case $|a_k| = |a_{k+1}| = \cdots = |a_{k+p}|$ needs only a trivial modification), the right-hand side is equal to $\int_0^R\frac{n(r)}{r}dr$, and thus the Jensen formula takes the form

$$\frac{1}{2\pi}\int_0^{2\pi}\text{Log}|f(Re^{i\theta})|\,d\theta = \text{Log}|f(0)| + \int_0^R\frac{n(r)}{r}dr.$$

Let $M(R) = \|f\|_{C(0,R)}$, and assume that $f(0) \neq 0$; then from the Jensen formula we have

$$\frac{1}{2\pi}\int_0^{2\pi}\text{Log}M(R)\,d\theta \geq \text{Log}|f(0)| + \sum_{|a_n|\leq R}\text{Log}\frac{R}{|a_k|};$$

hence

$$\text{Log}M(R) - \text{Log}|f(0)| \geq \sum_{|a_k|\leq R}\text{Log}\frac{R}{|a_k|}.$$

If we use only the roots inside or on the smaller circle $C(0, \frac{R}{2})$, then the sum on the right-hand side is decreased, and so

$$\text{Log}\frac{M(R)}{|f(0)|} \geq \sum_{|a_k| \leq \frac{R}{2}} \text{Log}\frac{R}{|a_k|} \geq \sum_{|a_k| \leq \frac{R}{2}} \text{Log } 2 = n\left(\frac{R}{2}\right) \cdot \text{Log } 2.$$

Thus, we obtain the following

COROLLARY 7.7.2 *Let $f(z)$ be analytic in the closed disc $\bar{D}(0, R)$, and $f(0) \neq 0$; then*

$$n\left(\frac{R}{2}\right) \leq \frac{1}{\text{Log } 2} \cdot \text{Log}\frac{M(R)}{|f(0)|}.$$

From this corollary, we see that if we have a bound on $|f(z)|$ on any circle and we know the value of the function at the center of the circle, then we have an upper bound on the number of roots inside and on the circle having half the radius of the original circle. From the proof it is obvious that similar upper bounds can be obtained for a circle whose radius is any fraction of R.

EXAMPLE. If $|f(0)| = 3$ and $M(R) \leq 180$, then

$$n\left(\frac{R}{2}\right) \leq \frac{1}{\text{Log } 2} \cdot \text{Log}\left(\frac{180}{3}\right) < \frac{\text{Log } 64}{\text{Log } 2} = 6,$$

and so there cannot be more than 5 roots inside or on the circle of radius $\frac{R}{2}$.

EXAMPLE. If $|f(0)| = 3$, and $M(R) \leq 5$, then

$$n\left(\frac{R}{2}\right) \leq \frac{1}{\text{Log } 2} \cdot \text{Log}\frac{5}{3} < 1,$$

and so there is *no* root inside or on the circle of radius $\frac{R}{2}$.

The Jensen formula can easily be extended to meromorphic functions.

THEOREM 7.7.3 (Jensen-Nevanlinna) *Let $f(z)$ be a function meromorphic in the closed disc $\bar{D}(0, R)$ and having neither root nor pole at the origin. Then denoting by $a_1, a_2, \ldots, a_n$ the roots and by $b_1, b_2, \ldots, b_m$*

the poles of $f(z)$ on $\bar{D}(0, R)$ (repeated as often as their multiplicities), we have

$$\frac{1}{2\pi}\int_0^{2\pi} \text{Log}\,|f(Re^{i\theta})|\,d\theta = \text{Log}\,|f(0)| + \sum_{k=1}^{n}\text{Log}\,\frac{R}{|a_k|} - \sum_{j=1}^{m}\text{Log}\,\frac{R}{|b_k|}$$

$$= \text{Log}\,|f(0)| + \int_0^R \frac{n(r)-m(r)}{r}\,dr,$$

where $n(r)$ and $m(r)$ denote the number of roots and poles, respectively, on the disc $\bar{D}(0, r)$.

If the origin is a root (or a pole) of multiplicity ℓ, then $\text{Log}\,|f(0)|$ must be replaced by

$$\text{Log}\,\left|\frac{f(z)}{z^\ell}\right|_{z=0} + \ell\,\text{Log}\,R \quad (\text{or by } \text{Log}\,\left|z^\ell\cdot f(z)\right|_{z=0} - \ell\,\text{Log}\,R)$$

in the formula above.

   Proof. Let

$$F(z) = f(z)\cdot\left\{\prod_{k=1}^{n}\frac{R(a_k-z)}{R^2-\bar{a}_kz}\right\}^{-1}\cdot\left\{\prod_{j=1}^{m}\frac{R(b_j-z)}{R^2-\bar{b}_jz}\right\}.$$

Then $F(z)$ is an analytic function on $\bar{D}(0, R)$ without roots.   □

   Remark. The Jensen-Nevanlinna formula can be further extended by applying the *Poisson kernel*. (See Exercise 11.16.)

7.8   Entire Functions of Finite Order

DEFINITION 7.8.1 *An entire function $f(z)$ is said to be **of finite order** if there exist positive constants k and r_0 such that*

$$|f(z)| \le e^{|z|^k} \quad \text{whenever } |z| \ge r_0.$$

*The **order** ρ of an entire function $f(z)$ of finite order is the infimum of all k such that*

$$|f(z)| \le e^{|z|^k} \quad \text{for all sufficiently large } |z|.$$

EXAMPLE. The function $f(z) = ze^z$ is of finite order; indeed $\rho = 1$, for

$$\begin{aligned} |f(z)| &= |ze^z| = |z| \cdot e^x \le |z| \cdot e^{|z|} \quad (\because x = \Re z \le |z|) \\ &\le e^{\epsilon|z|} \cdot e^{|z|} = e^{(1+\epsilon)|z|} \le e^{|z|^{1+\epsilon}}, \end{aligned}$$

because, for any $\epsilon > 0$, we have $|z| \le e^{\epsilon|z|}$ whenever $|z|$ is sufficiently large.

EXAMPLE. The function $f(z) = e^{e^z}$ is of infinite order, for

$$|f(z)| = \left| e^{e^z} \right| \not\le e^{|z|^k}$$

for any k as $z \to \infty$ along the positive real axis.

If $k \le 0$, then $|f(z)| \le e^{|z|^k} \le e$ $(|z| \ge 1)$ implies that $f(z)$ is a bounded entire function, hence a constant. We also note that if $k < h$ and $|f(z)| \le e^{|z|^k}$, then $|f(z)| \le e^{|z|^h}$.

EXAMPLE. The entire functions e^z, $\sin z$, $\cos z$ are all of order 1.

$$\cos \sqrt{z} = 1 - \frac{z}{2!} + \frac{z^2}{4!} - \frac{z^3}{6!} + - \cdots,$$

and

$$\frac{\sin \sqrt{z}}{\sqrt{z}} = 1 - \frac{z}{3!} + \frac{z^2}{5!} - \frac{z^3}{7!} + - \cdots$$

are entire functions of order $\frac{1}{2}$.

What is the order of a polynomial? Take $f(z) = z^n$. Then, for any $\epsilon > 0$,

$$|f(z)| = |z^n| = |z|^n \le \left(e^{|z|^\epsilon} \right)^n = e^{n|z|^\epsilon} \le e^{|z|^{\epsilon n}}$$

for sufficiently large $|z|$. Hence $f(z) = z^n$ is of order 0. Because the leading term dominates the rest of the terms in a polynomial for sufficiently large $|z|$, we see that the order of any polynomial is 0.

As before, set

$$M(r) = \|f\|_{C(0,r)} \ (= \max\{|f(z)| ; |z| = r\}).$$

Then $M(r) \le e^{r^k}$ implies

$$\frac{\text{Log Log } M(r)}{\text{Log } r} \le k. \quad \therefore \rho = \lim_{r \to \infty} \sup \frac{\text{Log Log } M(r)}{\text{Log } r}.$$

Suppose $f(z)$ is an entire function of finite order ρ. Then for any $\epsilon > 0$ we have

$$\text{Log } M(r) \le r^{\rho + \epsilon} \quad \text{whenever } r \text{ is sufficiently large.}$$

Thus, from Corollary 7.7.2, we have, for given $\epsilon > 0$, and large r,

$$n\left(\frac{r}{2}\right) \le \frac{1}{\text{Log } 2}\left\{r^{\rho + \frac{\epsilon}{2}} - \text{Log } |f(0)|\right\}$$

provided $f(0) \ne 0$, where, as before, $n(r)$ is the number of roots of $f(z)$ in the closed disc $\bar{D}(0, r)$.

Now we may assume, without loss of generality, that $|f(0)| = 1$. Then

$$n\left(\frac{r}{2}\right) \le \frac{r^{\rho + \frac{\epsilon}{2}}}{\text{Log } 2};$$

hence

$$n(r) \le \frac{2^{\rho + \frac{\epsilon}{2}}}{\text{Log } 2} \cdot r^{\rho + \frac{\epsilon}{2}} \quad \text{for sufficiently large } r.$$

Therefore if $0 < |a_1| \le |a_2| \le \cdots \le |a_k| \le \cdots$, where $\{a_k\}_{k=1}^{\infty}$ are the roots of $f(z)$, then, by taking $r = |a_k|$, we obtain

$$k \le n(|a_k|) \le A|a_k|^{\rho + \frac{\epsilon}{2}} \quad \text{for large } k,$$

where $A = \dfrac{2^{\rho + \frac{\epsilon}{2}}}{\text{Log } 2}$ is independent of k. Hence

$$\frac{1}{|a_k|^{\rho + \epsilon}} \le \left(\frac{A}{k}\right)^{\frac{\rho + \epsilon}{\rho + \frac{\epsilon}{2}}} \quad \text{for large } k.$$

Because $\dfrac{\rho + \epsilon}{\rho + \frac{\epsilon}{2}} > 1$, we obtain

$$\sum_{k=1}^{\infty} \frac{1}{|a_k|^{\rho + \epsilon}} < \infty.$$

Easy reasoning shows that if $f(z)$ is an entire function of finite order ρ, then so too is $cf(z)$, where c is a nonzero constant; the same holds for $\frac{f(z)}{z^m}$ if the origin is a root of multiplicity m for the entire function $f(z)$. Therefore, we have established the following

THEOREM 7.8.2 *If $f(z)$ is an entire function (not identically zero) of finite order ρ, and $a_1, a_2, \ldots, a_k, \ldots$ are the roots of $f(z)$ (other than the origin) repeated as often as their multiplicities, then, for any $\epsilon > 0$,*

$$\sum_{k=1}^{\infty} \frac{1}{|a_k|^{\rho+\epsilon}} < \infty.$$

The infimum, σ, of the positive numbers α for which

$$\sum_{k=1}^{\infty} \frac{1}{|a_k|^{\alpha}} < \infty$$

is called the *exponent of convergence* of the roots of $f(z)$. The theorem asserts that the exponent of convergence σ cannot be greater than the order ρ of the entire function; that is, $\sigma \le \rho$.

We may have $\sigma < \rho$. For example, if $f(z) = e^z$, then $\rho = 1$, but there is no root, so that $\sigma = 0$.

EXAMPLE. Let $f(z) = \cos z$. Then

$$|f(z)| \le \frac{1}{2} \left(\left| e^{iz} \right| + \left| e^{-iz} \right| \right) \le e^{|z|}. \quad \therefore \rho \le 1.$$

On the other hand, let $z = iy$; then

$$|f(z)| = \frac{1}{2} \left(e^{-y} + e^{y} \right) \ge \frac{1}{2} e^{|z|} \quad (y \to \infty).$$

Thus, ρ is at least 1 and hence $\rho = 1$.

Now the roots of $\cos z$ are all real, namely, they are the odd multiples of $\frac{\pi}{2}$. Therefore,

$$\sum_{k=1}^{\infty} \frac{1}{|a_k|^{\alpha}} = 2 \left\{ \frac{1}{\left(\frac{\pi}{2}\right)^{\alpha}} + \frac{1}{\left(\frac{3\pi}{2}\right)^{\alpha}} + \frac{1}{\left(\frac{5\pi}{2}\right)^{\alpha}} + \cdots \right\}$$

$$= \frac{2^{1+\alpha}}{\pi^{\alpha}} \sum_{k:\text{odd}} \frac{1}{k^{\alpha}} < \infty \quad \text{provided } \alpha > 1.$$

However, $\sum_{k=1}^{\infty} \frac{1}{|a_k|} = \infty$, and so in this case $\rho = \sigma = 1$.

This theorem enables us to show, as we now demonstrate, that for entire functions of finite order ρ there exists an integer $m \leq \rho$ that satisfies the condition of convergence of the canonical product. More precisely, we have the following

THEOREM 7.8.3 (Hadamard Factorization Theorem) *Suppose $f(z)$ is an entire function (not identically zero) of finite order ρ, and $a_1, a_2, \ldots, a_k, \ldots$ are its roots (other than the origin) repeated as often as their multiplicities; then there exists a polynomial $g(z)$ of degree $\leq \rho$ and an integer $m \leq \rho$ such that $f(z)$ can be expressed as*

$$f(z) = e^{g(z)} \cdot z^n \cdot \prod_{k=1}^{\infty} E_m \left(\frac{z}{a_k} \right),$$

where n is the multiplicity of the root at the origin ($n = 0$ if $f(0) \neq 0$).

We begin by rewriting the Cauchy Integral Formula 4.4.3 as follows: Let

$$g(z) = u(z) + iv(z) = \sum_{k=0}^{\infty} c_k z^k$$

be analytic in a closed disc $\bar{D}(0, R)$, where $u(z)$ and $v(z)$ are the real and imaginary parts of $g(z)$, respectively. Then

$$\frac{1}{2\pi i} \int_{C(0,R)} \frac{g(\zeta)}{\zeta^{k+1}} d\zeta = \begin{cases} c_k & (k = 0, 1, 2, \ldots), \\ 0 & (k = -1, -2, -3, \ldots); \end{cases}$$

hence

$$\frac{1}{2\pi R^k} \int_0^{2\pi} g\left(Re^{it}\right) \cdot e^{-ikt} dt = \begin{cases} c_k & (k = 0, 1, 2, \ldots), \\ 0 & (k = -1, -2, -3, \ldots). \end{cases}$$

Taking the complex conjugate for the case $k < 0$, we obtain

$$\frac{1}{2\pi R^k} \int_0^{2\pi} \overline{g\left(Re^{it}\right)} \cdot e^{-ikt} dt = 0 \quad (k = 1, 2, 3, \ldots).$$

Adding the last two equalities, we obtain

$$\frac{1}{\pi R^k} \int_0^{2\pi} u\left(Re^{it}\right) \cdot e^{-ikt} dt = c_k \quad (k = 1, 2, 3, \ldots).$$

For $k = 0$, we have

$$c_0 = \frac{1}{2\pi} \int_0^{2\pi} g(Re^{it}) \, dt = g(0),$$

$$\bar{c}_0 = \frac{1}{2\pi} \int_0^{2\pi} \overline{g(Re^{it})} \, dt = \overline{g(0)},$$

and so

$$c_0 + \bar{c}_0 = \frac{1}{\pi} \int_0^{2\pi} u\left(Re^{it}\right) \, dt = g(0) + \overline{g(0)}.$$

It follows that

$$c_0 = -\overline{g(0)} + \frac{1}{\pi} \int_0^{2\pi} u\left(Re^{it}\right) \, dt.$$

We can now write

$$g(z) = -\overline{g(0)} + \sum_{k=0}^{\infty} \frac{1}{\pi} \int_0^{2\pi} u\left(Re^{it}\right) \cdot \left(\frac{z}{Re^{it}}\right)^k \, dt$$

$$= -\overline{g(0)} + \frac{1}{\pi} \int_{C(0,R)} u\left(\zeta\right) \cdot \left\{\sum_{k=0}^{\infty} \left(\frac{z}{\zeta}\right)^k\right\} \cdot \frac{d\zeta}{i\zeta} \quad \left(\zeta = Re^{it}\right)$$

$$= -\overline{g(0)} + \frac{1}{\pi i} \int_{C(0,R)} u\left(\zeta\right) \cdot \frac{d\zeta}{\zeta - z}.$$

Differentiating under the integral sign, we obtain

$$g^{(p)}(z) = \frac{p!}{\pi i} \int_{C(0,R)} u\left(\zeta\right) \cdot \frac{d\zeta}{(\zeta - z)^{p+1}} \quad (p \in \mathbb{N}).$$

With the help of this formula, we now establish the following crucial step.

LEMMA 7.8.4 Let $f(z)$ be an entire function of finite order ρ, $f(0) \neq 0$, having roots $a_1, a_2, \ldots, a_k, \ldots$ (where multiple roots are repeated according to their multiplicities). Then, for $p + 1 > \rho$,

$$\frac{d^p}{dz^p} \left\{\frac{f'(z)}{f(z)}\right\} = -p! \sum_{k=1}^{\infty} \frac{1}{(a_k - z)^{p+1}}.$$

Proof. Let $a_1, a_2, \ldots, a_{n(R)}$ be the roots of $f(z)$ in $D(0, R)$, where $f(z)$ has no root on the circle $C(0, R)$. Then

$$f(z) \cdot \left\{\prod_{k=1}^{n(R)} \frac{R(a_k - z)}{(R^2 - \bar{a}_k z)}\right\}^{-1}$$

is analytic and has no root in $\bar{D}(0, R)$. Therefore, there exists an analytic function $g(z)$ such that

$$f(z) \cdot \left\{ \prod_{k=1}^{n(R)} \frac{R(a_k - z)}{(R^2 - \bar{a}_k z)} \right\}^{-1} = e^{g(z)} \qquad z \in D(0, R).$$

Taking the logarithmic derivative, we obtain

$$\frac{f'(z)}{f(z)} - \sum_{k=1}^{n(R)} \left\{ \frac{-1}{a_k - z} + \frac{\bar{a}_k}{R^2 - \bar{a}_k z} \right\} = g'(z).$$

Hence

$$\frac{d^p}{dz^p} \left\{ \frac{f'(z)}{f(z)} \right\} = p! \sum_{k=1}^{n(R)} \left\{ \frac{-1}{(a_k - z)^{p+1}} + \frac{\bar{a}_k^{p+1}}{(R^2 - \bar{a}_k z)^{p+1}} \right\} + g^{(p+1)}(z)$$

$$(p = 0, 1, 2, \ldots).$$

Using the Cauchy integral formula above for $g(z)$ and noting that the Blaschke product has modulus 1 on the circle $C(0, R)$, we have $\Re g(z) = \mathrm{Log}\,|f(z)|$ on that circle, and we obtain

$$\frac{d^p}{dz^p} \left\{ \frac{f'(z)}{f(z)} \right\} = p! \sum_{k=1}^{n(R)} \left\{ \frac{-1}{(a_k - z)^{p+1}} + \frac{\bar{a}_k^{p+1}}{(R^2 - \bar{a}_k z)^{p+1}} \right\}$$

$$+ \frac{(p+1)!}{\pi i} \int_{C(0,R)} \mathrm{Log}\,|f(\zeta)| \frac{d\zeta}{(\zeta - z)^{p+2}}.$$

Note that $\mathrm{Log}\,|f(\zeta)|$ could be very large in absolute value, and so we use the following device to evaluate the integral on the right: Because

$$-\mathrm{Log}\,M(R) \cdot \int_{C(0,R)} \frac{d\zeta}{(\zeta - z)^{p+2}} = 0 \quad (p = 0, 1, 2, \ldots),$$

where $M(R) = \|f\|_{C(0, R)}$, we have

$$\frac{(p+1)!}{\pi i} \int_{C(0,R)} \mathrm{Log}\,|f(\zeta)| \frac{d\zeta}{(\zeta - z)^{p+2}}$$

$$= \frac{(p+1)!}{\pi i} \int_{C(0,R)} \mathrm{Log}\frac{|f(\zeta)|}{M(R)} \cdot \frac{d\zeta}{(\zeta - z)^{p+2}}.$$

For R sufficiently large, the modulus of the right-hand side does not exceed

$$\frac{(p+1)!\ R^{\rho+\epsilon} - \text{Log}\,|f(0)|}{\pi} \cdot \frac{1}{(R-|z|)^{p+2}} \cdot 2\pi R \quad (\epsilon > 0),$$

on account of the *Jensen inequality*

$$\frac{1}{2\pi}\int_0^{2\pi} \text{Log}\,|f(Re^{it})|\,dt \geq \text{Log}\,|f(0)|.$$

Thus, for $p+1 > \rho$, the integral tends to zero as $R \to \infty$.
   Next, we estimate the term

$$p!\sum_{k=1}^{n(R)} \frac{\bar{a}_k^{p+1}}{(R^2 - \bar{a}_k z)^{p+1}}.$$

Its modulus does not exceed

$$p!\frac{R^{p+1}\cdot n(R)}{\{R(R-|z|)\}^{p+1}} \leq p!\frac{AR^{\rho+\epsilon} + B}{(R-|z|)^{p+1}} \longrightarrow 0 \quad (R \to \infty).$$

Here we have used the estimate

$$n(R) \leq AR^{\rho+\epsilon} + B \quad \text{(for some suitable constants } A \text{ and } B),$$

which is an easy consequence of the Jensen Formula 7.7.1. (See Exercise 7.39.) Therefore,

$$\frac{d^p}{dz^p}\left\{\frac{f'(z)}{f(z)}\right\} = -p!\sum_{k=1}^{\infty}\frac{1}{(a_k - z)^{p+1}} \quad (p+1 > \rho). \qquad \square$$

Remark. The lemma holds even if the condition "$f(0) \neq 0$" is replaced by "$f(z)$ *is not identically zero*." Simply apply the result above to $h(z) = \frac{f(z)}{z^n}$, where n is the multiplicity of the root of $f(z)$ at $z = 0$.
   Proof of the Hadamard factorization theorem. First assume that $f(0) \neq 0$. Let σ be the exponent of convergence of the roots and m the least non-negative integer such that $\sum_{k=1}^{\infty}\frac{1}{|a_k|^{m+1}} < \infty$. Then $m \leq \sigma$, and

$$P(z) = \prod_{k=1}^{\infty} E_m\left(\frac{z}{a_k}\right)$$

is the canonical product associated with the roots $\{a_k\}_{k=1}^{\infty}$. The function $\frac{f(z)}{P(z)}$ is entire with no root, and so there exists an entire function $g(z)$ such that

$$f(z) = e^{g(z)} \cdot P(z).$$

Therefore, by Theorem 7.4.10, we have

$$\frac{f'(z)}{f(z)} = g'(z) + \frac{P'(z)}{P(z)}$$

$$= g'(z) + \sum_{k=1}^{\infty} \left\{ \frac{-1}{a_k - z} + \frac{1}{a_k} + \frac{z}{a_k^2} + \cdots + \frac{z^{m-1}}{a_k^m} \right\}.$$

Differentiating p times ($p + 1 > \rho$), we obtain

$$g^{(p+1)}(z) = \frac{d^p}{dz^p} \left\{ \frac{f'(z)}{f(z)} \right\} + p! \sum_{k=1}^{\infty} \frac{1}{(a_k - z)^{p+1}} = 0,$$

by the preceding lemma. Thus, the entire function $g(z)$ is a polynomial of degree at most $p \leq \rho$.

If $f(0) = 0$, then let $f(z) = z^n \cdot h(z)$, where n is the multiplicity of the root of $f(z)$ at $z = 0$ and $h(z)$ is an entire function of order ρ having precisely the same roots as $f(z)$ except at the origin. By the preceding result, $h(z)$ can be factored as $h(z) = e^{g(z)} \cdot P(z)$, where $g(z)$ is a polynomial of degree at most ρ and $P(z)$ is the canonical product associated with the roots of $f(z)$ (other than the origin). Therefore,

$$f(z) = e^{g(z)} \cdot z^n \cdot \prod_{k=1}^{\infty} E_m \left(\frac{z}{a_k} \right). \qquad \square$$

Remark. $\rho = \max\{\sigma, \deg g\}$. (See Exercise 7.45.)

EXAMPLE. Consider the entire function $\frac{\sin \pi z}{\pi z}$. Its roots are all the non-zero integers (and are all simple), hence the exponent of convergence is 1. We also know its order ρ is 1. Therefore, it must be of the form

$$\frac{\sin \pi z}{\pi z} = e^{az+b} \cdot \prod_{-\infty}^{\infty}{}' \left\{ \left(1 - \frac{z}{k} \right) \cdot e^{\frac{z}{k}} \right\}$$

$$= e^{az+b} \cdot \prod_{k=1}^{\infty} \left(1 - \frac{z^2}{k^2} \right).$$

Letting $z \to 0$, we obtain $1 = e^b$, and so we may choose $b = 0$. Because $\frac{\sin \pi z}{\pi z}$ is an even function, we have $a = 0$. Therefore,

$$\sin \pi z = \pi z \cdot \prod_{k=1}^{\infty} \left(1 - \frac{z^2}{k^2} \right).$$

The Hadamard Factorization Theorem 7.8.3 for entire functions of finite order is very important and has many far-reaching consequences. Here we mention only specific cases of the "little" Picard Theorem 5.2.3 for entire functions of finite order.

COROLLARY 7.8.5 *A nonconstant entire function of finite order assumes every complex value with only one possible exception.*

Proof. Suppose $f(z)$ is an entire function of finite order with two distinct exceptional values α and β; that is,

$$f(z) \neq \alpha \quad \text{and} \quad f(z) \neq \beta \quad \text{for any } z \in \mathbb{C}.$$

Then $f(z) - \beta$ is an entire function having the same order as $f(z)$, and it possesses no root, so that $f(z) - \beta = e^{g(z)}$ for some nonconstant polynomial $g(z)$. Because $f(z) - \beta \neq \alpha - \beta$ for any $z \in \mathbb{C}$, it follows that $g(z) \neq \log(\alpha - \beta)$ for any $z \in \mathbb{C}$. However, since $g(z)$ is a nonconstant polynomial, it *must* assume every value without exception, by the fundamental theorem of algebra, so the proof is complete.   □

COROLLARY 7.8.6 *An entire function of finite noninteger order assumes every complex value infinitely many times.*

Proof. Clearly, the entire functions $f(z)$ and $f(z) - \alpha$ have the same order for any $\alpha \in \mathbb{C}$, and so we have only to show that $f(z)$ has infinitely many roots. If $f(z)$ has only finitely many roots $a_1, a_2, \ldots, a_n$, then

$$f(z) = e^{g(z)} \cdot \prod_{k=1}^{n} (z - a_k)$$

for some polynomial $g(z)$. But then the order of $f(z)$ is the degree of $g(z)$, which is an integer. This contradiction establishes the corollary.   □

7.9   The Runge Approximation Theorem

A function analytic in a disc can be expanded in a Taylor series. This implies that such a function can be approximated locally uniformly in the disc by polynomials, the partial sums of the Taylor series (and polynomials have singularities only at the point at infinity). A function analytic in an annulus can be expressed as a Laurent series, and so it can be approximated locally uniformly in the annulus by rational functions whose poles are at the center of the annulus and at the point at infinity. A function $f(z)$ meromorphic in the plane $\mathbb{C}$ with poles at $\{a_k\}_{k=1}^\infty$ can be expressed as a Mittag-Leffler series:

$$f(z) = g(z) + \sum_{k=1}^\infty \left\{ P_k \left(\frac{1}{z - a_k} \right) - Q_k(z) \right\},$$

where

(a) $g(z)$ is an entire function;

(b) $P_k(z)$ is a polynomial without the constant term, and $P_k \left(\frac{1}{z-a_k} \right)$, being the singular part of $f(z)$ at a_k, is a rational function whose only singularity is a pole at a_k; and

(c) $Q_k(z)$ is a suitable partial sum of the Taylor series expansion of $P_k \left(\frac{1}{z-a_k} \right)$ around the origin (assuming $a_k \neq 0$), hence is a polynomial.

It follows that the general term $P_k \left(\frac{1}{z-a_k} \right) - Q_k(z)$ of the Mittag-Leffler series of the meromorphic function $f(z)$ with poles at $\{a_k\}_{k=1}^\infty$ is a rational function with poles at a_k and the point at infinity; therefore $f(z)$ can be approximated locally uniformly in $\mathbb{C} \setminus \{a_k ; k \in \mathbb{N}\}$ by rational functions whose poles are outside the region of analyticity of $f(z)$. (To include $g(z)$ in this statement we can simply employ partial sums of its Taylor expansion around any point of $\mathbb{C}$, say the origin.)

Thus, it is natural to ask: Can a function that is analytic in an arbitrary region Ω be approximated locally uniformly by rational functions whose poles are outside the region Ω? It turns out that not only is the answer affirmative, but, remarkably, Ω need not even be connected.

THEOREM 7.9.1 (Runge) *A function $f(z)$ analytic in a nonempty open set $G \subset \mathbb{C}$ can be approximated locally uniformly by a sequence of rational functions whose poles are outside the set G. Moreover, given an arbitrary set $E \subset \hat{\mathbb{C}} \setminus G$ whose closure intersects every component of $\hat{\mathbb{C}} \setminus G$, then the approximating rational functions may be so constructed as to have all their poles in the set E.*

Note: The condition that the closure of the set E has nonempty intersection with every component of $\hat{\mathbb{C}} \setminus G$ cannot be relaxed.

EXAMPLE. Consider the function $f(z) = \frac{1}{z}$, which is analytic in the punctured disc $D'(0, 3)$. Does there exist a function $g(z)$ analytic in $D(0, 3)$ (not $D'(0, 3)$) such that

$$\|f - g\|_K < \frac{1}{4},$$

where K is the (closed) annulus $\{z \in \mathbb{C}; 1 \leq |z| \leq 2\}$? If so, for $z \in C(0, 2)$,

$$|g(z)| \leq |f(z)| + |g(z) - f(z)| < \frac{1}{2} + \frac{1}{4} = \frac{3}{4};$$

but for $z \in C(0, 1)$, we have

$$|g(z)| \geq |f(z)| - |g(z) - f(z)| > 1 - \frac{1}{4} = \frac{3}{4}.$$

These inequalities contradict the maximum modulus principle, since we have assumed that $g(z)$ is analytic in the whole disc $D(0, 3)$; that is, to obtain an approximation to $f(z)$ of the desired accuracy in the given annulus K we cannot require $g(z)$ to be analytic everywhere in $D(0, 3)$.

The proof of the Runge approximation theorem is carried out via a sequence of lemmas.

LEMMA 7.9.2 *For an arbitrary (nonempty) open set G of $\mathbb{C}$, there is a sequence $\{\Delta_n\}_{n=1}^{\infty}$ of open sets satisfying the following conditions:*

1^0 the boundary $\partial \Delta_n$ consists of a finite number of piecewise smooth curves;

2^0 the closure $\bar{\Delta}_n$ is compact, $\bar{\Delta}_n \subset \Delta_{n+1}$ $(n \in \mathbb{N})$, and $\cup_{n=1}^{\infty} \Delta_n = G$.

Proof. Let z_0 be an arbitrary point in the open set G, and choose $n_1 \in \mathbb{N}$ satisfying $\frac{\sqrt{2}}{n_1} < \text{dist}\{z_0, \partial G\}$. Place a "grid" of horizontal lines $\Im z = \Im z_0 + \frac{k}{n_1}$ and vertical lines $\Re z = \Re z_0 + \frac{k}{n_1}$ $(k \in \mathbb{Z})$. Let Δ_1 be the interior of the union of closed squares (with sides $\frac{1}{n_1}$) that are in $G \cap D(z_0, n_1)$. Clearly, $\bar{\Delta}_1$ is compact, and its boundary $\partial \Delta_1$ consists of a finite number of simple closed polygons. Next, choose $n_2 \in \mathbb{N}$ such that $\frac{\sqrt{2}}{n_2} < \text{dist}\{\bar{\Delta}_1, \partial G\}$, and carry out the same procedure to obtain Δ_2. Clearly, we have $\bar{\Delta}_1 \subset \Delta_2$ (for $n_1 < n_2$). Repeating this process, we obtain a sequence $\{\Delta_n\}_{n=1}^{\infty}$ of open sets that satisfies the required conditions. $\qquad\square$

Remark. If G is a region (i.e., if G is connected), then (by merely choosing Δ_n to be the component containing the point z_0), we can also make Δ_n connected. Moreover, if, in addition, G is simply connected, then we can make Δ_n simply connected.

Suppose $f(z)$ is analytic in a nonempty open set G and K is an arbitrary compact subset of G; then for n sufficiently large we have $K \subset \Delta_{n-1}$, and so, by the "Cauchy integral formula," we have

$$f(z) = \frac{1}{2\pi i} \int_{\partial \Delta_n} \frac{f(\zeta)}{\zeta - z} d\zeta \quad (z \in K).$$

LEMMA 7.9.3 *Suppose $f(z)$ is a function continuous on Γ, where Γ consists of a finite number of curves C_k ($k = 1, \ldots, m$). Let K be a compact set that does not intersect Γ. Then for any $\epsilon > 0$ there exists a rational function $\varphi(z)$ having all its poles on Γ and satisfying*

$$\left| \frac{1}{2\pi i} \int_{\Gamma} \frac{f(\zeta)}{\zeta - z} d\zeta - \varphi(z) \right| < \epsilon \quad (\text{for all } z \in K).$$

Proof. Because $K \cap \Gamma = \phi$ (and both K and Γ are compact), we may choose r satisfying $0 < r < \text{dist}\{K, \Gamma\}$. Let

$$\zeta_0^{(k)}, \zeta_1^{(k)}, \ldots, \zeta_{n-1}^{(k)} \quad \left(\zeta_n^{(k)} = \zeta_0^{(k)} \right)$$

be points on C_k such that the length of the arc $C_k^{(j)}$ joining $\zeta_{j-1}^{(k)}$ and $\zeta_j^{(k)}$ is $\frac{L_k}{n}$, where L_k is the length of C_k ($k = 1, \ldots, m$), and set

$$\varphi_n(z) = \frac{1}{2\pi i} \sum_{k=1}^{m} \sum_{j=1}^{n} \frac{f\left(\zeta_j^{(k)}\right)}{\zeta_j^{(k)} - z} \left(\zeta_j^{(k)} - \zeta_{j-1}^{(k)} \right).$$

Because $f(\zeta)$ is uniformly continuous on C_k, we can, given any $\epsilon > 0$, choose n so large that

$$|f(\zeta') - f(\zeta'')| < \epsilon \quad \text{for all } \zeta', \zeta'' \in C_k^{(j)} \quad (k = 1, \ldots, m; \; j = 1, \ldots, n).$$

Therefore, for $z \in K$, we have

$$\left| \frac{1}{2\pi i} \int_\Gamma \frac{f(\zeta)}{\zeta - z} d\zeta - \varphi_n(z) \right|$$

$$= \frac{1}{2\pi} \left| \sum_{k=1}^{m} \sum_{j=1}^{n} \int_{C_k^{(j)}} \left\{ \frac{f(\zeta)}{\zeta - z} - \frac{f\left(\zeta_j^{(k)}\right)}{\zeta_j^{(k)} - z} \right\} d\zeta \right|$$

$$\leq \frac{1}{2\pi} \sum_{k=1}^{m} \sum_{j=1}^{n} \int_{C_k^{(j)}} \left| \frac{f(\zeta) - f(\zeta_j^{(k)})}{\zeta - z} \right.$$

$$+ f\left(\zeta_j^{(k)}\right) \left\{ \frac{1}{\zeta - z} - \frac{1}{\zeta_j^{(k)} - z} \right\} \right| |d\zeta|$$

$$\leq \frac{1}{2\pi} \sum_{k=1}^{m} \sum_{j=1}^{n} \int_{C_k^{(j)}} \left\{ \left| \frac{f(\zeta) - f(\zeta_j^{(k)})}{\zeta - z} \right| \right.$$

$$+ \left| f\left(\zeta_j^{(k)}\right) \right| \cdot \left| \frac{\zeta_j^{(k)} - \zeta}{(\zeta - z)(\zeta_j^{(k)} - z)} \right| \right\} |d\zeta|$$

$$\leq \frac{1}{2\pi} \sum_{k=1}^{m} \sum_{j=1}^{n} \left\{ \frac{\epsilon}{r} + M \frac{L_k}{nr^2} \right\} \cdot \frac{L_k}{n}$$

$$\leq \frac{1}{2\pi} \left\{ \epsilon + \frac{mML}{nr} \right\} \cdot \frac{L}{r},$$

where $L = \sum_{k=1}^{m} L_k$ and $M = \|f\|_\Gamma$. It follows that $\{\varphi_n(z)\}_{n=1}^{\infty}$ converges to $\frac{1}{2\pi i} \int_\Gamma \frac{f(\zeta)}{\zeta - z} d\zeta$ uniformly on K (as $n \to \infty$). Clearly, the functions $\varphi_n(z)$ are rational functions having all their poles on Γ. $\qquad \square$

We have shown that if $f(z)$ is analytic in a nonempty open set G and if K is an arbitrary compact subset of G, then for n sufficiently large we have $K \subset \Delta_{n-1}$, and for any preassigned $\epsilon > 0$ there is a rational function $\varphi(z)$ having all its poles on $\partial \Delta_n$ such that

$$|f(z) - \varphi(z)| < \epsilon \quad (z \in K).$$

Now, because the poles of $\varphi(z)$ are in G, we must "transplant" these poles to the complement of G. We need the following

LEMMA 7.9.4 *Let K be a compact subset of $\mathbb{C}$ and a and b be two distinct arbitrary points that are in the same component of $\hat{\mathbb{C}} \setminus K$, $a \neq \infty$. Then*

any rational function $R_a(z)$ having its only pole at a can be uniformly approximated on K by rational functions whose only pole is at b; that is, given any $\epsilon > 0$, there exists a rational function $R_b(z)$ whose only pole is at b and is such that

$$|R_a(z) - R_b(z)| < \epsilon \quad (z \in K).$$

In particular, if $b = \infty$, then $R_b(z)$ becomes a polynomial.

Proof. (i) First, assume $b \neq \infty$. Because a and b are in the same component of $\hat{\mathbb{C}} \setminus K$, we can connect a and b by a broken line L in the component. Choose r satisfying $0 < 2r < \text{dist}\{K, L\}$, and let

$$a = a_0, a_1, \ldots, a_m = b$$

be points on L with $|a_j - a_{j-1}| < r$ $(j = 1, 2, \ldots, m)$. Clearly it is sufficient to restrict attention to the case $R_a(z)$ $(= Q_0(z))$ consisting of only one term, $\frac{1}{(z-a)^n}$. For $z \in K$ we have $\left|\frac{a_0 - a_1}{z - a_1}\right| < \frac{1}{2}$, and so, expanding in a Taylor series, we obtain

$$\frac{1}{(z - a_0)^n} = \frac{1}{\{(z - a_1) - (a_0 - a_1)\}^n}$$

$$= \frac{1}{(z - a_1)^n \left\{1 - \left(\dfrac{a_0 - a_1}{z - a_1}\right)\right\}^n}$$

$$= \frac{1}{(z - a_1)^n} \sum_{k=n}^{\infty} A_k \left(\frac{a_0 - a_1}{z - a_1}\right)^{k-n},$$

where $A_k = \frac{k!}{(k-n)!(n-1)!}$. Because the series $\sum_{k=n}^{\infty} A_k \left(\frac{1}{2}\right)^{k-n}$ is convergent (the radius of convergence of the power series $\sum_{k=n}^{\infty} A_k t^{k-n}$ is 1), the Weierstrass M-test 2.4.5 gives the uniform convergence on K of the series on the right. Hence, by choosing N sufficiently large we obtain a rational function

$$Q_1(z) = \frac{1}{(z - a_1)^n} \sum_{k=n}^{N} A_k \left(\frac{a_0 - a_1}{z - a_1}\right)^{k-n},$$

having a_1 as its only pole and such that

$$|R_a(z) - Q_1(z)| < \frac{\epsilon}{m} \quad (z \in K),$$

where ϵ is a preassigned positive number.

Suppose $Q_1(z)$, $Q_2(z)$, ... , $Q_{j-1}(z)$ $(1 \leq j < n)$ have been chosen such that each $Q_k(z)$ is a rational function having a_k as its only pole. Because $|a_j - a_{j-1}| < r$, $Q_{j-1}(z)$ is analytic in the complement of $\bar{D}(a_j, r)$, and so $Q_{j-1}(z)$ can be expanded in a Laurent series centered at a_j, which converges locally uniformly in the complement of $\bar{D}(a_j, r)$. Because the compact set K is in the complement of $\bar{D}(a_j, r)$, by taking sufficiently many terms of the said Laurent series we obtain a rational function $Q_j(z)$ having a_j as its only pole and such that

$$|Q_{j-1}(z) - Q_j(z)| < \frac{\epsilon}{m} \quad (z \in K).$$

Finally, by induction, we obtain a rational function $R_b(z)$ $(= Q_m(z))$ having b $(= a_m)$ as its only pole, and

$$|R_a(z) - R_b(z)| \leq \sum_{j=1}^{m} |Q_{j-1}(z) - Q_j(z)| < \epsilon \quad (z \in K).$$

(ii) If $b = \infty$, then choose b_0 such that $K \subset D(0, |b_0|)$, and, by what we have just shown, first approximate $R_a(z)$ on K by a rational function $Q(z)$ having b_0 as its only pole. Because $Q(z)$ is analytic in the disc $D(0, |b_0|)$ that contains the compact set K, we may expand $Q(z)$ in Taylor series around the origin and then take sufficiently many terms in this expansion to obtain a desired polynomial.   □

We can now complete the proof of the Runge approximation theorem. We have, for any $n \in \mathbb{N}$, a rational function $\varphi_n(z)$ that is the sum of terms of the form $\frac{k}{z-a}$, where $a \in \partial\Delta_n$ (k is a constant), and

$$|f(z) - \varphi_n(z)| < \frac{1}{2^{n+1}} \quad (z \in \Delta_{n-1}).$$

By the preceding lemma, there are rational functions $R_n(z)$ having poles only in E such that

$$|\varphi_n(z) - R_n(z)| < \frac{1}{2^{n+1}} \quad (z \in \Delta_{n-1}).$$
$$\therefore |f(z) - R_n(z)| < \frac{1}{2^n} \quad (z \in \Delta_{n-1}).$$

It follows that

$$f(z) = R_1(z) + \sum_{n=1}^{\infty} \{R_{n+1}(z) - R_n(z)\},$$

and the convergence is locally uniform in G. This completes the proof of the Runge approximation theorem.

COROLLARY 7.9.5 *Let G be an arbitrary (nonempty) open subset of $\mathbb{C}$, and let E be a set obtained by choosing one point from each component of $\hat{\mathbb{C}} \setminus G$. Then the space of all rational functions all of whose poles are in E is dense in the space of all analytic functions in G with the topology of local uniform convergence in G (i.e., the metric topology of Exercise 4.37).*

Extending our notation for regions to open sets, we denote by $H(G)$, where G is a nonempty open subset of the complex plane $\mathbb{C}$, the space of all functions analytic in G. Naturally, we equip $H(G)$ with the topology of local uniform convergence. (See Exercise 4.37.) Suppose G_1 and G_2 are two open subsets of the complex plane $\mathbb{C}$ with $G_1 \subset G_2$. Then every function in $H(G_2)$ can be considered as a function in $H(G_1)$; that is, $H(G_2)$ can be considered as a subspace of $H(G_1)$. We have the following

COROLLARY 7.9.6 (Runge) *Suppose G_1 and G_2 are two open subsets of the complex plane $\mathbb{C}$ with $G_1 \subset G_2$. Then $H(G_2)$ is dense in $H(G_1)$ if and only if every component of $\hat{\mathbb{C}} \setminus G_1$ intersects $\hat{\mathbb{C}} \setminus G_2$; that is, G_1 is simply connected in G_2.*

In other words, G_1 cannot have a "hole" unless this is forced upon it by a hole of G_2 (as seen from the preceding example).

COROLLARY 7.9.7 *If Ω is a simply connected region in the complex plane $\mathbb{C}$, then polynomials are dense in $H(\Omega)$; that is, for an arbitrary function $f(z) \in H(\Omega)$, any compact subset $K \subset \Omega$, and any $\epsilon > 0$, there exists a polynomial $p(z)$ satisfying*

$$|f(z) - p(z)| < \epsilon \quad (\text{for all } z \in K).$$

Proof. If Ω is simply connected, then $\hat{\mathbb{C}} \setminus \Omega$ has only one component that happens to be unbounded, and so we may choose $E = \{\infty\}$. □

Remark. If Ω is the disc $D(c, r)$, then its complement is connected; hence the theorem says that any function $f(z)$ analytic in the disc $D(c, r)$ can be approximated locally uniformly by polynomials. Of course this result is trivial from the Taylor-series expansion of $f(z)$:

$$f(z) = \sum_{k=0}^{\infty} a_k(z - c)^k \quad z \in D(c, r).$$

If we set $p_k(z) = (z - c)^k$, then we have

$$f(z) = \sum_{k=0}^{\infty} a_k p_k(z)$$

where $\{p_k(z)\}_{k=0}^{\infty}$ is independent of $f(z)$. In general, for a simply connected region Ω, there exists a sequence $\{p_k(z)\}_{k=0}^{\infty}$ of polynomials depending only on Ω such that every function $f(z) \in H(\Omega)$ can be expressed as

$$f(z) = \sum_{k=0}^{\infty} a_k p_k(z) \quad (z \in \Omega),$$

where $\{a_k\}_{k=0}^{\infty}$ depends on $f(z)$. The polynomials $\{p_k(z)\}_{k=0}^{\infty}$ are called the *Faber polynomials* of the region Ω.

As an application of the Runge Approximation Theorem 7.9.1, we prove the Mittag-Leffler Theorem 7.3.1 for a general region.

THEOREM 7.9.8 (Mittag-Leffler) *Let $\{a_k\}_{k=1}^{\infty}$ be a sequence of distinct points in a region $\Omega \subset \mathbb{C}$ having no accumulation point in Ω. Then, given a sequence $\{p_k(z)\}_{k=1}^{\infty}$ of polynomials without constant term, there exists a function $f(z)$ meromorphic in Ω having at each point a_k ($k \in \mathbb{N}$) the singular part $p_k\left(\frac{1}{z-a_k}\right)$ and no other singularity.*

Proof. The case $\Omega = \mathbb{C}$ has been established before, hence we assume that $\Omega \neq \mathbb{C}$. Let $d_n = \text{dist}\{a_n, \partial\Omega\}$. If there is a subsequence of $\{d_n\}_{n=1}^{\infty}$ that is bounded away from zero, then the corresponding poles a_n tend to the point at infinity, and so, by the original version of the Mittag-Leffler Theorem 7.1.2, we can construct a function $g(z)$ meromorphic in the complex plane $\mathbb{C}$ (hence also meromorphic in the region Ω) with the prescribed singularities. Hence, the desired meromorphic function in Ω can be obtained by simply adding $g(z)$ to the function that is constructed below. Therefore (by reordering, if necessary), it is sufficient to consider the case that $d_n \downarrow 0$ (as $n \to \infty$). Set

$$\Delta_n = \{z \in \Omega \,;\, |z| \leq n, \ \text{dist}(z, \partial\Omega) \geq 2d_n\}.$$

Then $\{\Delta_n\}_{n=1}^{\infty}$ is a sequence of compact sets in Ω with the properties:

(i) $\Delta_1 \subset \Delta_2 \subset \cdots \subset \Delta_n \subset \cdots$;

(ii) for each n, $\Omega \setminus \Delta_n$ has no relatively compact component in Ω;

(iii) every compact subset K of Ω is contained in some Δ_n;

(iv) $a_k \notin \Delta_n$ for all $k \geq n$.

By the Runge Approximation Theorem 7.9.1, there exist rational functions $R_n(z) \in H(\Omega)$ such that

$$\left| p_n \left(\frac{1}{z - a_n} \right) - R_n(z) \right| < \frac{1}{2^n} \quad (z \in \Delta_n).$$

Then

$$f(z) = \sum_{n=1}^{\infty} \left\{ p_n \left(\frac{1}{z - a_n} \right) - R_n(z) \right\}$$

satisfies the required conditions.   □

Exercises

1. Construct a meromorphic function that has a simple pole
   (a) at $z = n^2$ with residue n for every $n \in \mathbb{N}$;
   (b) at $z = n$ with residue n^2 for every $n \in \mathbb{N}$.

2. (a) Construct a meromorphic function that has a simple pole with residue 1 at each Gaussian integer $\omega_{mn} = m + in$ $(m, n \in \mathbb{Z})$.
   (b) Construct a meromorphic function that has a double pole with singular part $\frac{1}{(z - \omega_{mn})^2}$ at each Gaussian integer ω_{mn}.

3. Does there exist a function analytic in the complex plane $\mathbb{C}$ except at $n \in \mathbb{Z}$, where there is an isolated essential singularity with singular part

$$\sum_{k=1}^{\infty} \frac{1}{k^k (z - n)^k} \ ?$$

4. We have established

$$\frac{\pi^2}{8} = \frac{1}{1^2} + \frac{1}{3^2} + \frac{1}{5^2} + \frac{1}{7^2} + \cdots$$

and

$$\frac{\pi^2}{6} = \frac{1}{1^2} + \frac{1}{2^2} + \frac{1}{3^2} + \frac{1}{4^2} + \cdots$$

separately. Actually, either one implies the other. Prove this assertion.

5. Obtain the equality

$$\pi \cot \pi z = \frac{1}{z} + \sum_{n=1}^{\infty} \left(\frac{1}{z-n} + \frac{1}{z+n} \right),$$

by first evaluating

$$\frac{1}{2\pi i} \int_{\Gamma_R} \frac{\pi \cot \pi \zeta}{\zeta - z} d\zeta, \quad z \in D(0, R),$$

where Γ_R is the square with vertices at $\pm R \pm Ri$ ($R = n + \frac{1}{2}$) and then letting $n \to \infty$.

Hint: At one point, use the equality

$$\int_{\Gamma_R} \frac{\pi \cot \pi \zeta}{\zeta - z} d\zeta = \int_{\Gamma_R} \frac{\pi \cot \pi \zeta}{\zeta} d\zeta + z \int_{\Gamma_R} \frac{\pi \cot \pi \zeta}{\zeta(\zeta - z)} d\zeta$$

and observe that the first integral on the right is zero.

6. Set $z = i$ in

$$\pi \cot \pi z = \frac{1}{z} + 2z \sum_{n=1}^{\infty} \frac{1}{z^2 - n^2}$$

to evaluate

$$\sum_{n=1}^{\infty} \frac{1}{n^2 + 1}.$$

What if we set $z = ai$?

7. Evaluate $\sum_{n=1}^{\infty} \frac{1}{n^2 + a^2}$ ($a > 0$) by considering the integral

$$\int_{\Gamma_n} \frac{\pi \cot \pi z}{z^2 + a^2} dz,$$

where Γ_n are squares with vertices at $\pm \left(n + \frac{1}{2} \right) \pm i \left(n + \frac{1}{2} \right)$.

8. (a) Use the formula

$$\pi \cot \pi z = \frac{1}{z} + {\sum}' \left(\frac{1}{z-n} + \frac{1}{n} \right)$$

to prove the following striking result: For every *even* positive integer n the sum of the series

$$\zeta(n) = \frac{1}{1^n} + \frac{1}{2^n} + \frac{1}{3^n} + \cdots$$

is a rational multiple of π^n.

Hint: Expand both sides of the equation above in power series and match coefficients.

(b) Work out the values of $\zeta(2)$, $\zeta(4)$, $\zeta(6)$, $\zeta(8)$.

(c) Sum the series

$$\frac{1}{1^2} + \frac{1}{3^2} + \frac{1}{5^2} + \cdots$$

$$\frac{1}{1^4} + \frac{1}{3^4} + \frac{1}{5^4} + \cdots$$

$$\frac{1}{1^6} + \frac{1}{3^6} + \frac{1}{5^6} + \cdots$$

$$\frac{1}{1^8} + \frac{1}{3^8} + \frac{1}{5^8} + \cdots$$

9. (a) Show that the series

$$\zeta(s) = \frac{1}{1^s} + \frac{1}{2^s} + \frac{1}{3^s} + \cdots$$

converges absolutely and locally uniformly on the half-plane $\Re s > 1$. (See preceding problem. Of course, k^s is to be interpreted as $e^{s \operatorname{Log} k}$.)

(b) In the half-plane introduced in part (a), show that $\zeta(s)$ can be expressed as the infinite *product* $\prod_p \left(1 - \frac{1}{p^s}\right)^{-1}$, where p ranges over the primes 2, 3, 5, 7, ...

Hint: Use the fact that every integer greater than 1 is either a prime or a product of primes, and that the product is *unique* (except for the order of the factors).

(c) Accepting the fact that π^2 is known to be irrational, use (b) to show that the set of primes is infinite. (Of course, this fact was first proven with less sophisticated knowledge.)

10. Show that

$$f(z) = \sum_{n=-\infty}^{\infty} \frac{\sin z}{|n|!(z - n\pi)}$$

is an entire function, and evaluate

$$\sum_{n=0}^{\infty} f(n\pi).$$

11. Show that

(a)

$$\frac{\pi}{\sin \pi z} = \frac{1}{z} + \sum_{-\infty}^{\infty}{}'(-1)^n \left\{ \frac{1}{z-n} + \frac{1}{n} \right\}.$$

(b)

$$\cos \pi z = \prod_{n=1}^{\infty} \left\{ 1 - \frac{4z^2}{(2n-1)^2} \right\}.$$

12. Obtain a partial fraction expansion of $\frac{1}{e^z - 1}$.

13. Construct an entire function with simple roots at

(a) $a_n = n$ $(n \in \mathbb{N})$;

(b) Gaussian integers $\omega_{mn} = m + in$ $(m, n \in \mathbb{Z})$.

14. Construct an entire function $f(z)$ such that

$$f(n) = n! \quad (n = 0, 1, 2, \ldots).$$

15. Show that, given a function $f(z)$ analytic in an annulus $\{z \in \mathbb{C};\ a < |z| < b\}$, there exists an entire function $g(z)$ such that

$$f(z) = \sum_{n=-\infty}^{\infty} g(n)z^n \quad (a < |z| < b).$$

Is such an entire function $g(z)$ unique?

16. Let $\{a_k\}_{k=1}^{\infty}$ be a sequence of distinct points having no accumulation point in the finite complex plane $\mathbb{C}$, and $\{P_k(z)\}_{k=1}^{\infty}$ an arbitrary sequence of polynomials. Show that there exists an entire function $f(z)$ whose Taylor-series expansion at the point a_k starts with $P_k(z - a_k)$ $(k \in \mathbb{N})$. (See Corollary 7.2.3.)

17. (a) Let $\{a_j\}_{j=1}^{\infty}$ and $\{b_j\}_{j=1}^{\infty}$ be two sequences of distinct points having no finite accumulation point and also no point in common between the two sequences. Let $\{k_j\}_{j=1}^{\infty}$ and $\{m_j\}_{j=1}^{\infty}$ be two arbitrary sequences of positive integers. Find a meromorphic function with root of multiplicity k_j at each a_j and pole of multiplicity m_j at each b_j $(j \in \mathbb{N})$.

(b) Can you strengthen this result by prescribing the singular part at each b_j, instead of just its multiplicity?

(c) Let $\{a_j\}_{j=1}^{\infty}$ and $\{b_j\}_{j=1}^{\infty}$ be as in Part (a), and $\{P_k(z)\}_{k=1}^{\infty}$ and $\{Q_j(z)\}_{j=1}^{\infty}$ two arbitrary sequences of polynomials except that $Q_j(z)$ $(j \in \mathbb{N})$ have no constant terms. Construct a meromorphic function whose Taylor-series expansion at the point a_k starts with $P_k(z - a_k)$ $(k \in \mathbb{N})$ and whose singular part at the point b_j is $Q_j\left(\frac{1}{z-b_j}\right)$ $(j \in \mathbb{N})$.

18. Discuss the convergence of $\prod_{n=1}^{\infty}(1 + c_n)$ and that of $\sum_{n=1}^{\infty} c_n$, where

(a) $c_{2n-1} = \dfrac{1}{\sqrt{n}}, \quad c_{2n} = \dfrac{-1}{\sqrt{n}};$

(b) $c_{2n-1} = \dfrac{-1}{\sqrt{n}}, \quad c_{2n} = \dfrac{1}{\sqrt{n}} + \dfrac{1}{n}.$

19. Discuss the convergence of

(a) $\prod_{k=0}^{\infty} \dfrac{k(k-1)+(1+i)}{k(k-1)+(1-i)};$

(b) $\prod_{k=1}^{\infty} \left\{1 - \dfrac{1}{k(k+3)}\right\};$

(c) $\prod_{k=0}^{\infty} \dfrac{k^3+1}{k^3-i};$

(d) $\prod_{k=2}^{\infty} \left\{\dfrac{k}{\pi} \cdot \sin\dfrac{\pi}{k}\right\};$

20. Let $s_n = \int_0^{\frac{\pi}{2}} \sin^n \theta \, d\theta$. Prove that

$$\lim_{n \to \infty} \frac{s_{2n+1}}{s_{2n}} = 1,$$

and deduce the *Wallis formula*:

$$\frac{2}{\pi} = \prod_{n=1}^{\infty} \left(1 - \frac{1}{(2n)^2}\right).$$

21. Prove Theorem 7.4.9.

22. Find the region of convergence of the infinite products:

(a) $\prod_{k=1}^{\infty} \left(1 + z^k\right);$

(b) $\prod_{k=1}^{\infty} \cos\frac{z}{k};$

(c) $\prod_{k=1}^{\infty} \left\{ \dfrac{k}{z} \cdot \sin \dfrac{z}{k} \right\}$.

(d) $\prod_{k=1}^{\infty} \left(1 - e^{\frac{z}{k^2}} \right)$.

23. (a) Show that the infinite product

$$\prod_{k=1}^{\infty} \cos \frac{z}{2^k}$$

converges locally uniformly in the complex plane $\mathbb{C}$, and evaluate the product.

Hint: Apply the double-angle formula for the sine function repeatedly to

$$\sin \frac{z}{2^n} \cdot \prod_{k=1}^{n} \cos \frac{z}{2^k}.$$

(b) Show that

$$\frac{2}{\pi} = \frac{\sqrt{2}}{2} \cdot \frac{\sqrt{2 + \sqrt{2}}}{2} \cdot \frac{\sqrt{2 + \sqrt{2 + \sqrt{2}}}}{2} \cdots.$$

(c) How about

$$\prod_{k=1}^{\infty} \cosh \frac{z}{2^k} \ ?$$

24. Obtain infinite product representations of $\sinh \pi z$ and $\cosh \pi z$.

25. Show that

$$\pi z \cdot \prod_{k=1}^{\infty} \left\{ \left(1 + \frac{z}{2k-1} \right) \left(1 + \frac{z}{2k} \right) \left(1 - \frac{z}{k} \right) \right\}$$

$$= \sin \pi z \cdot \exp \{ z \mathrm{Log}\, 2 \}.$$

26. Show that

$$\prod_{k=0}^{\infty} \left(1 + z^{2^k} \right) = \frac{1}{1 - z} \quad (z \in D),$$

where the convergence is locally uniform in the unit disc D.

27. Suppose $R > 0$ is the radius of convergence of the power series $\sum_{k=0}^{\infty} a_k z^k$. Show that the infinite product $\prod_{k=0}^{\infty} \left(1 + a_k z^k \right)$ is analytic in $D(0, R)$.

28. Show that the following infinite products define entire functions:

    (a) $\displaystyle\prod_{k=0}^{\infty}\left(1+\frac{z^k}{k!}\right)$;

    (b) $\displaystyle\prod_{k=2}^{\infty}\left\{1+\frac{z}{k(\log k)^2}\right\}$.

29. Show that $f(z) = \prod_{k=1}^{\infty}\left(1-\pi z^{k^2}\right)$ defines an analytic function in the unit disc D. Further show that $\sum_{k=1}^{\infty}(1-|a_k|) = \infty$, but $\sum_{k=1}^{\infty}(1-|a_k|)^2 < \infty$, where $\{a_k\}_{k=1}^{\infty}$ are the roots of $f(z)$.

30. Show that $f(z) = \prod_{k=1}^{\infty}\left(1-\pi z^{k}\right)$ defines an analytic function in the unit disc D. Further show that $\sum_{k=1}^{\infty}(1-|a_k|)^2 = \infty$, but $\sum_{k=1}^{\infty}(1-|a_k|)^3 < \infty$, where $\{a_k\}_{k=1}^{\infty}$ are the roots of $f(z)$.

31. Show that $f(z) = \prod_{k=0}^{\infty}\left(1-a^k z\right)$ $(|a| < 1)$ defines an entire function. Using the identity $f(z) = (1-z)f(az)$, find its Taylor coefficients (around the origin). Show also that the order of this entire function is zero.

32. Suppose $\{a_k\}_{k=1}^{\infty}$ and $\{b_k\}_{k=1}^{\infty}$ are sequences of complex numbers with $\sum_{k=1}^{\infty}|a_k - b_k| < \infty$. Discuss the convergence of

$$\prod_{k=1}^{\infty}\frac{z-a_k}{z-b_k}.$$

33. Prove the Weierstrass Factorization Theorem 7.6.3 for a general region Ω by applying

    (a) the Mittag-Leffler Theorem 7.3.1;

    (b) the Weierstrass primary factors.

    Hint: Given a sequence $\{a_k\}_{k=1}^{\infty}$ of points in the region Ω (having no accumulation point in Ω), choose a sequence of points $\{c_k\}_{k=1}^{\infty}$ on the boundary $\partial\Omega$ satisfying

$$|a_k - c_k| = \inf\{|a_k - z| \;;\; z \in \partial\Omega\},$$

    and determine as precisely as possible the behavior of

$$\prod_{k=1}^{\infty}E_k\left(\frac{a_k - c_k}{z - c_k}\right).$$

34. Give an example of a sequence $\{a_k\}_{k=1}^{\infty}$ of points in the unit disc D for which $\prod_{k=1}^{\infty} |a_k|$ is convergent, yet the infinite product

$$\prod_{k=1}^{\infty} \frac{a_k - z}{1 - \bar{a}_k z}$$

does not converge in the unit disc D.

35. Show that a bounded analytic function in the right half-plane that has a root at each positive integer is identically zero.
    Hint: Use a Möbius transformation to map the right half-plane onto the unit disc.

36. (Ostrowski) Suppose $f(z)$ is analytic in the unit disc D, and $a_1, a_2,$ ..., $a_k,$ ... are the roots of $f(z)$ (repeated as often as their multiplicities). Show that the product $\prod_{k=1}^{\infty} |a_k|$ converges (i.e., $\sum_{k=1}^{\infty} (1 - |a_k|) < \infty$) if and only if the integral

$$\int_0^{2\pi} \text{Log} \, |f(re^{i\theta})| \, d\theta$$

is bounded above as $r \to 1-$. In particular, $\sum_{k=1}^{\infty}(1 - |a_k|) < \infty$ if $f(z)$ belongs to the *Nevanlinna class*; that is, if $f(z)$ is analytic in the unit disc and

$$\int_0^{2\pi} \log^+ |f(re^{i\theta})| \, d\theta$$

is bounded for $0 \le r < 1$, where

$$\log^+ x = \begin{cases} \text{Log} \, x, & (x \ge 1); \\ 0, & (0 \le x < 1). \end{cases}$$

Hint: The Jensen Formula 7.7.1.

37. Give an example of a sequence $\{a_k\}_{k=1}^{\infty}$ of nonzero complex numbers such that $\sum_{k=1}^{\infty} \frac{1}{|a_k|^\sigma} < \infty$ where σ is the exponent of convergence.

38. Show that if $f(z)$ is entire, with nonzero roots $\{a_k\}_{k=1}^{\infty}$ (each repeated according to its multiplicity), and

$$\int_1^{\infty} \frac{\text{Log} \, M(r)}{r^{\lambda+1}} dr < \infty \quad \text{for some } \lambda > 0,$$

then $\sum_{k=1}^{\infty} \frac{1}{|a_k|^\lambda} < \infty$.

39. Suppose $f(z)$ is an entire function of finite order ρ. Show that for any $\epsilon > 0$, there exist positive real numbers A and B such that

$$n(r) \leq Ar^{\rho+\epsilon} + B \quad \text{for all } r \geq 0.$$

40. Suppose $M(r) \leq ce^{r^k}$ or $M(r) \leq e^{cr^k}$ for r sufficiently large, where c is a positive constant. Show that $\rho \leq k$.

41. Let $f(z) = \sum_{k=0}^{\infty} c_k z^k$ be an entire function. Show that the order ρ of the function $f(z)$ is given by

$$\rho = \limsup_{k \to \infty} \left\{ \frac{k \operatorname{Log} k}{\operatorname{Log} \frac{1}{|c_k|}} \right\},$$

where the quantity inside the parentheses { } is considered to be 0 if $c_k = 0$.

Hint: The Cauchy Estimate 4.4.7.

42. Suppose that, for $f(z) = \sum_{k=0}^{\infty} c_k z^k$,

$$\varphi = \limsup_{k \to \infty} \left\{ \frac{\operatorname{Log} k}{\operatorname{Log} \left| \frac{c_k}{c_{k+1}} \right|} \right\} < \infty.$$

Show that $f(z)$ is an entire function, and its order $\rho \leq \varphi$. Moreover, if "lim sup" can be replaced by "lim," then $\rho = \varphi$.

43. Find the order of

(a) $\sum_{k=1}^{\infty} \frac{z^k}{k^k}$;

(b) $\sum_{k=1}^{\infty} \frac{z^k}{2^{k^2}}$.

44. Suppose $f_1(z)$ and $f_2(z)$ are entire functions of order ρ_1 and ρ_2, respectively. Show that the order of the sum $f_1(z) + f_2(z)$ cannot be greater than $\max\{\rho_1, \rho_2\}$. How about $f_1'(z)$, $cf_1(z)$ $(c \neq 0)$, $f_1(z) \cdot f_2(z)$ and the quotient $\frac{f_1(z)}{f_2(z)}$ (assuming that the latter is entire)? Also $(f_1 \circ f_2)(z)$?

45. Let $g(z)$ be a polynomial, and

$$f(z) = e^{g(z)} \cdot z^n \cdot \prod_{k=1}^{\infty} E_m \left(\frac{z}{a_k} \right).$$

Show that $\rho \le \max\{\sigma, \deg g\}$.

46. Let $f(z)$ be an entire function whose only roots are at $a_k = k^\alpha$ ($k \in \mathbb{N}$, $\alpha > 0$). What is the lowest possible order of the function $f(z)$?

47. Show that for any $\lambda \ne 0$ and any polynomial $p(z) \not\equiv 0$, $e^{\lambda z} - p(z)$ has infinitely many roots. Indeed, $e^{\lambda z} - p(z)$ assumes *every* value in $\mathbb{C}$ infinitely often.

48. Let K be a proper closed subset of the unit circle; that is, $K \subset C(0, 1)$. Show that every function continuous on K can be uniformly approximated by polynomials on K.

49. Let Ω_1 and Ω_2 be two open sets in $\mathbb{C}$, and $f(z)$ a function analytic in $\Omega_1 \cap \Omega_2$. Show that there are functions $g_1(z) \in H(\Omega_1)$ and $g_2(z) \in H(\Omega_2)$ such that $f(z) = g_1(z) - g_2(z)$ ($z \in \Omega_1 \cap \Omega_2$).

50. Construct a sequence of functions analytic in the unit disc that converges pointwise to zero, but does not converge locally uniformly in the unit disc.

Chapter 8

Analytic Continuation

Our treatment of analytic continuation will be in the manner the subject was treated 50 to 100 years ago and lacks a certain amount of rigor compared to today's more abstract treatment. However, it is easier to understand in the older context and is therefore better suited to an introductory course in complex analysis. In this chapter we shall use the notation $(f(z), \Omega)$ to express $f(z) \in H(\Omega)$ when it is desired to indicate the domain of definition of the function $f(z)$.

8.1 The Power Series Method

The problem of analytic continuation arises naturally when considering power series expansions.

EXAMPLE. Consider

$$\frac{1}{1-z} = 1 + z + z^2 + \cdots + z^n + \cdots.$$

The right-hand side represents an analytic function in its disc of convergence; that is, in the unit disc D. However, the left-hand side is analytic everywhere in the complex plane $\mathbb{C}$ except at $z = 1$. Thus, it appears that a function analytic globally is represented by a power series only locally.

EXAMPLE. Suppose we are given a power series

$$z + \frac{z^2}{2} + \frac{z^3}{3} + \cdots + \frac{z^n}{n} + \cdots.$$

This power series converges for all z in the closed unit disc except at $z = 1$. We could guess that it too may represent a function that is analytic in a region larger than the unit disc.

These examples lead to the question: When does a power series represent a function that is defined and analytic beyond the disc of convergence? Given a function $f(z) \in H(\Omega_0)$, suppose there exists a function $g(z) \in H(\Omega)$, where $\Omega \supset \Omega_0$, and $g(z) = f(z)$ for $z \in \Omega_0$. Then, by the uniqueness theorem (Corollary 2.7.5), such a function $g(z)$ must be unique in Ω. This uniquely determined function $(g(z), \Omega)$ is called an *analytic continuation* of $(f(z), \Omega_0)$. Thus the analytic continuation can be regarded as an extension of the definition of $f(z)$ from Ω_0 to Ω.

EXAMPLE. Consider

$$f(z) = 1 + z + z^2 + \cdots + z^n + \cdots \quad (z \in D).$$

Then $f(z) = \frac{1}{1-z}$ $(z \in D)$. However, for $z \in D(-1, 2)$,

$$\frac{1}{2}\left\{1 + \left(\frac{z+1}{2}\right) + \left(\frac{z+1}{2}\right)^2 + \cdots + \left(\frac{z+1}{2}\right)^n + \cdots\right\}$$

$$= \frac{1}{2} \cdot \frac{1}{1 - \dfrac{z+1}{2}} = \frac{1}{1-z}.$$

Thus, $\frac{1}{2}\sum_{k=0}^{\infty}\left(\frac{z+1}{2}\right)^k$ is an analytic continuation of $\sum_{k=0}^{\infty} z^k$ in the unit disc D to the disc $D(-1, 2)$.

Now assume we are given a power series $\sum_{k=0}^{\infty} a_k(z-a)^k$ with radius of convergence r_a $(0 < r_a < \infty)$. Then in the disc of convergence $D_a = D(a, r_a)$ it defines an analytic function, $f(z)$. To extend $f(z)$, choose a point b other than a in the disc D_a. Then, because $f(z)$ is analytic at b, it can be expanded into a power series $\sum_{k=0}^{\infty} b_k(z-b)^k$ around b, which converges at least in the disc with center at b and tangent (internally) to the circle $C_a = C(a, r_a)$; that is, in the disc $D(b, r_a - |a - b|)$. However, it may happen that the disc of convergence $D_b = D(b, r_b)$ for this new series extends outside the disc D_a (i.e., it is possible that $r_b > r_a - |a-b|$). In this case, the function $f(z)$ can be continued analytically to the union of the two discs.

Again, choose a point c inside the disc D_b of convergence and reexpand $f(z)$ around the point c. If the disc D_c of convergence breaks out of D_b, then $f(z)$ can be further continued to D_c (Figure 8.1).

Restated: Given an arc Γ : $z = z(t)$ $(0 \leq t \leq 1)$, with $z(0) = a$, $z(1) = b$, and a power series $f(z)$ centered at $z = a$, suppose that, for each point $z(t)$ on the arc Γ, there is a power series $P(z, t)$ satisfying

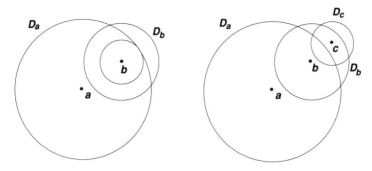

Figure 8.1

1^0 $P(z, 0) = f(z)$;

2^0 for each point $z(t_0)$ on the arc Γ, there exists a positive number $\delta(t_0)$ such that $P(z, t)$ is a reexpansion of $P(z, t_0)$ whenever $|t - t_0| < \delta(t_0)$.

Then we say $f(z)$ can be continued analytically along the arc Γ, and $P(z, 1)$ is called an *analytic continuation of $f(z)$ at the point b along the arc Γ*. Clearly, $f(z) = P(z, 0)$ is the analytic continuation of $P(z, 1)$ at the point a along the arc $-\Gamma$.

THEOREM 8.1.1 *Given a power series $f(z)$ at a point a and an arc Γ joining a to b. Then an analytic continuation of $f(z)$ at the point b along the arc Γ is uniquely determined (provided $f(z)$ can be continued along the arc Γ).*

Proof. Suppose there are two continuations $P(z, t)$ and $Q(z, t)$ where $P(z, 0) = Q(z, 0) = f(z)$. Consider the set

$$\{t \in [0, 1];\ P(z, t) \neq Q(z, t)\}.$$

If this set is empty, then there is nothing to prove. If not, let its infimum be t_0. Suppose $P(z, t_0) = Q(z, t_0)$. Then for all t sufficiently close to t_0 ($t > t_0$), both $P(z, t)$ and $Q(z, t)$ are reexpansions of $P(z, t_0) = Q(z, t_0)$. Hence $P(z, t)$ and $Q(z, t)$ must be the same for such t, which contradicts the definition of t_0. If $P(z, t_0) \neq Q(z, t_0)$, then $t_0 > 0$, and so choose $t_1 \in (0, t_0)$ so that $|z(t_1) - z(t_0)| < \frac{1}{2} \min\{r_P, r_Q\}$, where r_P and r_Q are the radii of convergence of $P(z, t_0)$ and $Q(z, t_0)$, respectively. Then both $P(z, t_0)$ and $Q(z, t_0)$ are reexpansions of $P(z, t_1) = Q(z, t_1)$; hence we must have $P(z, t_0) = Q(z, t_0)$, a contradiction. $\square$

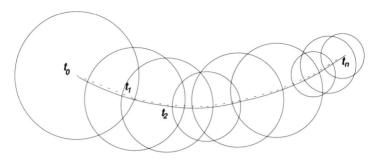

Figure 8.2

Now, suppose $f(z) = P(z, 0)$ can be continued along an arc Γ : $z = z(t)$ $(0 \leq t \leq 1)$, and let $r(t)$ be the radius of convergence of the power series $P(z, t)$. For $t_0 \in [0, 1]$, choose $\delta_0 > 0$ such that

$$|z(t) - z(t_0)| < \frac{r(t_0)}{2} \quad \text{whenever} \quad |t - t_0| < \delta_0 \quad (0 \leq t \leq 1).$$

Then, for such t, $P(z, t)$ and $P(z, t_0)$ are reexpansions of each other, hence

$$r(t) \geq r(t_0) - |z(t_0) - z(t)|, \quad r(t_0) \geq r(t) - |z(t) - z(t_0)|.$$

$$\therefore \; |r(t) - r(t_0)| \leq |z(t) - z(t_0)|.$$

This inequality means that $r(t)$ is a continuous function of $t \in [0, 1]$.

Let $r_0 = \inf\{r(t); 0 \leq t \leq 1\}$. Then, because $r(t)$ is a continuous function on a compact set $[0, 1]$, there must exist $t' \in [0, 1]$ for which $r(t') = r_0$. Because we are assuming that the analytic continuation along the arc Γ is possible, we obtain $r_0 = r(t') > 0$.

Consider a partition

$$0 = t_0 < t_1 < t_2 < \cdots < t_{n-1} < t_n = 1$$

of the interval $[0, 1]$ such that

$$|z(t) - z(t_k)| < r_0 \quad \text{whenever} \; t_{k-1} \leq t \leq t_k \quad (k = 1, 2, \ldots, n).$$

Then $P(z, t_k)$ is a reexpansion of $P(z, t_{k-1})$ (Figure 8.2).

We have shown that the analytic continuation $\{P(z, t)\}_{0 \leq t \leq 1}$ along an arc Γ can be obtained by a finite chain of reexpansions $\{P(z, t_k)\}_{k=0}^{n}$, which, in turn, can be considered an analytic continuation along a broken line having vertices $\{z(t_k)\}_{k=0}^n$ on the arc Γ. Moreover, it is easy to see that the broken line can be replaced by another broken line with rational vertices (except perhaps at one or both endpoints) arbitrarily close to the arc Γ. This argument shows that the set of analytic continuations of a given power series is "*essentially*" countably infinite.

8.2   Natural Boundaries

Starting from a given power series and performing analytic continuation along all possible arcs, we may arrive (in principle) at a stage where it is impossible to extend the original series by any means beyond a certain region. We then say that the boundary of this region constitutes the *natural boundary* of the function.

When we extend a function in the manner described in the preceding section, we get a new (power series) representation in each succeeding region. We speak of these representations as *function elements*; that is, these are analytic functions defined only in a certain region. The underlying function is then the collection of all these elements. Weierstrass spoke of *Funktionelemente* and *Funktion im Grossen.*[1] The Funktion im Grossen is the total collection of function elements obtained by extending as far as possible.

Actually, the above concept of analytic continuation can be carried out independently of power series. Let $f(z)$ and $g(z)$ be functions analytic in the regions Ω_f and Ω_g, respectively. Suppose $\Omega_f \cap \Omega_g \neq \phi$ and $f(z) = g(z)$ for $z \in \Omega_f \cap \Omega_g$. We define

$$
h(z) = \begin{cases} f(z) & (z \in \Omega_f); \\ g(z) & (z \in \Omega_g). \end{cases}
$$

Thus we obtain a function $h(z)$ that is analytic in $\Omega = \Omega_f \cup \Omega_g$. Because $\Omega_f \cap \Omega_g \neq \phi$, we say that the function elements $(f(z), \Omega_f)$ and $(g(z), \Omega_g)$ are *direct* analytic continuations of each other. Two function elements $(f(z), \Omega_f)$ and $(g(z), \Omega_g)$ are said to be *equivalent*, $(f(z), \Omega_f) \sim (g(z), \Omega_g)$, if

(a) there exists a finite chain $\{(f_k(z), \Omega_k)\}_{k=1}^n$ of function elements such that

$$
(f_1(z), \Omega_1) = (f(z), \Omega_f), \quad (f_n(z), \Omega_n) = (g(z), \Omega_g); \quad \text{and}
$$

---

[1] "*im Grossen*" is rendered in English as "*in the large.*"

(b) $(f_{k+1}(z), \Omega_{k+1})$ is a direct analytic continuation of $(f_k(z), \Omega_k)$ $(k = 1, 2, 3, \ldots, n - 1)$.

This relation clearly satisfies the equivalence axioms:

(a) *Reflexivity*: $(f(z), \Omega_f) \sim (f(z), \Omega_f)$;

(b) *Symmetry*: $(f(z), \Omega_f) \sim (g(z), \Omega_g) \Longrightarrow (g(z), \Omega_g) \sim (f(z), \Omega_f)$;

(c) *Transitivity*: $\left.\begin{matrix} (f(z), \Omega_f) \sim (g(z), \Omega_g) \\ (g(z), \Omega_g) \sim (h(z), \Omega_h) \end{matrix}\right\} \Longrightarrow (f(z), \Omega_f) \sim (h(z), \Omega_h).$

This equivalence relation partitions the set of all function elements into equivalence classes. Each equivalence class consists of all function elements equivalent to each other, and is determined by any one of its elements. Thus, we may define a Funktion im Grossen as an equivalence class of function elements.

When a function is continued analytically along an arc Γ, it may happen that we can go up to a point of Γ, but cannot get beyond the point. Such a point is defined to be a singular point of the Funktion im Grossen. On the circle of convergence of a power series there must exist at least one singular point of the function, and such a point is also a singular point of the Funktion im Grossen. However, there are cases in which *every* point on the circle of convergence is a singular point of the function. If this situation occurs, it is impossible to continue the function analytically to a point outside the circle of convergence, and hence the circle is the natural boundary of the function.

To present such an example, we need the following

LEMMA 8.2.1 (Pringsheim) *Suppose the radius of convergence of the power series* $f(z) = \sum_{k=0}^{\infty} c_k z^k$ *is R and that all the coefficients* c_k *are nonnegative real numbers. Then the point* $z = R$ *is a singularity of the function* $f(z)$.

Proof. Without loss of generality we may assume that $R = 1$. Let $z_0 \in D$. Then, by reexpanding $f(z)$ in powers of $(z - z_0)$, $|z_0| = r_0 < 1$ and in powers of $(z - r_0)$, then comparing the corresponding terms in the two expansions, keeping in mind that all the coefficients c_k are nonnegative, we see that if the function $f(z)$ could be continued analytically along the positive real axis beyond the point $z = 1$ then it could be continued along any other radius as well. Now this situation is impossible, for there exists at least one singular point on the circle of convergence.   □

We are now ready to present an example of a power series whose circle of convergence is its natural boundary.

EXAMPLE. Let

$$f(z) = z + z^2 + z^4 + z^8 + \cdots + z^{2^k} + \cdots.$$

Clearly, the circle of convergence is the unit circle. Let m, n be two arbitrary positive integers and consider the function $f(z)$ under the transformation $z = \zeta e^{i\frac{m\pi}{2^n}}$; then

$$f(z) = f\left(\zeta e^{i\frac{m\pi}{2^n}}\right) = \sum_{k=0}^{n} \left(\zeta e^{i\frac{m\pi}{2^n}}\right)^{2^k} + \sum_{k=n+1}^{\infty} \left(\zeta e^{i\frac{m\pi}{2^n}}\right)^{2^k}$$

$$= \sum_{k=0}^{n} \left(\zeta e^{i\frac{m\pi}{2^n}}\right)^{2^k} + \sum_{k=n+1}^{\infty} \zeta^{2^k}.$$

By the above lemma, $\zeta = 1$ is a singular point of the second term on the right, which implies that the points $z = e^{i\frac{m\pi}{2^n}}$ are singular points of $f(z)$. Because such points are dense on the unit circle and the set of singular points is a closed set (why?), we have proven that the unit circle is the natural boundary of this power series.

Actually, the above example is a particular case of the *Hadamard gap theorem*. A sequence $\{\lambda_k\}_{k=1}^{\infty}$ of positive integers is said to be *Hadamard-lacunary*, or simply *lacunary*, if there exists a constant $q > 1$ such that

$$\lambda_{k+1} > q\lambda_k \quad \text{(for all } k \in \mathbb{N}).$$

A power series $\sum_{k=1}^{\infty} a_k z^{\lambda_k}$ is *lacunary* if the sequence $\{\lambda_k\}_{k=1}^{\infty}$ is.

THEOREM 8.2.2 (Hadamard 1892) *A lacunary power series has its circle of convergence as a natural boundary.*
    Proof. (Mordell 1927) Without loss of generality we may assume that the radius of convergence is 1. If the theorem were false, then there would exist a point $z_0 = e^{i\theta}$ such that $f(z) = \sum_{k=1}^{\infty} a_k z^{\lambda_k}$ can be continued analytically to a neighborhood $D(z_0, \epsilon)$ of z_0 for some $\epsilon > 0$. By rotation, this implies that

$$g(z) = f(e^{i\theta} z) = \sum_{k=1}^{\infty} a_k e^{i\lambda_k \theta} z^{\lambda_k}$$

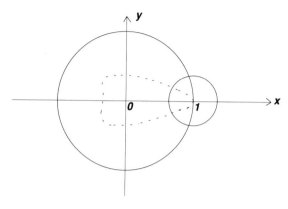

Figure 8.3

can be continued analytically to $D(1, \epsilon)$. Because the power series defining $f(z)$ and $g(z)$ have the same gaps, we may assume from the beginning that $z_0 = 1$.

Now consider the mapping

$$z = \frac{1}{2}\left(w^p + w^{p+1}\right),$$

where p is a positive integer to be chosen later. The closed unit disc $\bar{D}_w$ in the w-plane is mapped to a proper subset of the closed unit disc $\bar{D}_z$ in the z-plane, and the only point in $\bar{D}_w$ that is mapped to the boundary point of $\bar{D}_z$ is the point $w = 1$, which is mapped to $z = 1$ (Figure 8.3). Therefore, given $\epsilon > 0$, there corresponds $r > 1$ such that the image of the disc $D_w(0, r)$ is in $D_z \cup D(1, \epsilon)$ (because the image of $\bar{D}_w$ has a positive distance from the boundary of $D_z \cup D(1, \epsilon)$).

If $w \in D_w$, then $z \in D_z$, and so

$$
\begin{aligned}
F(w) \;=\; f(z) &= \sum_{k=1}^{\infty} a_k \left(\frac{w^p + w^{p+1}}{2}\right)^{\lambda_k} \\
&= \sum_{k=1}^{\infty}\left\{\frac{a_k}{2^{\lambda_k}} \sum_{\lambda=0}^{\lambda_k} \binom{\lambda_k}{\lambda} w^{p(\lambda_k - \lambda)} \cdot w^{(p+1)\lambda}\right\} \\
&= \sum_{k=1}^{\infty}\left\{\frac{a_k}{2^{\lambda_k}} \sum_{\lambda=0}^{\lambda_k} \binom{\lambda_k}{\lambda} w^{p\lambda_k + \lambda}\right\}
\end{aligned}
$$

is analytic in D_w. Recall that we have assumed that there exists a constant $q > 1$ such that $q\lambda_k < \lambda_{k+1}$ (for all $k \in \mathbb{N}$). If we choose p so large that

$\frac{p+1}{p} < q$, then

$$(p+1)\lambda_k < pq\lambda_k < p\lambda_{k+1} \quad \text{for all } k \in \mathbb{N}.$$

Thus the highest power in one of the finite sums is less than the least power in the next finite sum. It follows that for such a value of p every power of w appears at most once. If $f(z)$ can be continued analytically to $D_z(1, \epsilon)$, then the power series for $F(w)$ will have its radius of convergence greater than 1. This implies that the series $F(w)$ is absolutely convergent for some w real and greater than 1, which we call ρ. But then, because there is no duplication of powers,

$$\sum_{k=1}^{\infty} \left\{ \frac{|a_k|}{2^{\lambda_k}} \sum_{\lambda=0}^{\lambda_k} \binom{\lambda_k}{\lambda} \rho^{p\lambda_k + \lambda} \right\} < \infty;$$

that is,

$$\sum_{k=1}^{\infty} |a_k| \cdot \left(\frac{\rho^p + \rho^{p+1}}{2} \right)^{\lambda_k} < \infty.$$

However, because $\frac{1}{2}(\rho^p + \rho^{p+1}) > 1$, this implies that the Taylor expansion of $f(z)$ converges at a point outside the circle of convergence; thus we obtain a contradiction.   □

Remarks. (a) The Hadamard Gap Theorem 8.2.2 can also be proven as a corollary of the Ostrowski theorem on *overconvergence*. A power series is said to be *overconvergent* if it has a subsequence of partial sums converging locally uniformly outside the circle of convergence. Ostrowski showed that a power series "*with sufficiently many gaps*" is overconvergent in a neighborhood of every nonsingular point on the circle of convergence of the power series.

(b) Fabry showed that if $\frac{\lambda_k}{k} \longrightarrow \infty$ $(k \to \infty)$, then the circle of convergence of the power series $\sum_{k=1}^{\infty} a_k z^{\lambda_k}$ is its natural boundary. Note that Fabry's hypothesis is *weaker* than Hadamard's.

In passing, we state without proof two interesting results concerning power series. Let a power series having finite radius of convergence be given. (It must have an infinite number of nonzero coefficients, for otherwise it would be a polynomial and the radius of convergence would be infinite.) We can derive from this power series an infinite family of distinct power series by changing the signs of the coefficients. For example, if

$$f(z) = 1 + z + z^2 + z^3 + \cdots + z^k + \cdots,$$

then

$$f_1(z) = -1 + z + z^2 + z^3 + \cdots + z^k + \cdots,$$
$$f_2(z) = 1 + z - z^2 + z^3 + \cdots + z^k + \cdots,$$
$$f_3(z) = 1 - z - z^2 + z^3 + \cdots + z^k + \cdots,$$
$$\cdots \qquad \cdots \qquad \cdots$$

are all distinct, and we can obviously obtain uncountably many in this manner. All the power series obtained in this way have the same radius of convergence, for the radius of convergence is determined only by the absolute values of the coefficients.

The first result is attributed to Fatou. *Given any power series with finite radius of convergence, it is always possible, by changing signs, to obtain a power series for which the circle of convergence is the natural boundary.*

The second result is attributed to Pólya. *Given any power series with a finite radius of convergence, form the derived family of power series by making all possible sign changes. Then, with probability 1, an arbitrary power series selected from this family has the circle of convergence as its natural boundary.*

In concluding this section we prove the following theorem mentioned at the end of Section 7.3.

THEOREM 8.2.3 *For any region $\Omega \subset \hat{\mathbb{C}}$ there exists a function $f(z) \in H(\Omega)$ having the boundary $\partial\Omega$ of the region Ω as its natural boundary.*

Proof. If the region Ω has no boundary; that is, when $\Omega = \hat{\mathbb{C}}$, then clearly the only analytic functions in Ω are the constant functions. Therefore, we may assume that $\partial\Omega \neq \phi$. Choose a sequence $\{a_k\}_{k=1}^{\infty}$ of points in the region Ω such that every point on the boundary $\partial\Omega$, but *no* point of Ω, is an accumulation point of this sequence. This can be achieved as follows: By a simple application of a Möbius transformation (if necessary), we may assume, without loss of generality, that the point at infinity is not on the boundary $\partial\Omega$. Then $\partial\Omega$ is a closed bounded set. Thus, by the Heine-Borel theorem, for each $n \in \mathbb{N}$ there must exist a finite number of discs with radius $\frac{1}{n}$ and centers on the boundary $\partial\Omega$ that cover $\partial\Omega$. Now from each covering disc choose a point in the intersection of the disc and the region Ω. As n ranges over $\mathbb{N}$ we obtain a countable set of points in Ω such that every point on the boundary $\partial\Omega$ is an accumulation point of this countable set of points in Ω.

By the Weierstrass Theorem 7.3.2, there exists a function $f(z) \in H(\Omega)$ having roots precisely at $\{a_k\}_{k=1}^{\infty}$. (Because, in the construction described above, each point appearing in the sequence $\{a_k\}_{k=1}^{\infty}$ appears only *finitely* many times, we may prescribe that at each of these points $f(z)$ has a root whose order equals the number of times this point appears; or we could choose distinct points from the beginning.) If $f(z)$ could be continued analytically to include a point on the boundary, then $f(z)$ has roots that accumulate in the region of analyticity, and so we must have $f(z) \equiv 0$; but $f(z)$ was supposed to vanish only at $\{a_k\}_{k=1}^{\infty}$; this contradiction establishes the theorem.   □

Remark. This theorem is in sharp contrast to the case of several complex variables, where there exist pairs of regions, Ω_0 and Ω ($\Omega_0 \subset \Omega, \Omega_0 \neq \Omega$), with the property that every function analytic in Ω_0 is automatically analytic in Ω.

8.3   Multiple-Valued Functions

So far we have dealt only with single-valued functions. However, analytic continuation may lead to a loss of single-valuedness. For example, let a function $f(z)$ be continued from a point a to a point b along arcs Γ and Γ' to regions Ω and Ω', respectively. If the union $\Omega \cup \Omega'$ is not simply connected, then it is possible that the values of the function $f(z)$ at the point b do not agree (Figure 8.4). Alternatively, we can express the same idea without specifically mentioning analytic continuation, in the following way. Let Ω, Ω_1, Ω_2 be three regions, where $\Omega \subset \Omega_1 \cap \Omega_2$, and let $f(z)$, $f_1(z)$, $f_2(z)$ be defined and analytic in Ω, Ω_1, Ω_2, respectively. If $f(z) \equiv f_1(z) \equiv f_2(z)$ everywhere in Ω, can we conclude that $f_1(z) \equiv f_2(z)$ everywhere in $\Omega_1 \cap \Omega_2$? The answer is "No."

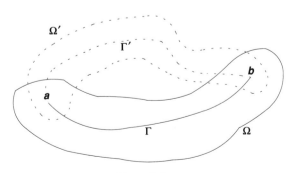

Figure 8.4

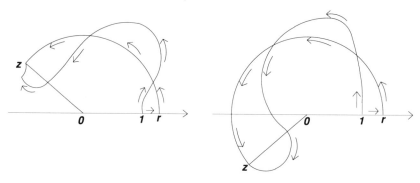

Figure 8.5

EXAMPLE. Consider the definite integral

$$\log z = \int_1^z \frac{d\zeta}{\zeta}.$$

This equality holds if z is a positive real number. Because the integrand $\frac{1}{\zeta}$ is analytic except at the origin, the integral defines an analytic function of z in a simply connected region Ω containing the point 1 but not the origin; viz., the integral $\int_1^z \frac{d\zeta}{\zeta}$ is the unique analytic continuation of $\log z$ ($z \in \mathbb{R}_+$) to such a region Ω.

Thus, if we cut the complex plane along the negative real axis and let Ω be the region so obtained, then the analytic continuation of $\log z$ ($z \in \mathbb{R}_+$) is the principal value $\mathrm{Log}\, z$. In fact, let Γ be a rectifiable arc joining 1 to $z = re^{i\theta}$ ($-\pi < \theta < \pi$) in the region Ω. Then

$$\int_\Gamma \frac{d\zeta}{\zeta} = \int_{\Gamma_1} \frac{d\zeta}{\zeta} + \int_{\Gamma_2} \frac{d\zeta}{\zeta},$$

where Γ_1 is the segment of the real axis joining 1 to r, and Γ_2 is the circular arc from r to z (Figure 8.5).

$$\therefore \int_\Gamma \frac{d\zeta}{\zeta} = \int_1^r \frac{dx}{x} + i \int_0^\theta dt = \mathrm{Log}\, r + i\theta$$
$$= \mathrm{Log}\, |z| + i\mathrm{Arg}\, z \qquad (-\pi < \mathrm{Arg}\, z \le \pi)$$
$$= \mathrm{Log}\, z.$$

We can certainly continue the integral across the cut as in Figure 8.5 on the right; then

$$\log z = \int_\Gamma \frac{d\zeta}{\zeta} = \int_1^r \frac{dx}{x} + i \int_0^\theta dt = \mathrm{Log}\, r + i\theta.$$

However, in this case $\theta > \pi$, and so this is not the principal value of $\log z$. Because the imaginary part of the principal value is $(\theta - 2\pi)$, we have $\log z = \text{Log } z + 2\pi i$. If z continues to encircles the origin, so that $(2n - 1)\pi < \theta \le (2n + 1)\pi$, then we obtain $\log z = \text{Log } z + 2n\pi i$. Similarly, if z encircles the origin in the negative sense (i.e., clockwise) and crosses the cut n times, then $\log z = \text{Log } z - 2n\pi i$. In general, if we determine the value of $\log z_0$ at the point z_0 inside a simply connected region Ω not containing the origin, then $\log z$ becomes a single-valued analytic function, and we call it a *branch* of $\log z$ (in Ω). If the region Ω contains a punctured neighborhood of the origin, then, depending on whether the point z encircles the origin in the positive or the negative sense, the value of $\log z$ changes by an integer multiple of $2\pi i$, and so we see that $\log z$ is not a single-valued function in a neighborhood of the origin; the origin is a *branch point* of $\log z$. Note that to obtain a single-valued branch of $\log z$ it suffices to cut along any simple arc joining the origin and the point at infinity.

EXAMPLE. Powers z^α of z ($z \ne 0$, $\alpha \in \mathbb{C}$) are defined by

$$z^\alpha = e^{\alpha \log z}.$$

Due to the multiple-valuedness of $\log z$, this is generally a multiple-valued function. However, if we restrict $\log z$ to its principal value, then a branch of z^α that we call the *principal value* of z^α is uniquely determined. If α is real, then the principal value of z^α is an analytic continuation of z^α ($z \in \mathbb{R}_+$). In general, because $\log z = \text{Log } z + 2n\pi i$, we have

$$z^\alpha = e^{\alpha \text{Log } z} \cdot e^{2n\alpha\pi i} = e^{\alpha \text{Log } z} \cdot \left(e^{2\alpha\pi i}\right)^n.$$

Thus, the factor $e^{2\alpha\pi i}$ is the cause of the multiple-valuedness. If z moves around the origin once and returns to the starting point, then the value of z^α is multiplied by the factor $e^{2\alpha\pi i}$. The origin is called a *branch point* of z^α. For example, the principal value of i^i is

$$i^i = e^{i\frac{\pi}{2}i} = e^{-\frac{\pi}{2}},$$

and, in general,

$$i^i = e^{-\frac{\pi}{2}} \cdot \left(e^{2i\pi i}\right)^n = e^{-\frac{\pi}{2}(1+4n)}.$$

Of course, if α is an integer, then $e^{2\alpha\pi i} = 1$, so the function z^α is single-valued; and if $\alpha = \frac{p}{q}$ is rational ($p, q \in \mathbb{Z}$, $q > 0$, $(p, q) = 1$), then z^α is q-valued.

For example, consider the case $\alpha = \frac{1}{2}$. If z moves around the origin once and returns to the starting point, the value of $z^{\frac{1}{2}}$ is multiplied by the factor $e^{\pi i} = -1$; that is, it changes sign, and if z moves around the origin once more, the value of $z^{\frac{1}{2}}$ returns to the original value. Thus $z^{\frac{1}{2}}$ is a double-valued function; in particular, for each real z ($\neq 0$), there are two values of $z^{\frac{1}{2}}$; for example, $\sqrt{1} = \pm 1$. The usual convention employed in elementary courses that $\sqrt{x} > 0$ if $x > 0$ does not bring us happiness (= analyticity).

If $z_0 \neq 0$, then z^α is analytic in a neighborhood of z_0, and we have (as expected)

$$\frac{dz^\alpha}{dz} = \frac{d}{dz} e^{\alpha \log z} = e^{\alpha \log z} \cdot \frac{\alpha}{z} = \alpha z^{\alpha - 1}.$$

Naturally, $\log z$ in $z^{\alpha - 1} = e^{(\alpha - 1) \log z}$ is the same as the one in $z^\alpha = e^{\alpha \log z}$. Similarly,

$$\frac{d^n z^\alpha}{dz^n} = \alpha(\alpha - 1) \cdots (\alpha - n + 1) z^{\alpha - n}.$$

As we mentioned earlier, an analytic function (*Funktion im Grossen*) is uniquely determined, independent of the choice of a representative function element. For any point in the complex plane $\mathbb{C}$, if there are at most n distinct values of a function $f(z)$ (and there are actually n distinct values at some point), then the function is said to be n-valued. We ask: How large can the cardinal number n be? The answer is given by the following

THEOREM 8.3.1 (Poincaré-Volterra) *An analytic function can take at most countably many distinct values at any point in the complex plane $\mathbb{C}$.*

Proof. Let $P(z, a)$ be a power series representing the analytic function $f(z)$ in a neighborhood of a point a. Then any power series $P(z, b)$ representing the analytic function $f(z)$ at a point b can be obtained from $P(z, a)$ along a broken line with vertices at $\{a_1, a_2, \ldots, a_n\}$ ($a_1 = a$, $a_n = b$), where $\{a_k\}_{k=2}^{n-1}$ are suitably chosen *rational* points. (Cf. the concluding remark in Section 8.1.) Thus, to every function element $P(z, b)$ there corresponds a finite ordered set of rational points. Clearly, no two distinct function elements correspond to the same finite ordered sets of rational points. Because the set of all rational points is countable, the collection of all finite ordered sets of rational points is countable. Therefore, there can be at most countably many distinct function elements at the point b. $\square$

8.4   Riemann Surfaces

From the examples above, we know that multiple-valuedness is an intrinsic property of analytic functions, and to understand it we used two devices. The first is the so-called *branch-cut method*, which gives a simply connected region in which the function under consideration is single-valued. The second is to observe the manner in which the value of the function varies along certain arcs. This method provides a way to see how different branches of the function are connected with each other and suggests an idea that leads to the *Riemann surface* of a function.

We take the function $f(z) = \sqrt[3]{z}$ to illustrate the *scissors-and-paste* construction of a Riemann surface. This function has three different values for each z (except for $z = 0$). However, in a simply connected region not containing the origin, its value is unambiguously determined once a value is given at any one point in the region. For example, consider the region obtained by cutting along the negative real axis (although any cut along a simple arc joining the origin to the point at infinity will do). This is a simply connected region, and the value of the function in this region is completely determined once we choose $\sqrt[3]{1} = 1, e^{\frac{2\pi i}{3}}$ or $e^{\frac{4\pi i}{3}}$. Each of these three "branches" is single-valued if z is restricted to this region. However, there is really no bad behavior on the cut except at $z = 0$, and in fact, as z moves continuously across the cut from the second quadrant to the third we have $f(-1) = e^{i\frac{\pi}{3}}$ (assuming $f(1) = 1$) and $f(-1 - \epsilon i) \approx e^{i\frac{\pi}{3}}$, but had we moved around clockwise we would have had $f(-1) = e^{-i\frac{\pi}{3}}$ and $f(-1 - \epsilon i) \approx e^{-i\frac{\pi}{3}}$.

Now, to each of the three branches of $f(z)$, Riemann assigned a replica of the cut plane in which the function is single-valued. Each of these replicas is called a *sheet* and they are ordered according to the value of $f(1)$—a replica of the cut plane is the kth sheet if $f(1) = e^{\frac{2k\pi i}{3}}$ ($k = 0, 1, 2$) (Figure 8.6).

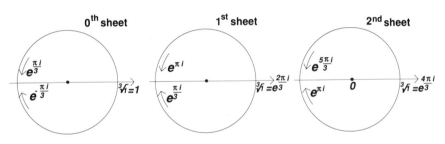

Figure 8.6

As we have just explained, on the 0th sheet of the cut plane, if z starts from $z = 1$ and moves along the unit circle in the positive sense to -1, then we obtain $\sqrt[3]{-1} = e^{\frac{\pi i}{3}}$; if, instead, z moves along the unit circle in the negative sense, then $\sqrt[3]{-1} = e^{-\frac{\pi i}{3}}$.

Similarly, on the 1st sheet, because $\sqrt[3]{1} = e^{\frac{2\pi i}{3}}$, we have

$$\sqrt[3]{-1} = e^{\pi i} \quad \text{on the upper edge of the cut, and}$$
$$\sqrt[3]{-1} = e^{\frac{\pi i}{3}} \quad \text{on the lower edge of the cut.}$$

Finally, on the 2nd sheet, because $\sqrt[3]{1} = e^{\frac{4\pi i}{3}}$, we have

$$\sqrt[3]{-1} = e^{\frac{5\pi i}{3}} \quad \text{on the upper edge of the cut, and}$$
$$\sqrt[3]{-1} = e^{\pi i} \quad \text{on the lower edge of the cut.}$$

These observations suggest that we stack the sheets together so that the 1st sheet is immediately above the 0th sheet and the 2nd sheet above the 1st one. Then the upper edge of the cut in the 0th sheet is *"pasted"* together with the lower edge of the cut in the 1st sheet; the upper edge of the cut in the 1st sheet with the lower edge of the cut of the 2nd sheet; and finally the upper edge of the cut in the 2nd sheet is pasted together with the lower edge of the cut in the 0th sheet—because $e^{\frac{5\pi i}{3}} = e^{-\frac{\pi i}{3}}$. Of course the last step is impossible without self-intersection in our physical world. This fact, however, in no way affects the validity of our construction of the idealized model. The result of the construction is the *Riemann surface*[2] on which the function $f(z) = \sqrt[3]{z}$ is single-valued and continuous, which is to be regarded as the true domain of definition of the function $z^{\frac{1}{3}}$.

Unlike all the other points, each of which is on only one sheet (assuming that the negative real numbers belong to the upper edge, say), the origin is on all the sheets of the surface, because the origin is the cause of the difficulty—if we make a circuit around the origin, we arrive at new values but this occurs for no other point in the (finite) plane. Such a point is called a *branch point*. If it belongs to k sheets ($k > 1$), then we say it is a branch point of *order* $k - 1$.

The Riemann surface of $f(z) = \log z$ can be constructed in a similar way, the sole difference being that it now requires an infinite number of sheets, and the origin is a branch point of order ∞.

---

[2] Once again we emphasize that the *"scissors-and-paste"* idea is not mathematically rigorous, and it has been modified to meet current standards of rigor and abstraction. However, we believe that the method which we have sketched is preferable for a beginning course. (After all, it was good enough for Riemann!)

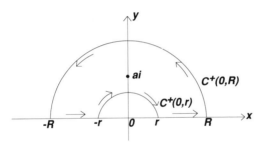

Figure 8.7

It is also possible to have more than one branch point. For example, the function $f(z) = \sqrt{z(z-1)}$ has branch points at $z = 0$ and at $z = 1$. The Riemann surface of this function can be constructed by taking two copies of the complex plane that are cut along the line segment joining the origin to the point $z = 1$ and pasting the upper edge of the cut in each of the sheets to the lower edge of the cut in the other sheet.

Even for a multiple-valued function, the Cauchy Residue Theorem 5.4.1 can be applied to a single-valued branch (or rather, on the Riemann surface). We give two examples.

EXAMPLE.
$$\int_0^\infty \frac{\log x}{x^2 + a^2}\,dx = \frac{\pi}{2a}\log a \quad (a > 0).$$

The function $f(z) = \frac{\log z}{z^2 + a^2}$ has a branch point at the origin, but if we choose the contour as in Figure 8.7, then the contour lies on a sheet and the function is single-valued there. There is only one singularity inside the contour, a simple pole at $z = ai$. The residue there is
$$\frac{\log(ai)}{2ai} = \frac{1}{2ai}\left(\log a + i\frac{\pi}{2}\right).$$

Therefore, we have
$$\int_r^R + \int_{C^+(0,R)} + \int_{-R}^{-r} - \int_{C^+(0,r)} = \frac{\pi}{a}\left(\log a + i\frac{\pi}{2}\right).$$

On $C^+(0,\ R)$, we have $z = Re^{i\theta}$, and so
$$\left|\int_{C^+(0,R)} \frac{\log z}{z^2 + a^2}\,dz\right| = \left|\int_0^\pi \frac{\log R + i\theta}{(Re^{i\theta})^2 + a^2} \cdot iRe^{i\theta}\,d\theta\right|$$
$$\leq \frac{\pi R(\log R + \pi)}{R^2 - a^2} \longrightarrow 0 \quad (R \to \infty).$$

Similarly, on $C^+(0, r)$ we have

$$\left| \int_{C^+(0,r)} \frac{\log z}{z^2 + a^2} dz \right| \le \frac{\pi r(\log r + \pi)}{a^2 - r^2} \longrightarrow 0 \quad (r \to 0).$$

Now, for the interval $[r, R]$ on the positive real axis we have $z = x$ $(r \le x \le R)$, and so

$$\int_r^R \frac{\log z}{z^2 + a^2} dz = \int_r^R \frac{\log x}{x^2 + a^2} dx;$$

but for the interval $[-R, -r]$ on the negative real axis, we have $z = x e^{i\pi}$ $(R \ge x \ge r)$, $dz = e^{i\pi} dx = -dx$, and so

$$\int_{-R}^{-r} \frac{\log z}{z^2 + a^2} dz = - \int_R^r \frac{(\log x) + i\pi}{x^2 + a^2} dx = \int_r^R \frac{(\log x) + i\pi}{x^2 + a^2} dx.$$

Thus, letting $R \to \infty, r \to 0$, we obtain

$$\int_0^\infty \frac{2(\log x) + i\pi}{x^2 + a^2} dx = \frac{\pi}{a} \left(\log a + i\frac{\pi}{2} \right).$$

Taking the real part of both sides, we obtain the desired result.

Remark. The case $a = 1$:

$$\int_0^\infty \frac{\log x}{x^2 + 1} dx = 0$$

is an easy exercise in calculus, from which the general case follows immediately. Note that for any $\alpha > 0$, we have

$$\lim_{x \to \infty} \frac{\log x}{x^\alpha} = 0 \quad \text{and} \quad \lim_{x \to 0+} x^\alpha \log x = 0;$$

thus the convergence of the improper integral is obvious. (The case $\alpha = 1$ is used in the above computation.)

EXAMPLE.

$$\int_0^\infty \frac{x^{a-1}}{1 + x} dx = \frac{\pi}{\sin a\pi} \quad (0 < a < 1).$$

The function

$$f(z) = \frac{z^{a-1}}{1 + z} = \frac{e^{(a-1)\log z}}{1 + z}$$

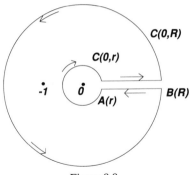

Figure 8.8

has a branch point at the origin, but in the annulus

$$\{z \in \mathbb{C};\ r < |z| < R\} \quad (0 < r < 1 < R)$$

with a slit along the interval $[r,\ R]$ on the positive real axis, it is single-valued and has only one singularity, which is a simple pole at $z = -1$ (Figure 8.8).

$$\left(\int_{AB} + \int_{C(0,R)} + \int_{BA} - \int_{C(0,r)}\right) \frac{z^{a-1}}{1+z}dz = 2\pi i \cdot e^{(a-1)\pi i} = -2\pi i e^{a\pi i}.$$

We have

$$\left|\int_{C(0,R)}\right| \leq \frac{R^{a-1}}{R-1} \cdot 2\pi R \longrightarrow 0 \quad (R \to \infty,\ \because a < 1),$$

and

$$\left|\int_{C(0,r)}\right| \leq \frac{r^{a-1}}{1-r} \cdot 2\pi r \longrightarrow 0 \quad (r \to 0,\ \because a > 0).$$

On BA, we have $z^{a-1} = e^{(a-1)(\log x + 2\pi i)}$ $(R \geq x \geq r)$, and so

$$\int_{AB} + \int_{BA} = \left(1 - e^{(a-1)2\pi i}\right) \int_r^R \frac{x^{a-1}}{1+x}dx.$$

$$\therefore \int_0^\infty \frac{x^{a-1}}{1+x}dx = \frac{-2\pi i e^{a\pi i}}{1 - e^{2a\pi i}} = \frac{2i}{e^{a\pi i} - e^{-a\pi i}} \cdot \pi = \frac{\pi}{\sin a\pi}.$$

Remark. This integral is $B(a,\ 1-a)$, where

$$B(p,\ q) = \int_0^1 x^{p-1}(1-x)^{q-1}dx \quad (p > 0,\ q > 0)$$

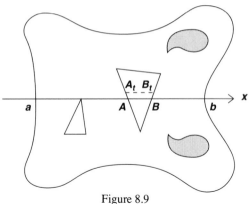

Figure 8.9

is the *Euler beta function*. (See Section 8.8, *The Euler Gamma Function*.)

8.5   The Schwarz Symmetry Principle

There are ways other than using power series to continue a function analytically. Here we mention the important *Schwarz symmetry (reflection) principle*. Our discussion of the Euler gamma function (Section 8.8) provides an example of another method, that of using a functional equation.

THEOREM 8.5.1 (Schwarz) *Let Ω be a region above the real axis having an interval (a, b) on the real axis as part of its boundary. Suppose that $f(z) \in H(\Omega)$ has real continuous boundary values on the interval (a, b), so that $f(z)$ is continuous in $\Omega \cup (a, b)$. Then $f(z)$ can be continued analytically across the interval (a, b) into the region Ω' symmetric to Ω with respect to the real axis by setting*

$$f(z) = \overline{f(\bar{z})} \quad (z \in \Omega').$$

Proof. Let

$$g(z) = \begin{cases} f(z), & z \in \Omega \cup (a, b); \\ \\ \overline{f(\bar{z})}, & z \in \Omega'. \end{cases}$$

The assumptions on $f(z)$ imply that $g(z)$ is continuous in $\Omega \cup \Omega' \cup (a, b)$ (Figure 8.9). By the Morera Theorem 4.5.4, it is sufficient to prove that

$$\int_{\Gamma} g(z) \, dz = 0$$

for every boundary Γ of a triangle such that $[\Gamma] \subset \Omega \cup \Omega' \cup (a, b)$. If Γ is in Ω, then the vanishing of the integral follows from the Cauchy-Goursat Theorem 4.3.1. If Γ is in Ω', then, by changing the variable $z \rightsquigarrow \bar{z}$ we obtain the mirror image Γ' (of Γ) in Ω; hence

$$\int_\Gamma g(z)\,dz = \int_\Gamma \overline{f(\bar{z})}\,dz = \int_{\Gamma'} \overline{f(z)}\,d\bar{z} = \overline{\int_{\Gamma'} f(z)\,dz} = 0,$$

again by the Cauchy-Goursat Theorem 4.3.1. Thus $g(z)$ is analytic in Ω'. If the boundary Γ of the triangle $[\Gamma]$ intersects the interval (a, b) at two points A and B (the case that the triangle has one side on the interval (a, b), or that Γ intersects (a, b) at only one point can be treated similarly), then

$$\int_\Gamma = \int_{\Gamma_\Omega} + \int_{\Gamma_{\Omega'}},$$

where Γ_Ω and $\Gamma_{\Omega'}$ are portions of Γ that are in Ω and Ω', respectively, plus the line segment $[A, B]$ with appropriate sense in each case. Let Γ_ϵ be the portion of the boundary of the triangle $[\Gamma]$ that is above the line $y = \epsilon$ ($\epsilon > 0$) plus the line segment $A_\epsilon B_\epsilon$, where A_ϵ and B_ϵ are the intersections of Γ with the line $y = \epsilon$. By the Cauchy-Goursat Theorem 4.3.1,

$$\int_{\Gamma_\epsilon} g(z)\,dz = 0.$$

Now, because $g(z)\,(= f(z))$ is uniformly continuous on the compact set $[\Gamma_\Omega]$, the left-hand side is a continuous function of ϵ, and as $\epsilon \to 0$ we obtain

$$\int_{\Gamma_\Omega} g(z)\,dz = \lim_{\epsilon \to 0} \int_{\Gamma_\epsilon} g(z)\,dz = 0.$$

Similarly,

$$\int_{\Gamma_{\Omega'}} g(z)\,dz = 0. \quad \therefore \quad \int_\Gamma g(z)\,dz = 0. \qquad \square$$

Note that the above argument can be employed to prove the following result.

THEOREM 8.5.2 (Painlevé) *Suppose Ω_1 and Ω_2 are two regions such that $\Omega_1 \cap \Omega_2 = \phi$ and $\bar{\Omega}_1 \cap \bar{\Omega}_2 = \Gamma$ is a simple rectifiable arc (Fig-*

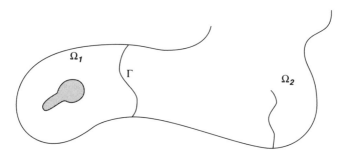

Figure 8.10

*ure 8.10). Let $f_k(z)$ be a function analytic on Ω_k and continuous on $\Omega_k \cup \Gamma$
$(k = 1, 2)$. If $f_1(z) = f_2(z)$ for $z \in \Gamma$, then*

$$g(z) = \begin{cases} f_1(z), & z \in \Omega_1 \cup \Gamma; \\ f_2(z), & z \in \Omega_2, \end{cases}$$

is a function analytic on $\Omega_1 \cup \Omega_2 \cup \Gamma$.

8.6   The Monodromy Theorem

Next, we sketch a proof of the following important result.

THEOREM 8.6.1 *Let Ω be a simply connected region. Suppose a function $f(z)$ defined and analytic on a subregion of Ω can be continued analytically throughout Ω. Then all analytic continuations of the original function element lead to the same result.*

Proof. We have to show that, given a power-series expansion of $f(z)$ centered at a and two arcs Γ_1 and Γ_2 (in Ω) joining a to b, the analytic continuations along Γ_1 and Γ_2 give the same results at b. (We may assume that the original function element and all its successors are power series, even though in practice other types of representations may be employed.) But this is equivalent to the assertion that the analytic continuation of $f(z)$ along any (closed) curve Γ in Ω starting and terminating at the same point results in the same function we started with. Furthermore, by the observation at the end of Section 8.1, the curve Γ may be taken as a polygon. Because a polygon can be triangulated, it is sufficient to consider the case in which Γ is the boundary of a triangle (Figure 8.11).

Suppose the analytic continuation of $f(z)$ along the boundary of a triangle $[\Gamma_0]$ does not result in the same function $f(z)$ we started with; then, by dividing $[\Gamma_0]$ into four congruent triangles we find that there is at least one

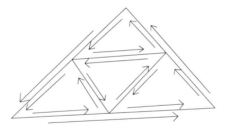

Figure 8.11

triangle $[\Gamma_1]$ with the same property as $[\Gamma_0]$. Now, divide $[\Gamma_1]$ further into four congruent triangles and choose a triangle $[\Gamma_2]$ with the same property. Repeating this procedure, we obtain a sequence $\{[\Gamma_k]\}_{k=0}^{\infty}$ of nested triangles. By the Cantor Intersection Theorem 2.3.2, this sequence of triangles converges to a point $z_0 \in [\Gamma_0] \subset \Omega$. Consider the power-series expansion $P(z, z_0)$ around z_0. For n sufficiently large, the triangle $[\Gamma_n]$ is entirely in the circle of convergence of $P(z, z_0)$. Then each analytic continuation along Γ_n is a reexpansion of $P(z, z_0)$, and so the analytic continuation along Γ_n and back to the initial point results in the same function we started with, which contradicts the construction of Γ_n.   □

Remark. Even if the region is not simply connected, the results of analytic continuations along two different arcs in Ω are the same if the paths of continuations are *homotopic* in Ω; that is, if the paths can be deformed continuously into each other *within* Ω and with the endpoints held fixed.

8.7   Law of Permanence

Recall that we defined the exponential function by the equation

$$e^z = 1 + \frac{z}{1!} + \frac{z^2}{2!} + \frac{z^3}{3!} + \cdots + \frac{z^k}{k!} + \cdots,$$

which is simply a formal extension of the function from the real domain to the complex domain. However, from our discussion it is seen that there is no other way to extend the exponential function analytically. The same is true for $\sin z$, $\cos z$, and so on.

We now consider a slightly more intricate example of analytic continuation. If z and w are real numbers, then it is known that

$$\sin(z + w) = \sin z \cdot \cos w + \cos z \cdot \sin w.$$

We also know that there exist entire functions—also called *sine* and *cosine*—that coincide with the real trigonometric functions sin z and cos z (on $\mathbb{R}$). If w is a real constant, then the two sides of the equality above are both entire functions of z that coincide for real values of z, hence they must be identical; that is, the equality holds for all $z \in \mathbb{C}$ and $w \in \mathbb{R}$. If we now keep z fixed and vary w, the same reasoning extends the validity of the equality to arbitrary complex values z and w.

The reasoning employed in this case may be considered an informal proof of the following theorem.

Law of Permanence. *An analytic relationship that holds in a subregion of the region Ω holds everywhere in Ω (assuming that all the functions involved in the relationship are defined and analytic in Ω).*

In fact, "*in a subregion*" may be replaced by "*on an arc.*"

8.8   The Euler Gamma Function

The *Euler integral of the second kind* is defined by the equation

$$\Gamma(x) = \int_0^\infty e^{-t} t^{x-1} \, dx.$$

The reader has probably encountered this integral in an elementary calculus course, the parameter x being considered real and positive; we shall begin with this restriction, but later we shall (of course!) replace x by the complex parameter $z = x + iy$. First we show that the integral is convergent for all values of x (satisfying the above restriction) and defines $\Gamma(x)$ as a continuous function of x. Let x_0 and x_1 be two arbitrary real numbers, $x_0 < x_1$. Then for all x in the interval $[x_0, x_1]$ the integrand is dominated by t^{x_0-1} when $0 < t \leq 1$ and by $e^{-t} t^{x_1-1}$ when $t > 1$. Since the dominating function is integrable from 0 to ∞, the existence of $\Gamma(x)$ is established and the continuous dependence of $\Gamma(x)$ on x now follows from the integral analogue of the Weierstrass M-test 2.4.5 (or the Lebesgue dominated convergence theorem). Since x_0 and x_1 are arbitrary, $\Gamma(x)$ is continuous for all positive x.

Now, if in the above argument we replace the real variable x by the complex variable z and interpret $t^{z-1} = e^{(z-1)\log t}$ as the principal value, then, because

$$\left| e^{-t} t^{z-1} \right| = e^{-t} t^{x-1} \quad (\Re z = x),$$

we obtain the local uniform convergence of the improper integral

$$\Gamma(z) = \int_0^\infty e^{-t} t^{z-1} \, dt$$

in the right half-plane $\Re z > 0$, and so $\Gamma(z)$ is continuous there. In fact, more is true—$\Gamma(z)$ is analytic in the right half-plane $\Re z > 0$. To prove this assertion, we appeal to the Morera Theorem 4.5.4. Let C be a curve in the right half-plane; then

$$\int_C \Gamma(z) \, dz = \int_C dz \int_0^\infty e^{-t} t^{z-1} \, dt = \int_0^\infty e^{-t} dt \int_C t^{z-1} \, dz.$$

Change of the order of integration is justified by the absolute convergence of the iterated integral. Now, for each fixed t the function $t^{z-1} = e^{(z-1)\operatorname{Log} t}$ is analytic for z in the right half-plane $\Re z > 0$. Therefore, by the Cauchy-Goursat Theorem 4.3.1,

$$\int_C t^{z-1} \, dz = 0 \quad \text{and hence} \quad \int_C \Gamma(z) \, dz = 0.$$

Thus we have established that the function $\Gamma(z)$ is analytic in the right half-plane.

The functional equation

$$\Gamma(z+1) = z \cdot \Gamma(z)$$

follows from integration by parts:

$$\begin{aligned}
\Gamma(z+1) &= \int_0^\infty e^{-t} t^z \, dt \\
&= \left[-e^{-t} t^z \right]_0^\infty + z \int_0^\infty e^{-t} t^{z-1} \, dt = z \cdot \Gamma(z).
\end{aligned}$$

In particular,

$$\begin{aligned}
\Gamma(n+1) &= n \cdot \Gamma(n) = n \cdot (n-1) \cdot \Gamma(n-1) = \cdots = n! \cdot \Gamma(1) \\
&= n! \int_0^\infty e^{-t} \, dt = n!.
\end{aligned}$$

Using this functional equation, we can continue $\Gamma(z)$ analytically to the left half-plane. If $\Re z > -1$, then $\Re(z+1) > 0$, and so the right-hand side of the equation $\Gamma(z) = \frac{\Gamma(z+1)}{z}$ is analytic for $\Re z > -1$ except at the origin, where it has a simple pole. The singular part of $\Gamma(z)$ at 0 is seen to be $\frac{1}{z}$, because, as $z \to 0$,

$$\Gamma(z+1) \longrightarrow \Gamma(1) = 1.$$

If $\Re z > -2$, then $\frac{\Gamma(z+1)}{z}$ is now a well-defined analytic function except at $z = 0$ and at -1. It has a simple pole at $z = -1$ with the singular part $\frac{-1}{z+1}$. Repeating this process, or using the relation

$$\Gamma(z) = \frac{\Gamma(n+1+z)}{z(1+z)(2+z)\cdots(n+z)} \quad \text{for } \Re z > -(n+1),$$

which can be obtained by repeated application of the functional equation, we see that $\Gamma(z)$ is a meromorphic function with simple poles at $z = 0$, $-1, -2, \ldots$, having the singular part $\frac{(-1)^n}{n!(n+z)}$ at $z = -n$.

Remarks: (a) From Corollary 7.2.3, there exist (uncountably many) *entire* functions $f(z)$ that interpolate $f(n) = n!$ (or $f(n+1) = n!$) for all nonnegative integers n; yet the Euler gamma function $\Gamma(z)$ is only meromorphic.

(b) The functional equation $f(z+1) = z \cdot f(z)$ is not sufficient in itself to characterize the gamma function. For a long time after the discovery of the gamma function by Euler, mathematicians searched for a simple characterization. The difficulty lies in the fact that attempts were made to derive such a characterization from an analysis of a differential equation satisfied by $\Gamma(z)$. No suitable equation was found; the impossibility of this task was shown long after Euler's work, in 1887, by Hölder, who showed that $\Gamma(z)$ is "*transcendentally transcendental*"; that is, it does not satisfy any linear differential equation with coefficients drawn from the class of *algebraic functions*, which, in turn, are defined as roots of a polynomial whose coefficients are themselves polynomials.

We have succeeded in continuing the gamma function $\Gamma(z)$ to the whole plane $\mathbb{C}$ except at the nonpositive integers; however, the Euler integral

$$\Gamma(z) = \int_0^\infty e^{-t} t^{z-1} \, dt$$

does not converge for $\Re z \leq 0$. Therefore, we seek a representation of $\Gamma(z)$ that is valid in the plane. For this purpose, we need to establish one more property of the gamma function,[3] namely that for x real and positive,

---

[3] We follow an elegant treatment in E. Artin, *Einführung in die Theorie der Gamma-funktion*, Hamburg, 1931.

Log $\Gamma(x)$ is a convex function of x; that is,

$$\frac{d^2}{dx^2}\text{Log}\,\Gamma(x) = \frac{\Gamma(x)\cdot\Gamma''(x) - \{\Gamma'(x)\}^2}{\{\Gamma(x)\}^2} > 0.$$

This inequality is established as follows: Beginning with the definition

$$\Gamma(x) = \int_0^\infty e^{-t}t^{x-1}\,dt$$

we differentiate twice under the integral sign (justified by the uniform convergence of the resulting integrals), obtaining

$$\Gamma'(x) = \int_0^\infty e^{-t}t^{x-1}(\text{Log}\,t)\,dt$$

and

$$\Gamma''(x) = \int_0^\infty e^{-t}t^{x-1}(\text{Log}\,t)^2\,dt.$$

Therefore,

$$u^2\Gamma(x) + 2u\Gamma'(x) + \Gamma''(x)$$
$$= \int_0^\infty e^{-t}t^{x-1}\left\{u^2 + 2u(\text{Log}\,t) + (\text{Log}\,t)^2\right\}dt$$
$$= \int_0^\infty e^{-t}t^{x-1}\,(u + \text{Log}\,t)^2\,dt > 0 \quad \text{for all } u \in \mathbb{R};$$

hence the discriminant $\{\Gamma'(x)\}^2 - \Gamma(x)\cdot\Gamma''(x) < 0$.

   We have shown that the gamma function $\Gamma(x)$ $(x > 0)$ satisfies the conditions

1^0 $f(1) = 1$;
2^0 $f(x + 1) = x \cdot f(x)$;
3^0 $f(x)$ is logarithmically convex (i.e., Log $f(x)$ is convex).

Now, we shall show that any function $f(x)$ defined and positive for positive real x satisfying these three conditions is necessarily given by

$$f(x) = \lim_{n\to\infty}\frac{n!\,n^x}{x(1 + x)\cdots(n + x)}.$$

In particular, such a function is unique. This beautiful characterization of the gamma function, which we now prove, was discovered by Bohr and Møllerup in 1922.

Let $n > 1$ be a positive integer and $0 < x \leq 1$. (By 2^0, the function is completely determined for all $x > 0$ if its values in the interval $(0, 1]$ are determined.) Then $n - 1 < n < n + x \leq n + 1$; therefore, by 3^0, we have

$$\frac{\mathrm{Log}\, f(n) - \mathrm{Log}\, f(n-1)}{n - (n-1)} < \frac{\mathrm{Log}\, f(n+x) - \mathrm{Log}\, f(n)}{(n+x) - n}$$

$$\leq \frac{\mathrm{Log}\, f(n+1) - \mathrm{Log}\, f(n)}{(n+1) - n}.$$

Using 2^0 to rewrite the above inequalities, we obtain

$$\mathrm{Log}\,(n-1) < \frac{\mathrm{Log}\, f(n+x) - \mathrm{Log}\, f(n)}{x} \leq \mathrm{Log}\, n;$$

that is,

$$x\mathrm{Log}\,(n-1) < \mathrm{Log}\frac{f(n+x)}{f(n)} \leq x\mathrm{Log}\, n,$$

viz.,

$$(n-1)^x \cdot f(n) < f(n+x) \leq n^x \cdot f(n).$$

By 2^0,

$$f(n+x) = (n-1+x)(n-2+x)\cdots(1+x)x\, f(x).$$

$$\therefore \quad \frac{(n-1)^x f(n)}{x(1+x)\cdots(n-1+x)} < f(x) \leq \frac{n^x f(n)}{x(1+x)\cdots(n-1+x)}.$$

Because the middle term is independent of n, and n is arbitrary, we may change n to $n+1$ on the left-hand side, obtaining

$$\frac{n^x f(n+1)}{x(1+x)\cdots(n+x)} < f(x) \leq \frac{n^x f(n+1)}{x(1+x)\cdots(n+x)} \cdot \frac{n+x}{n};$$

that is,

$$f(x) \cdot \frac{n}{n+x} \leq \frac{n^x n!}{x(1+x)\cdots(n+x)} < f(x).$$

Letting $n \to \infty$, we obtain

$$f(x) = \lim_{n \to \infty} \frac{n^x n!}{x(1+x)\cdots(n+x)}.$$

We have established the *Gauss formula*

$$\Gamma(x) = \lim_{n \to \infty} \frac{n^x n!}{x(1+x)\cdots(n+x)},$$

and so we have proven

THEOREM 8.8.1 (Bohr-Møllerup) *Suppose a function* $f(x)$ *is defined and positive for* $x > 0$ *and satisfies the conditions*

1^0 $f(1) = 1$;
2^0 $f(x+1) = x \cdot f(x)$;
3^0 $f(x)$ *is logarithmically convex.*

Then $f(x)$ must be the Euler gamma function $\Gamma(x)$.

Remark. If condition 1^0 is replaced by $f(1) = k$, then

$$f(n+1) = n! \cdot f(1) = n! \cdot k.$$

Thus, we have $f(x) = k \cdot \Gamma(x)$.

Because

$$\frac{z(1+z)\cdots(n+z)}{n!n^z} = e^{(1+\frac{1}{2}+\cdots+\frac{1}{n}-\mathrm{Log}\,n)z} \cdot z \cdot \left(1 + \frac{z}{1}\right)e^{-z}$$
$$\cdot \left(1 + \frac{z}{2}\right)e^{-\frac{z}{2}} \cdots \left(1 + \frac{z}{n}\right)e^{-\frac{z}{n}},$$

we have

$$\frac{1}{\Gamma(z)} = e^{\gamma z} \cdot z \cdot \prod_{n=1}^{\infty}\left\{\left(1 + \frac{z}{n}\right) \cdot e^{-\frac{z}{n}}\right\} \qquad \text{(Weierstrass)},$$

where $\gamma = \lim_{n \to \infty}\left(1 + \frac{1}{2} + \cdots + \frac{1}{n} - \mathrm{Log}\,n\right) = 0.57721\cdots$ is the *Euler constant*. (The question of whether γ is rational or irrational seems to be still open.)

Note that we have established the Weierstrass formula only for z real and positive; however, both sides are analytic (in fact, entire), and so, by the law of permanence (Section 8.7), it is valid throughout the complex plane $\mathbb{C}$. Note also that this is the canonical product of the entire function $\frac{1}{\Gamma(z)}$ and that $\Gamma(z)$ never vanishes in the complex plane $\mathbb{C}$.

Now, if in the Weierstrass formula we replace z by $-z$ and use $\Gamma(1-z) = -z \cdot \Gamma(-z)$, we obtain

$$\frac{1}{\Gamma(1-z)} = e^{-\gamma z} \cdot \prod_{n=1}^{\infty} \left\{ \left(1 - \frac{z}{n} \right) \cdot e^{\frac{z}{n}} \right\}.$$

$$\therefore \quad \frac{1}{\Gamma(z) \cdot \Gamma(1-z)} = z \cdot \prod_{n=1}^{\infty} \left(1 - \frac{z^2}{n^2} \right) = \frac{\sin \pi z}{\pi},$$

by a result in Section 7.6. We have thus established the identity

$$\Gamma(z) \cdot \Gamma(1-z) = \frac{\pi}{\sin \pi z}.$$

Substituting $z = \frac{1}{2}$, we then obtain

$$\Gamma \left(\frac{1}{2} \right) = \sqrt{\pi}.$$

Because

$$\Gamma \left(\frac{1}{2} \right) = \int_0^{\infty} e^{-t} \frac{dt}{\sqrt{t}} = 2 \int_0^{\infty} e^{-x^2} dx, \quad \text{we obtain} \quad \int_0^{\infty} e^{-x^2} dx = \frac{\sqrt{\pi}}{2}.$$

We also obtain from the Gauss formula the remarkable formula

$$\sqrt{\pi} = \Gamma \left(\frac{1}{2} \right) = \lim_{n \to \infty} \frac{n! \sqrt{n}}{\frac{1}{2} \left(1 + \frac{1}{2} \right) \cdots \left(n + \frac{1}{2} \right)}$$

$$= \lim_{n \to \infty} \frac{2^{n+1} n! \sqrt{n}}{1 \cdot 3 \cdot 5 \cdots (2n+1)}$$

$$= \lim_{n \to \infty} \frac{2^{2n+1} (n!)^2}{(2n)!} \cdot \frac{\sqrt{n}}{2n+1}$$

$$= \lim_{n \to \infty} \frac{2^{2n} (n!)^2}{(2n)! \sqrt{n}}.$$

This is known as the *Wallis formula*.

We now consider the *Euler integral of the first kind*,

$$B(p, q) = \int_0^1 t^{p-1} (1 - t)^{q-1} dt,$$

where the real parts of p and q are positive, and (for $0 < t < 1$)

$$t^{p-1} = e^{(p-1)\mathrm{Log}t}, \quad (1-t)^{q-1} = e^{(q-1)\mathrm{Log}(1-t)}$$

are the principal values. It is simple to check that the integral exists; it is known as the *Euler beta function*. Suppose $p > 0$, $q > 0$, and set

$$f(p) = B(p, q) \cdot \Gamma(p+q).$$

(We are considering q as a parameter.) We want to show that $f(p)$ satisfies the conditions 2^0 and 3^0 in the Bohr-Møllerup Theorem 8.8.1. Applying integration by parts, we obtain

$$
\begin{aligned}
B(p+1, q) &= \int_0^1 t^p (1-t)^{q-1} dt \\
&= \left[-\frac{t^p (1-t)^q}{q} \right]_0^1 + \int_0^1 \frac{p}{q} t^{p-1} (1-t)^q dt \\
&= \frac{p}{q} B(p, q+1).
\end{aligned}
$$

On the other hand, because

$$t^p (1-t)^{q-1} + t^{p-1}(1-t)^q = t^{p-1}(1-t)^{q-1},$$

we have

$$B(p+1, q) + B(p, q+1) = B(p, q).$$

Hence

$$\left(1 + \frac{q}{p}\right) B(p+1, q) = B(p, q), \quad \text{and so } B(p+1, q) = \frac{p}{p+q} B(p, q).$$

Therefore,

$$
\begin{aligned}
f(p+1) &= B(p+1, q) \cdot \Gamma(p+q+1) \\
&= \frac{p}{p+q} B(p, q) \cdot (p+q)\Gamma(p+q) \\
&= p \cdot f(p),
\end{aligned}
$$

and so the condition 2^0 is satisfied. The logarithmic convexity of $B(p, q)$ with respect to p is shown exactly as in the case of $\Gamma(x)$. The proof of the logarithmic convexity of $f(p)$ now follows from the fact that the product

310

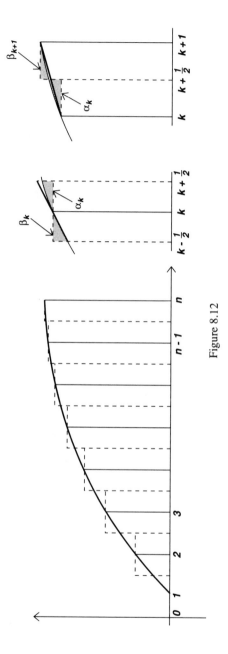

Figure 8.12

of logarithmically convex functions is logarithmically convex. (Why?)
Now

$$f(1) = B(1, q) \cdot \Gamma(q + 1)$$

and

$$B(1, q) = \int_0^1 (1 - t)^{q-1} dt = \frac{1}{q},$$

and so

$$f(1) = \frac{1}{q} \cdot q\Gamma(q) = \Gamma(q).$$

It follows that $f(p) = \Gamma(p) \cdot \Gamma(q)$, and therefore

$$B(p, q) = \frac{\Gamma(p) \cdot \Gamma(q)}{\Gamma(p + q)}.$$

Although we began with the restriction that p and q are positive real
numbers, we note that, by the law of permanence, this equality holds for
complex p and q with positive real parts.

Next we shall derive the important *Stirling formula*, which gives an
accurate estimate of $n!$ for large n. To estimate $n!$, we approximate $\text{Log } n$
by $\int_{n-\frac{1}{2}}^{n+\frac{1}{2}} \text{Log } x \, dx$ (Figure 8.12); then we obtain

$$\int_1^n \text{Log } x \, dx = \text{Log } 2 + \text{Log } 3 + \cdots + \text{Log } (n - 1) + \frac{1}{2}\text{Log } n + \delta_n.$$

On the other hand,

$$\int_1^n \text{Log } x \, dx = n\text{Log } n - n + 1,$$

$$\therefore \text{Log } \{(n - 1)!\} = \left(n - \frac{1}{2}\right) \text{Log } n - n + 1 - \delta_n;$$

that is,

$$\Gamma(n) = (n - 1)! = n^{n-\frac{1}{2}} \cdot e^{-n} \cdot e^{1-\delta_n}.$$

As can be seen from Figure 8.12, the error $\delta_n = \alpha_1 - \beta_2 + \alpha_2 - \beta_3 + \cdots -$
β_n is given by the partial sum of an alternating series. Because the function
$\text{Log } x$ is concave, the terms are monotone decreasing: $\beta_k > \alpha_k > \beta_{k+1}$.
Clearly,

$$\alpha_k \longrightarrow 0, \quad \beta_k \longrightarrow 0 \quad (\text{as } k \to \infty),$$

and so δ_n approaches a limit, which we call δ (as $n \to \infty$). Set $\mu(n) = \delta - \delta_n$ and $e^{1-\delta} = c$; then

$$\Gamma(n) = cn^{n-\frac{1}{2}} \cdot e^{-n} \cdot e^{\mu(n)}, \quad \text{with} \quad \lim_{n \to \infty} \mu(n) = 0.$$

To determine the constant c, we use the Wallis formula,

$$\begin{aligned}
\sqrt{\pi} &= \lim_{n \to \infty} \frac{(n!)^2 \cdot 2^{2n}}{(2n)!\sqrt{n}} \\
&= \lim_{n \to \infty} \frac{\left\{c(n+1)^{n+\frac{1}{2}} \cdot e^{-(n+1)}\right\}^2 \cdot 2^{2n}}{\left\{c(2n+1)^{2n+\frac{1}{2}} \cdot e^{-(2n+1)}\right\}\sqrt{n}} \\
&= \lim_{n \to \infty} c\left(\frac{n+1}{2n+1}\right)^{2n+1} \cdot \left(\frac{2n+1}{n}\right)^{\frac{1}{2}} \cdot \frac{2^{2n}}{e} \\
&= \frac{c}{\sqrt{2e}} \lim_{n \to \infty} \left\{\left(1 + \frac{\frac{1}{2}}{n+\frac{1}{2}}\right)^{n+\frac{1}{2}}\right\}^2 = \frac{c}{\sqrt{2}}.
\end{aligned}$$

Thus

$$c = \sqrt{2\pi}$$

and (using the notation $a_n \sim b_n$ to mean $\lim_{n \to \infty} \frac{a_n}{b_n} = 1$) we obtain

$$\Gamma(n) = \sqrt{2\pi} \cdot n^{n-\frac{1}{2}} \cdot e^{-n} \cdot e^{\mu(n)} \sim \sqrt{2\pi} \cdot n^{n-\frac{1}{2}} \cdot e^{-n}.$$

Hence

$$\Gamma(n+1) = n! \sim \sqrt{2\pi} \cdot n^{n+\frac{1}{2}} \cdot e^{-n}.$$

Remark.

$$\mu(n) = \frac{1}{12n} - \frac{1}{360n^3} + \frac{1}{1260n^5} - \frac{\theta}{1680n^7} \quad (0 < \theta < 1).$$

Finally, we give two examples to illustrate how the gamma function can be used to evaluate some definite integrals that arise frequently in both practical and theoretical problems. We start from the last example in Section 8.4.

EXAMPLE.

$$\int_0^\infty \frac{x^{a-1}}{1+x}dx \quad (0 < a < 1).$$

Set $t = \frac{1}{1+x}$; then

$$\int_0^\infty \frac{x^{a-1}}{1+x} dx = -\int_1^0 t \cdot \left(\frac{1-t}{t}\right)^{a-1} \cdot \frac{dt}{t^2}$$

$$= \int_0^1 t^{-a} \cdot (1-t)^{a-1} dt = B(1-a, a)$$

$$= \frac{\Gamma(1-a) \cdot \Gamma(a)}{\Gamma(1)} = \frac{\pi}{\sin \pi a}.$$

EXAMPLE.

$$\int_0^\infty x^{s-1} \cdot \cos x \, dx = \Gamma(s) \cdot \cos \frac{s\pi}{2},$$

$$\int_0^\infty x^{s-1} \cdot \sin x \, dx = \Gamma(s) \cdot \sin \frac{s\pi}{2} \qquad (0 < s < 1).$$

These two equalities can be obtained by splitting the equality

$$\int_0^\infty e^{-ix} x^{s-1} dx = \Gamma(s) e^{-\frac{s\pi}{2} i} = \frac{\Gamma(s)}{i^s},$$

into its real and imaginary parts (for real s). We therefore proceed to prove the last equality.

Let C be the curve shown in Figure 8.13. Then

$$\int_C e^{-z} z^{s-1} dz = 0; \quad \text{that is,} \quad \int_{AB} + \int_{C_R} - \int_{A'B'} - \int_{C_r} = 0.$$

$$\left|\int_{C_R}\right| \le R^s \int_0^{\frac{\pi}{2}} e^{-R\cos\theta} d\theta = R^s \int_0^{\frac{\pi}{2}} e^{-R\sin\theta} d\theta$$

$$< R^s \int_0^{\frac{\pi}{2}} e^{-\frac{2R}{\pi}\theta} d\theta < \frac{\pi}{2R^{1-s}} \longrightarrow 0 \quad (\text{as } R \to \infty, \; \because s < 1);$$

$$\left|\int_{C_r}\right| \le r^s \int_0^{\frac{\pi}{2}} e^{-r\cos\theta} d\theta < r^s \cdot \frac{\pi}{2} \longrightarrow 0 \quad (\text{as } r \to 0, \; \because s > 0).$$

$$\int_{AB} = \int_r^R e^{-x} x^{s-1} dx \longrightarrow \Gamma(s) \quad (R \to \infty, \; r \to 0),$$

$$\int_{A'B'} = i \int_r^R e^{-iy} (iy)^{s-1} dy$$

$$= i^s \int_r^R e^{-ix} x^{s-1} dx \longrightarrow i^s \int_0^\infty e^{-ix} x^{s-1} dx \quad (R \to \infty, \; r \to 0).$$

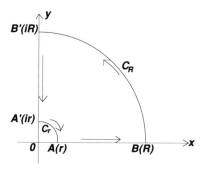

Figure 8.13

Thus we conclude that

$$i^s \int_0^\infty e^{-ix} x^{s-1} dx = \Gamma(s),$$

as was to be proven.

Exercises

1. Show that

$$\frac{1}{1-i} \sum_{k=0}^\infty \left(\frac{z-i}{1-i}\right)^k$$

is a direct analytic continuation of $\sum_{k=0}^\infty z^k$.

2. Let

$$f(x) = \begin{cases} \exp\left\{-\dfrac{1}{x^2}\right\}, & (x \neq 0); \\ 0, & (x = 0). \end{cases}$$

Show that $f(x) \in C^\infty(\mathbb{R})$. Can it be continued analytically to a complex region containing the origin?

3. Classify the singularities at $z = 1$ of the functions defined by the following power series:

(a) $\displaystyle\sum_{k=1}^\infty k z^k$;   (b) $\displaystyle\sum_{k=1}^\infty z^k$;   (c) $\displaystyle\sum_{k=1}^\infty \frac{z^k}{k}$;   (d) $\displaystyle\sum_{k=1}^\infty \frac{z^k}{k(k+1)}$.

4. Show that the series

$$f(z) = \frac{z}{1-z^2} + \frac{z^2}{1-z^4} + \frac{z^4}{1-z^8} + \cdots + \frac{z^{2^k}}{1-z^{2^{k+1}}} + \cdots$$

converges locally uniformly to

$$f(z) = \begin{cases} \dfrac{z}{1-z} & \text{if } z \in D; \\[2ex] \dfrac{1}{1-z} & \text{if } z \notin \bar{D}. \end{cases}$$

Are they analytic continuations of each other?

5. (a) Show that $f(z) = \sum_{k=0}^{\infty} \frac{z^{k!}}{1-z^{k!}}$ is analytic in the unit disc and the unit circle is its natural boundary.

   (b) How about $f(z) = \sum_{k=1}^{\infty} \frac{z^k}{1-z^k}$?

6. Show that the unit circle is the natural boundary of

$$f(z) = z + z^2 + z^4 + z^8 + \cdots + z^{2^k} + \cdots,$$

by using the functional equation $f(z) = z + f(z^2)$.

7. Show that the following power series have their circles of convergence as their natural boundaries (without appealing to the Hadamard Gap Theorem 8.2.2).

   (a) $\sum_{k=0}^{\infty} z^{k!}$;

   (b) $\sum_{k=0}^{\infty} \dfrac{z^{k!}}{k^3}$;

   (c) $\sum_{k=0}^{\infty} \dfrac{z^{2^k}}{k^3}$;

   (d) $\sum_{k=0}^{\infty} \dfrac{z^{2^k+2}}{(2^k+1)(2^k+2)}$.

   Note that the last three series converge absolutely and uniformly on their respective circles of convergence.

8. For any $a, b \in \mathbb{C}$, with $|a - b| \geq 2$, show that

$$\sum_{k=0}^{\infty} \{(z-a)(z-b)\}^k$$

represents an analytic function in neighborhoods of the points a and b and that they are analytic continuations of each other. Classify the singularities of this function.

9. (a) Let $\{a_k\}_{k=1}^{\infty}$ be a sequence of (distinct) points dense on the real axis, and $\{c_k\}_{k=1}^{\infty}$ a sequence of positive real numbers such that $\sum_{k=1}^{\infty} c_k < \infty$. Show that the functions

$$f(z) = \sum_{k=1}^{\infty} \frac{c_k}{z - a_k} \quad (\Im z > 0),$$

$$g(z) = \sum_{k=1}^{\infty} \frac{c_k}{z - a_k} \quad (\Im z < 0)$$

are analytic on their respective regions, but that they are *not* analytic continuations of each other.

(b) What if $\{a_k\}_{k=1}^{\infty}$ is a sequence of (distinct) points dense on the unit circle, and the series $\sum_{k=1}^{\infty} c_k$ is absolutely convergent (but c_k are not necessarily real numbers)? That is, show that the functions

$$f(z) = \sum_{k=1}^{\infty} \frac{c_k}{z - a_k} \quad (z \in D),$$

$$g(z) = \sum_{k=1}^{\infty} \frac{c_k}{z - a_k} \quad (z \notin \bar{D})$$

are analytic in their respective regions, but that they are *not* analytic continuations of each other.

(c) Can you generalize?

10. Suppose $f(z)$ is continuous in the whole plane $\mathbb{C}$, and analytic in the upper half-plane $\{z \in \mathbb{C};\ \Im z > 0\}$ and in the lower half-plane $\{z \in \mathbb{C};\ \Im z < 0\}$. Show that $f(z)$ is entire.

11. (Schwarz) Let Ω be a region in the unit disc having an arc Γ of the unit circle as a part of its boundary. Suppose $f(z)$ is analytic in the region Ω and is continuous in $\Omega \cup \Gamma$.

(a) If $f(z)$ takes real values on the arc Γ, show that $f(z)$ can be continued analytically across the arc Γ by setting

$$f(z) = \overline{f\left(\frac{1}{\bar{z}}\right)}.$$

(b) If $|f(z)| = 1$ on the arc Γ, show that $f(z)$ can be continued analytically across the arc Γ by setting

$$f(z) = \frac{1}{f\left(\dfrac{1}{\bar{z}}\right)}.$$

12. Let $f(z) \in H(\Omega)$, where Ω is a region symmetric with respect to the real axis. Show that there exist two functions $g(z), h(z) \in H(\Omega)$ such that

$$f(z) = g(z) + ih(z),$$

and $g(z)$ and $h(z)$ assume real values on the portion of the real axis in Ω. (For example, $e^{iz} = \cos z + i \sin z$.) Is this decomposition of $f(z)$ unique?

13. Let $f(z)$ be analytic in the (open) first quadrant, continuous in the closed first quadrant, and assume real values on the boundary (of the first quadrant). Show that $f(z)$ can be continued analytically to the whole complex plane $\mathbb{C}$ (i.e., $f(z)$ is an entire function). Further, show that $f(z)$ is even; that is, $f(-z) = f(z)$ for all $z \in \mathbb{C}$.

14. Suppose $f(z)$ is analytic in a neighborhood of the origin where it satisfies $f(2z) = \{f(z)\}^2$. Show that $f(z)$ is an entire function. Can you find, by inspiration rather than calculation, all functions (other than $f(z) \equiv 0$) which satisfy this relation?

15. Describe the Riemann surfaces for the following functions; that is, state the number of sheets, the branch points, the cuts, and the way cuts in different sheets are pasted together.

(a) $\sqrt[3]{(z-a)^2}$;

(b) $\sqrt[3]{\dfrac{z-a}{z-b}}$;

(c) $\sqrt{z} + \sqrt[3]{z-1}$;

(d) $\log(1 + z^2)$.

16. Evaluate

(a) $\int_{C_1} \sqrt{z}\, dz$, where $C_1 : z(t) = e^{it}$ $(0 \leq t \leq 2\pi)$, and the value of $\sqrt{z}$ at $t = 0$ is 1.

(b) $\int_{C_2} \sqrt{z}\, dz$, where $C_2 : z(t) = e^{it}$ $(-\pi \leq t \leq \pi)$, and the value of $\sqrt{z}$ at $t = -\pi$ is i.

17. Evaluate

(a) $\displaystyle\int_0^\infty \frac{\log x}{(1+x^2)^2}dx.$ $\left[-\dfrac{\pi}{4}\right]$

(b) $\displaystyle\int_0^\infty \frac{\log(1+x^2)}{1+x^2}dx.$ $[\pi \log 2]$

(c) $\displaystyle\int_0^\infty \frac{(\log x)^2}{x^2+a^2}dx$ $(a>0).$ $\left[\dfrac{\pi}{8a}\left\{\pi^2+4(\log a)^2\right\}\right]$

(d) $\displaystyle\int_0^\infty \frac{x^p}{1+x^2}dx$ $(-1<p<1).$ $\left[\pi\dfrac{\sin\frac{p\pi}{2}}{\sin p\pi}\right]$

(e) $\displaystyle\int_0^\infty \frac{x^{1+p}}{(1+x^2)^2}dx$ $(-2<p<2).$ $\left[\dfrac{p\pi}{4\sin\frac{p\pi}{2}}\right]$

18. Show that
$$e^{z+w} = e^z \cdot e^w \quad (z,\, w \in \mathbb{C}),$$

by the law of permanence.

19. (a) Suppose $f(z)$ is analytic in the strip $-\frac{\pi}{2} < \Re z < \frac{\pi}{2}$, continuous in the strip $-\frac{\pi}{2} \le \Re z \le \frac{\pi}{2}$, and real for $\Re z = \pm\frac{\pi}{2}$, and for $\Im z = 0$. Show that $f(z)$ can be continued analytically to the whole complex plane $\mathbb{C}$, and that
$$f(\pi - z) = f(z), \quad f(z + 2\pi) = f(z) \quad (z \in \mathbb{C}).$$

(b) Show also that
$$f(z) = \sum_{k=-\infty}^{\infty} a_k e^{ikz},$$

where $a_{-k} = \bar{a}_k$, and the series converges for all $z \in \mathbb{C}$ and uniformly in every strip $-R \le \Im z \le M$ ($M > 0, R > 0$).

20. Show that $\Gamma(x)$ has only one minimum for $x > 0$, which is in the interval $(1, 2)$.

21. Show that

(a)
$$\Gamma(p) = 2\int_0^\infty t^{2p-1}e^{-t^2}dt$$
$$= \frac{1}{p}\int_0^\infty e^{-t^{1/p}}dt = \int_0^1 \left(\text{Log}\,\frac{1}{t}\right)^{p-1}dt \quad (p>0).$$

(b)

$$B(p, q) = B(q, p) = \frac{1}{a^{p+q-1}} \int_0^a t^{p-1}(a-t)^{q-1}dt \quad (a > 0)$$

$$= 2 \int_0^{\frac{\pi}{2}} (\sin t)^{2p-1}(\cos t)^{2q-1}dt = \int_0^\infty \frac{t^{p-1}}{(1+t)^{p+q}}dt$$

$$= \frac{1}{2^{p+q-1}} \int_{-1}^1 (1+t)^{p-1}(1-t)^{q-1}dt \quad (p > 0, \, q > 0).$$

22. Verify

(a) $\displaystyle\int_0^\infty \frac{x^2}{(1+x)^5}dx = \frac{1}{12}$;

(b) $\displaystyle\int_{-1}^1 (1-x^2)^n dx = \frac{2^{2n+1}(n!)^2}{(2n+1)!}$;

(c) $\displaystyle\int_0^1 \frac{dx}{\sqrt{1-x^4}} = \frac{\{\Gamma(\frac{1}{4})\}^2}{\sqrt{32\pi}}$;

(d) $\displaystyle\int_0^1 \frac{x\,dx}{\sqrt{1-x^4}} = \frac{\pi}{4}$;

(e) $\displaystyle\int_0^1 \frac{x^2\,dx}{\sqrt{1-x^4}} = \frac{\pi\sqrt{2\pi}}{\{\Gamma(\frac{1}{4})\}^2}$;

(f) $\displaystyle\int_0^1 \frac{x^{m-1}\,dx}{\sqrt{1-x^n}} = \frac{\sqrt{\pi}}{n} \frac{\Gamma\left(\frac{m}{n}\right)}{\Gamma\left(\frac{m}{n} + \frac{1}{2}\right)}$;

(g) $\displaystyle\int_0^\pi \sin^n x\,dx = \int_{-\frac{\pi}{2}}^{\frac{\pi}{2}} \cos^n x\,dx = 2 \int_0^{\frac{\pi}{2}} \sin^n x\,dx$

$$= 2 \int_0^{\frac{\pi}{2}} \cos^n x\,dx = \sqrt{\pi}\frac{\Gamma\left(\frac{n+1}{2}\right)}{\Gamma\left(\frac{n+2}{2}\right)};$$

(h) $\displaystyle\int_0^{\frac{\pi}{2}} \sin^{2n+1}\theta\,d\theta = \int_0^{\frac{\pi}{2}} \cos^{2n+1}\theta\,d\theta = \frac{1}{2}B(n+1, \frac{1}{2})$;

(i) $\displaystyle\int_0^{\frac{\pi}{2}} \sin^{2n}\theta\,d\theta = \int_0^{\frac{\pi}{2}} \cos^{2n}\theta\,d\theta = \frac{1}{2}B(n+\frac{1}{2}, \frac{1}{2})$;

(j) $\displaystyle\int_0^{\frac{\pi}{2}} \sqrt{\tan\theta}\,d\theta = \frac{\pi}{\sqrt{2}}$.

Can you express the results in (g), (h), and (i) without using the Euler gamma function or the beta function (for the cases in which n is a positive integer)?

23. Show that the function $f(z) = e^{2\pi i z} \cdot \Gamma(z)$ also satisfies the functional equation

$$f(z+1) = z \cdot f(z).$$

Can you find other examples?

24. Show that if y is a nonzero real number, then

$$|\Gamma(iy)| = \sqrt{\frac{\pi}{y \sinh \pi y}}.$$

Hint: $\Gamma(\bar{z}) = \overline{\Gamma(z)}$.

25. Show that

(a) $\lim_{n\to\infty} \dfrac{n^z \Gamma(n)}{\Gamma(n+z)} = 1$;

(b) $\dfrac{\Gamma'(z)}{\Gamma(z)} = -\dfrac{1}{z} - \gamma + \displaystyle\sum_{k=1}^{\infty} \left(\dfrac{1}{k} - \dfrac{1}{k+z} \right)$;

(c) $\Gamma'(1) + \gamma = 0$;

(d) $\dfrac{\Gamma'(n)}{\Gamma(n)} + \gamma = 1 + \dfrac{1}{2} + \cdots + \dfrac{1}{n-1}$;

(e) $\dfrac{\Gamma'(1+z)}{\Gamma(1+z)} - \dfrac{\Gamma'(z)}{\Gamma(z)} = \dfrac{1}{z}$;

(f) $\dfrac{\Gamma'(1-z)}{\Gamma(1-z)} - \dfrac{\Gamma'(z)}{\Gamma(z)} = \pi \cot \pi z$;

(g) $\dfrac{d}{dz} \left\{ \dfrac{\Gamma'(z)}{\Gamma(z)} \right\} = \displaystyle\sum_{k=0}^{\infty} \dfrac{1}{(k+z)^2}$.

(From this equality we can also see that $\Gamma(x)$ is logarithmically convex for $x > 0$.)

26. Show that

$$\Gamma(z) = \sum_{k=0}^{\infty} \frac{(-1)^k}{k!(k+z)} + \int_1^{\infty} e^{-t} t^{z-1}\, dt.$$

Note that the first term is the sum of all singular parts of $\Gamma(z)$, and the second term is an entire function.

27. Prove the *Gauss multiplication formula*:

$$(2\pi)^{\frac{n-1}{2}} \Gamma(z) = n^{z-\frac{1}{2}} \cdot \Gamma\left(\frac{z}{n}\right) \cdot \Gamma\left(\frac{z+1}{n}\right) \cdots \Gamma\left(\frac{z+n-1}{n}\right).$$

The case $n = 2$ is known as the *Legendre duplication formula*:

$$\sqrt{\pi}\Gamma(2z) = 2^{2z-1} \cdot \Gamma(z) \cdot \Gamma\left(z + \frac{1}{2}\right).$$

Hint: The Bohr-Møllerup Theorem 8.8.1.

28. Show that $\frac{1}{\Gamma(z)}$ is an entire function of order 1.

29. Show that $\sum_{k=1}^{\infty} \frac{z^k}{\Gamma(\alpha k+1)}$ $(\alpha > 0)$ is an entire function of order $\frac{1}{\alpha}$.

30. (a) Suppose $\alpha, \beta, \gamma, \delta$ are complex numbers, where none of them is a negative integer and $\alpha + \beta = \gamma + \delta$. Show that the infinite product

$$\prod_{k=1}^{\infty} \frac{(\alpha + k)(\beta + k)}{(\gamma + k)(\delta + k)}$$

converges absolutely, and express its value in terms of the gamma function.

   (b) Can you generalize?

31. Show that for $s > 0, p > 0$,

$$\int_0^{\infty} e^{-px} x^{s-1} \cos qx \, dx = \frac{\Gamma(s)}{(p^2 + q^2)^{\frac{s}{2}}} \cos s\varphi,$$

$$\int_0^{\infty} e^{-px} x^{s-1} \sin qx \, dx = \frac{\Gamma(s)}{(p^2 + q^2)^{\frac{s}{2}}} \sin s\varphi,$$

where $\varphi = \text{Arctan} \frac{q}{p}, -\frac{\pi}{2} < \varphi < \frac{\pi}{2}$.

32. (a) Show that, in n-dimensional Euclidean space, the volume of the solid

$$\left\{(\xi_1, \xi_2, \ldots, \xi_n) \in \mathbb{R}^n ; \sum_{k=1}^{n} |\xi_k|^{\alpha} \leq r^{\alpha}\right\} \quad (r > 0, \alpha > 0)$$

is given by

$$V = \frac{(2r)^n}{\alpha^{n-1}} \cdot \frac{\left\{\Gamma\left(\frac{1}{\alpha}\right)\right\}^n}{n\Gamma\left(\frac{n}{\alpha}\right)}.$$

In particular, setting $\alpha = 2$ and $r = 1$, we obtain the volume of the unit ball in the n-dimensional Euclidean space:

$$V = \frac{\pi^{\frac{n}{2}}}{\Gamma\left(\dfrac{n}{2} + 1\right)} = \begin{cases} \dfrac{(2\pi)^{\frac{n}{2}}}{2 \cdot 4 \cdot 6 \cdots n} & (n \text{ even}); \\[3ex] \dfrac{2^{\frac{n+1}{2}} \pi^{\frac{n-1}{2}}}{1 \cdot 3 \cdot 5 \cdots n} & (n \text{ odd}). \end{cases}$$

(b) How about the surface area (for the case $\alpha = 2$)?

Chapter 9

Normal Families

9.1 The Montel Selection Theorem

It is of fundamental importance in mathematical analysis to have a condition guaranteeing that a certain set of elements has an accumulation point. For example, the Bolzano-Weierstrass property is equivalent to the completeness of the real number field, without which hardly a trace of mathematical analysis would be left. We have seen (Exercise 4.37) that in the vector space $H(\Omega)$ of all analytic functions in a region Ω a metric can be defined so that $H(\Omega)$ becomes a complete metric space, and convergence in this metric is equivalent to local uniform convergence in Ω. In this section we discuss a condition that guarantees the existence of an accumulation point (that is, a member of $H(\Omega)$) of a family of analytic functions in $H(\Omega)$. We start with the definition of equicontinuity of a family of functions.

DEFINITION 9.1.1 *Let $\mathcal{F}$ be a family of functions (not necessarily analytic) defined in a region Ω, and S a subset of Ω. Suppose that for every $\epsilon > 0$ there exists $\delta > 0$ such that*

$$|f(z) - f(z')| < \epsilon \quad \text{whenever} \quad f(z) \in \mathcal{F} \text{ and } |z - z'| < \delta, \ \ z, z' \in S.$$

Then we say that the family $\mathcal{F}$ is **equicontinuous** *on S.*

Clearly, every function $f(z)$ in $\mathcal{F}$ is uniformly continuous on S. Note carefully, however, that the choice of δ does not depend on the choice of $f(z)$.

We now formulate the Ascoli-Arzelà theorem in a way that suits our purpose.

THEOREM 9.1.2 (Ascoli-Arzelà) *Let $\mathcal{F}$ be a family of functions defined in a region $\Omega \subset \mathbb{C}$ and pointwise bounded in Ω (i.e., for each $z \in \Omega$, there* *exists a positive number M_z satisfying $|f(z)| \leq M_z$ for every function $f(z)$ in $\mathcal{F}$) and equicontinuous on each compact subset of Ω. Then every sequence $\{f_k(z)\}_{k=1}^{\infty}$ in $\mathcal{F}$ contains a subsequence that converges locally uniformly on Ω.*

   Proof. Let $\{z_k\}_{k=1}^{\infty}$ be a sequence of points that is dense in Ω. (Why does such a sequence exist?) Because the sequence $\{f_k(z_1)\}_{k=1}^{\infty}$ of complex numbers is bounded, it contains, by the Bolzano-Weierstrass Property 2.1.3, a convergent subsequence $\{f_{k1}(z_1)\}_{k=1}^{\infty}$. Now consider the sequence $\{f_{k1}(z_2)\}_{k=1}^{\infty}$. This is also a bounded sequence of complex numbers, hence, again by the Bolzano-Weierstrass Property 2.1.3, it contains a convergent subsequence $\{f_{k2}(z_2)\}_{k=1}^{\infty}$. Continuing this procedure for each positive integer m, we obtain a subsequence $\{f_{km}(z)\}_{k=1}^{\infty}$ of $\{f_k(z)\}_{k=1}^{\infty}$ such that

(a) $\{f_{km}(z)\}_{k=1}^{\infty}$ is a subsequence of $\{f_{kn}(z)\}_{k=1}^{\infty}$ if $m > n$;

(b) $\{f_{km}(z_j)\}_{k=1}^{\infty}$ converges for each $j = 1, 2, \ldots, m$.

Set $g_k(z) = f_{kk}(z) \, (k \in \mathbb{N})$. Then, except possibly for the first $m-1$ terms, $\{g_k(z)\}_{k=1}^{\infty}$ is a subsequence of $\{f_{km}(z)\}_{k=1}^{\infty}$; therefore, $\{g_k(z_m)\}_{k=1}^{\infty}$ converges for each $m \in \mathbb{N}$.

   We claim that the sequence $\{g_k(z)\}_{k=1}^{\infty}$ satisfies the required condition; that is, it converges locally uniformly in Ω. To prove this assertion, it is sufficient, by Lemma 2.4.4, to show that each point in Ω has a neighborhood in which the sequence converges uniformly.

   Let z_0 be an arbitrary point in Ω and choose $r > 0$ so that the closed disc $\bar{D}(z_0, r)$ is entirely in Ω. Then, because the closed disc is compact and the sequence $\{f_k(z)\}_{k=1}^{\infty}$, hence $\{g_k(z)\}_{k=1}^{\infty}$ also, is equicontinuous in the disc $\bar{D}(z_0, r)$, for any given $\epsilon > 0$ there exists $\delta > 0$ such that

$$|g_k(z) - g_k(z')| < \frac{\epsilon}{3} \quad \text{for all } k \in \mathbb{N} \text{ and } |z - z'| < \delta, \ z, z' \in \bar{D}(z_0, r).$$

Because $\{z_k\}_{k=1}^{\infty}$ is dense in Ω, there exists a point z_{j_0} in the sequence such that $z_{j_0} \in D(z_0, \delta_0)$, where $\delta_0 = \min\{r, \frac{\delta}{2}\}$ (Figure 9.1).

   As $\{g_k(z_j)\}_{k=1}^{\infty}$ converges for all $j \in \mathbb{N}$, it converges, in particular, for $j = j_0$; hence there exists $N = N(j_0)$ such that

$$|g_m(z_{j_0}) - g_n(z_{j_0})| < \frac{\epsilon}{3} \quad \text{provided } m, n \geq N.$$

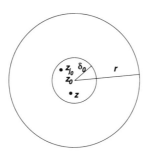

Figure 9.1

Then for any $z \in D(z_0, \delta_0)$ we have $|z - z_{j_0}| < 2\delta_0 \leq \delta$; therefore, by equicontinuity,

$$|g_k(z) - g_k(z_{j_0})| < \frac{\epsilon}{3} \quad \text{for all } k \in \mathbb{N}.$$

It follows that for all points z in the disc $D(z_0, \delta_0)$,

$$
\begin{aligned}
|g_m(z) - g_n(z)| &\leq |g_m(z) - g_m(z_{j_0})| + |g_m(z_{j_0}) - g_n(z_{j_0})| \\
&\quad + |g_n(z_{j_0}) - g_n(z)| \\
&< \frac{\epsilon}{3} + \frac{\epsilon}{3} + \frac{\epsilon}{3} = \epsilon \quad \text{if } m, n \geq N,
\end{aligned}
$$

and this completes the proof.   □

So far we have not assumed that the functions are analytic; if $\mathcal{F}$ is a family of analytic functions, then the assumption that it is equicontinuous in each compact subset can be removed provided the *pointwise* boundedness is replaced by the *local uniform boundedness* in Ω; that is, for each point $z_0 \in \Omega$ there exist positive numbers $\delta(z_0)$ and $M(z_0)$ such that $\|f\|_{\bar{D}(z_0, \delta(z_0))} \leq M(z_0)$ for all $f(z) \in \mathcal{F}$.

THEOREM 9.1.3 (Montel) *Let $\mathcal{F} \subset H(\Omega)$, and suppose it is locally uniformly bounded in the region Ω. Then every sequence $\{f_k(z)\}_{k=1}^{\infty}$ in $\mathcal{F}$ contains a subsequence that converges locally uniformly on Ω.*

Proof. By the Ascoli-Arzelà Theorem 9.1.2 it is sufficient to establish that analyticity plus local uniform boundedness implies "local" equicontinuity; that is, each point in Ω has a neighborhood in which the sequence $\{f_k(z)\}_{k=1}^{\infty}$ is equicontinuous. (Readers are urged to carry out the detailed argument here.)

Let z_0 be an arbitrary point in Ω and $0 < \delta < \text{dist}\,(z_0,\ \partial\Omega)$. By the
Cauchy Integral Formula 4.4.3,

$$f_k'(z) = \frac{1}{2\pi i} \int_{C(z_0,\delta)} \frac{f_k(\zeta)}{(\zeta - z)^2} d\zeta \quad k \in \mathbb{N}, \ z \in D\left(z_0, \frac{\delta}{2}\right).$$

Replacing δ by a smaller (positive) value, if necessary, we may assume
that $\{f_k(z)\}_{k=1}^{\infty}$ is uniformly bounded by some constant M in the disc
$\bar{D}\,(z_0,\ \delta)$. Then

$$\|f_k'\|_{D(z_0, \frac{\delta}{2})} \leq \frac{M}{2\pi \left(\frac{\delta}{2}\right)^2} \cdot 2\pi\delta = \frac{4M}{\delta} \quad (k \in \mathbb{N});$$

(i.e., $\{f_k'(z)\}_{k=1}^{\infty}$ is also locally uniformly bounded, hence uniformly
bounded on each compact subset). Therefore,

$$|f_k(z) - f_k(z')| = \left| \int_{z'}^{z} f_k'(\zeta)\,d\zeta \right| \leq \frac{4M}{\delta}|z - z'| \quad (k \in \mathbb{N}),$$

whenever z and z' are in the disc $D\left(z_0, \frac{\delta}{2}\right)$, which is what we want to
show. $\square$

DEFINITION 9.1.4 *A family $\mathcal{F}$ of analytic functions defined in a region
Ω is called* **normal** *if every sequence of functions chosen from $\mathcal{F}$ contains
a subsequence that converges locally uniformly in Ω.*

Thus, the Montel Theorem 9.1.3 asserts that *a locally uniformly bounded
family of analytic functions constitutes a normal family.*

THEOREM 9.1.5 (Vitali) *Let $\{f_k(z)\}_{k=1}^{\infty}$ be a sequence of functions in
$H(\Omega)$ that is locally uniformly bounded in Ω (hence it is a normal family of
analytic functions) and converges on a set of distinct points $\{z_k\}_{k=1}^{\infty}$ having
an accumulation point in Ω. Then the sequence $\{f_k(z)\}_{k=1}^{\infty}$ converges
locally uniformly in Ω.*

Proof. First we show that the sequence $\{f_k(z)\}_{k=1}^{\infty}$ converges pointwise everywhere in Ω. Suppose the sequence does not converge at some
point $z_0 \in \Omega$. Then the sequence $\{f_k(z_0)\}_{k=1}^{\infty}$, being bounded, must
have at least two subsequences $\{g_k(z_0)\}_{k=1}^{\infty}$ and $\{h_k(z_0)\}_{k=1}^{\infty}$ that converge to *different* values a and b, respectively, by the Bolzano-Weierstrass
Property 2.1.3. By the Montel Theorem 9.1.3, the sequence $\{g_k(z)\}_{k=1}^{\infty}$
contains a subsequence that converges locally uniformly to an analytic

function $g(z) \in H(\Omega)$. Similarly, $\{h_k(z)\}_{k=1}^{\infty}$ contains a subsequence that converges locally uniformly to an analytic function $h(z) \in H(\Omega)$. Because $\{f_k(z)\}_{k=1}^{\infty}$ converges at $\{z_k\}_{k=1}^{\infty}$, its subsequences $\{g_k(z)\}_{k=1}^{\infty}$ and $\{h_k(z)\}_{k=1}^{\infty}$ must also converge to the same limits at these points. Thus

$$g(z_k) = h(z_k) \quad (k \in \mathbb{N}).$$

As the sequence $\{z_k\}_{k=1}^{\infty}$ has an accumulation point in Ω, the two analytic functions $g(z)$ and $h(z)$ must be identical. But then

$$a = g(z_0) = h(z_0) = b.$$

This contradiction establishes the *pointwise* convergence of $\{f_k(z)\}_{k=1}^{\infty}$ everywhere in Ω. Let the limit function be $f(z)$.

Next we show that the convergence which has just been established is locally uniform in Ω. Assume that the sequence fails to converge uniformly on some compact subset K of Ω. Then there exists a positive number ϵ and an increasing sequence $\{n_k\}_{k=1}^{\infty}$ of positive integers such that

$$\|f - f_{n_k}\|_K \geq \epsilon \quad (k \in \mathbb{N}).$$

But this result contradicts the Montel Theorem 9.1.3, which asserts that $\{f_{n_k}(z)\}_{k=1}^{\infty}$ contains a subsequence $\{f_{n_{k_j}}(z)\}_{j=1}^{\infty}$ that converges locally uniformly in Ω and, in particular, uniformly on K; that is,

$$\|f - f_{n_{k_j}}\|_K < \epsilon \quad \text{for all } j \text{ sufficiently large.} \qquad \square$$

Remark. The last half of the proof can be replaced by the argument used in the proof of the Ascoli-Arzelà Theorem 9.1.2.

We shall use the Montel Selection Theorem 9.1.3 when we prove the Riemann Mapping Theorem 10.2.1. Here we illustrate the technique by proving the following

THEOREM 9.1.6 *Suppose $f(z)$ is a bounded analytic function in the unit disc D and that θ_0 is a value for which the radial limit*

$$\lim_{r \to 1-} f(re^{i\theta_0}) = \ell$$

exists; then $f(z)$ tends to the same limit ℓ as $z \to e^{i\theta_0}$ nontangentially.

Proof. If z approaches $e^{i\theta_0}$ nontangentially, then there exists α ($0 < \alpha < \frac{\pi}{2}$) such that z stays inside the sector with vertex at $e^{i\theta_0}$, of angle 2α,

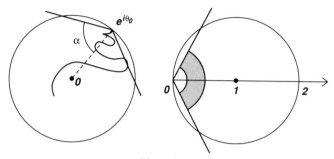

Figure 9.2

symmetric with respect to the radius reaching the point $e^{i\theta_0}$ (Figure 9.2). However, it is easier to handle this problem by changing the variable so that $f(z)$ is a bounded analytic function in the disc $D(1,1)$ having a limit ℓ as $z \to 0$ along the positive real axis.

Now consider the sequence of functions

$$f_k(z) = f\left(\frac{z}{k}\right) \quad (k \in \mathbb{N}).$$

Because the functions $\{f_k(z)\}_{k=1}^{\infty}$ are well defined and uniformly bounded in the disc $D(1,1)$, and because for every real z in the interval $0 < z < 2$

$$\lim_{k \to \infty} f_k(z) = \ell,$$

it follows from the Vitali Theorem 9.1.5 that $\{f_k(z)\}_{k=1}^{\infty}$ converges to the function $f_0(z) \equiv \ell$ locally uniformly in the disc $D(1, 1)$. In particular,

$$\lim_{k \to \infty} f_k(z) = \ell$$

uniformly on the compact set

$$K = \left\{ z \,;\, |\arg z| \leq \alpha < \frac{\pi}{2}, \; \frac{\cos\alpha}{2} \leq |z| \leq \cos\alpha \right\}.$$

This result implies that $f(z)$ approaches ℓ (as z approaches 0) from inside the sector $|\arg z| \leq \alpha$.   □

Remark. As a matter of fact, Fatou proved that for a function analytic and bounded in the disc the radial limit exists almost everywhere; thus it has a nontangential limit almost everywhere. (Cf. Hoffman, *Banach Spaces of Analytic Functions*, Prentice-Hall, 1962.)

Note. The Vitali Theorem 9.1.5 can also be used to prove the Jentzsch theorem mentioned in the remark after Corollary 5.6.7.

9.2 Univalent Functions

DEFINITION 9.2.1 *Let $f(z)$ be a function defined in a region Ω. Suppose $f(z_1) \neq f(z_2)$ whenever $z_1 \neq z_2$, $z_1, z_2 \in \Omega$; then we say that the function $f(z)$ is **univalent** (schlicht) in Ω.*

We start from the following

THEOREM 9.2.2 *If $f(z)$ is a univalent analytic function in a region Ω, then $f'(z)$ never vanishes in Ω.*
    Proof. Suppose $f'(z_0) = 0$, $z_0 \in \Omega$. Then

$$f(z) = a_0 + a_k(z - z_0)^k + a_{k+1}z^{k+1} + \cdots \quad (k \geq 2,\ a_k \neq 0),$$

and $z = z_0$ is a root of multiplicity k for the equation $f(z) - a_0 = 0$. Now we take $\delta > 0$ so small that

1^0 the closed disc $\bar{D}(z_0,\ \delta)$ is in the region Ω;

2^0 z_0 is the only root (which turns out to be of multiplicity k) of the equation $f(z) - a_0 = 0$ in the closed disc $\bar{D}(z_0,\ \delta)$; and

3^0 $f'(z) \neq 0$ for $z \in \bar{D}'(z_0,\ \delta) = \{z \in \mathbb{C}\,;\ 0 < |z - z_0| \leq \delta\}$.

(Why is this possible?) Then we have

$$\frac{1}{2\pi i} \int_{C(z_0,\delta)} \frac{f'(z)}{f(z) - a_0} dz = k.$$

Because the left-hand side is a continuous function of a_0 (as long as it stays inside the image of the circle $C(z_0,\ \delta)$ under the mapping $f(z)$—which is a simple closed curve) and is integer-valued, we conclude that, if a_0' is sufficiently close to a_0, then

$$\frac{1}{2\pi i} \int_{C(z_0,\delta)} \frac{f'(z)}{f(z) - a_0'} dz = k.$$

Thus the equation $f(z) - a_0' = 0$ has precisely k roots in the disc $D(z_0,\ \delta)$. (This part of the proof can be replaced by an argument using the Rouché Theorem 5.6.8. See also the proof of the next theorem.)

Because $f'(z) \neq 0$ for $z \in D'(z_0, \delta)$, all the roots of the equation $f(z) - a_0' = 0$ $(a_0' \neq a_0)$ are simple. Let them be $z_1, z_2, \ldots, z_k$. Then $f(z_1) = f(z_2) = \cdots = f(z_k) = a_0'$. This result contradicts the assumption that $f(z)$ is univalent, for $z_1, z_2, \ldots, z_k$ are distinct (and $k \geq 2$). $\square$

THEOREM 9.2.3 *Let $\{f_k(z)\}_{k=1}^{\infty}$ be a sequence of univalent analytic functions that converges locally uniformly to a nonconstant function $f(z)$ in a region Ω. Then $f(z)$ is (analytic and) univalent in Ω.*

Proof. Suppose $f(z_1) = f(z_2) = w_0$, where z_1 and z_2 are distinct points of Ω. Let $r > 0$ be so small that the closed discs $\bar{D}(z_1, r)$ and $\bar{D}(z_2, r)$ are mutually disjoint and are entirely in Ω. Moreover, we may assume that $f(z) \neq w_0$ for z on the circles $C(z_1, r)$ and $C(z_2, r)$. This is possible, since, by assumption, $f(z)$ is not constant. Set

$$m_1 = \min\{|f(z) - w_0| ; z \in C(z_1, r)\},$$
$$m_2 = \min\{|f(z) - w_0| ; z \in C(z_2, r)\};$$

then $m_1 > 0$, $m_2 > 0$. As the sequence $\{f_k(z)\}_{k=1}^{\infty}$ converges uniformly to $f(z)$ on the circles $C(z_1, r)$ and $C(z_2, r)$, for $0 < \epsilon < \min\{m_1, m_2\}$ there exists an integer N such that

$$\|f_k - f\|_{C(z_1,r)} < \epsilon, \quad \|f_k - f\|_{C(z_2,r)} < \epsilon, \quad \text{for } k \geq N.$$

Thus
$$|f(z) - w_0| \geq \min\{m_1, m_2\} > \epsilon > |f_k(z) - f(z)|$$

for z on either one of the circles $C(z_1, r)$ or $C(z_2, r)$ and $k \geq N$. Because

$$f_k(z) - w_0 = \{f_k(z) - f(z)\} + \{f(z) - w_0\},$$

by the Rouché Theorem 5.6.8 the equations

$$f_k(z) - w_0 = 0 \quad \text{and} \quad f(z) - w_0 = 0 \quad (k \geq N)$$

have the same number of roots inside either of the circles. In particular, $f_k(z) = w_0$ has roots inside both $C(z_1, r)$ and $C(z_2, r)$, which contradicts the assumption that the functions $f_k(z)$ are univalent. $\square$

Remark. Actually, this theorem is an immediate consequence of (the proof of) the Hurwitz Theorem 5.6.4.

THEOREM 9.2.4 (Bieberbach Area Theorem) *Let*

$$w = f(z) = z + b_0 + \frac{b_1}{z} + \frac{b_2}{z^2} + \cdots + \frac{b_k}{z^k} + \cdots$$

be a function analytic and univalent in the region $\{z \in \mathbb{C}; 1 < |z| < \infty\}$. *Then*

$$\sum_{k=1}^{\infty} k\,|b_k|^2 \le 1.$$

Proof. Let $A(r)$ be the area enclosed by the image of the circle $C(0, r), r > 1$ (which is a simple closed curve in the w-plane). Then

$$A(r) = \int_{C(0,r)} u\left(re^{i\theta}\right) dv\left(re^{i\theta}\right),$$

where $f(z) = u(z) + iv(z)$. If we use the subscript θ to denote differentiation with respect to θ, then we have

$$
\begin{aligned}
A(r) &= \int_{C(0,r)} u\left(re^{i\theta}\right) \cdot v_\theta\left(re^{i\theta}\right) d\theta \\
&= \int_0^{2\pi} \frac{1}{2}\left\{f\left(re^{i\theta}\right) + \overline{f\left(re^{i\theta}\right)}\right\} \cdot \frac{1}{2i}\left\{f_\theta\left(re^{i\theta}\right) - \overline{f_\theta\left(re^{i\theta}\right)}\right\} d\theta \\
&= \frac{1}{4}\int_0^{2\pi}\left\{re^{i\theta} + re^{-i\theta} + b_0 + \bar{b}_0 + \sum_{k=1}^{\infty}\frac{b_k e^{-ik\theta} + \bar{b}_k e^{ik\theta}}{r^k}\right\} \\
&\quad \times \left\{re^{i\theta} + re^{-i\theta} - \sum_{k=1}^{\infty}\frac{k}{r^k}\left(b_k e^{-ik\theta} + \bar{b}_k e^{ik\theta}\right)\right\} d\theta \\
&= \pi\left(r^2 - \sum_{k=1}^{\infty}\frac{k\,|b_k|^2}{r^{2k}}\right),
\end{aligned}
$$

where we have used the facts that the series converges absolutely and uniformly on the circle $C(0, r), r > 1$ and that the only terms of the product that have nonzero integrals are the constant terms. Because the orientation of the path $C(0, r)$ along which we have integrated is positive, the integral assigns a positive value to the area $A(r)$, and therefore

$$r^2 \ge \sum_{k=1}^{\infty}\frac{k\,|b_k|^2}{r^{2k}}.$$

Letting $r \to 1$, we obtain the desired result. $\qquad\square$

Let us now consider the family of functions

$$f(z) = a_0 + a_1 z + a_2 z^2 + \cdots + a_k z^k + \cdots$$

analytic and univalent in the unit disc. Because the constant term affects neither analyticity nor univalence, we may assume without loss of generality that $a_0 = 0$. Also, because $f'(z) \neq 0$, we may normalize the function by dividing by the nonzero constant $f'(0)$, so that $a_1 = 1$.

THEOREM 9.2.5 (Bieberbach 1916) *Let*

$$f(z) = z + a_2 z^2 + \cdots + a_k z^k + \cdots$$

be analytic and univalent in the unit disc D. Then

$$|a_2| \leq 2.$$

Proof. Set

$$f\left(z^2\right) = z^2 + a_2 z^4 + a_3 z^6 + \cdots = z^2 g(z).$$

Then

$$g(z) = 1 + a_2 z^2 + a_3 z^4 + a_4 z^6 + \cdots$$

is analytic and never vanishes in the unit disc D. Hence, choosing the branch of $\{g(z)\}^{\frac{1}{2}}$ satisfying $\{g(0)\}^{\frac{1}{2}} = 1$, we see that

$$\begin{aligned} F(z) &= \sqrt{f\left(z^2\right)} = z\{g(z)\}^{\frac{1}{2}} \\ &= z + \frac{a_2}{2} z^3 + \cdots \end{aligned}$$

is analytic in D, obviously odd, and vanishes only at $z = 0$. Moreover, $F(z)$ is also univalent in the unit disc D. For, if $F(z_1) = F(z_2)$, then $f\left(z_1^2\right) = f\left(z_2^2\right)$, and so $z_1^2 = z_2^2$ (because $f(z)$ is univalent). Then $z_1 = \pm z_2$, but $z_1 = -z_2$ contradicts the fact that $F(z)$ is an odd function, unless $z_1 = z_2 = 0$.

Now let

$$\varphi(z) = \frac{1}{F\left(\dfrac{1}{z}\right)} = z - \frac{a_2}{2} \frac{1}{z} + \cdots ;$$

then $\varphi(z)$ is analytic and univalent for $1 < |z| < \infty$. Thus, by the Area Theorem 9.2.4,

$$\left|-\frac{a_2}{2}\right| \leq 1, \quad \text{or} \quad |a_2| \leq 2. \qquad \square$$

Theorems 9.2.4 and 9.2.5, together with an elementary calculation, show that the equality $|a_2| = 2$ can occur only with the *Koebe function*

$$f_0(z) = \frac{z}{(1 - \epsilon z)^2} = \sum_{k=1}^{\infty} k \epsilon^{k-1} z^k \quad (|\epsilon| = 1).$$

Impressed by the fact that the Koebe function appears in many problems concerning univalent functions, Bieberbach conjectured in 1916 that, if

$$f(z) = z + \sum_{k=2}^{\infty} a_k z^k$$

is analytic and univalent in the unit disc D, then

$$|a_n| \leq n \quad (\text{for all } n = 2, 3, \ldots).$$

This conjecture was proven, over a period of six decades, for the specific values $n = 3, 4, 5, 6$ (by some of the most powerful analysts), and finally the entire conjecture was proven in 1984 by deBranges. (The original proof, long and difficult, has subsequently been simplified and shortened, both by deBranges himself and a number of other investigators.) For all n the maximization of $|a_n|$ is achieved only by the Koebe function.

A vast amount of important and interesting literature is centered on the Bieberbach conjecture. Here we present only the following result, known as the "$\frac{1}{4}$-theorem":

THOEREM 9.2.6 *Let*

$$w = f(z) = z + a_2 z^2 + \cdots$$

be a univalent analytic function in the unit disc D. Then the image of the unit disc contains the disc $D\left(0, \frac{1}{4}\right)$.

Proof. Suppose ζ is a point in the w-plane that is not in the image of the unit disc D. Then the function

$$F(z) = \frac{f(z)}{1 - \frac{f(z)}{\zeta}} = z + \left(a_2 + \frac{1}{\zeta}\right) z^2 + \cdots$$

is analytic and univalent in the unit disc D, because the composition of a Möbius transformation and a univalent function is again univalent. By the Bieberbach Theorem 9.2.5 (and the triangle inequality), we have

$$\left| a_2 + \frac{1}{\zeta} \right| \le 2, \quad \therefore \quad \left| \frac{1}{\zeta} \right| \le 2 + |a_2| \le 4, \quad \text{hence } |\zeta| \ge \frac{1}{4}. \qquad \square$$

Note that, once again, the extremal case is furnished by the Koebe function. It should be remarked that the mapping effected by this function is very simple and striking; if we choose, for convenience, $\epsilon = 1$, then the image of the unit disc consists of $\mathbb{C}$ minus the slit from $-\frac{1}{4}$ to infinity along the negative real axis. (See Example 2 in Section 10.3.)

Exercises

1. Show that in the conclusion of the Ascoli-Arzelà Theorem 9.1.2 the *local uniform convergence* on Ω cannot be replaced by the stronger assertion of *uniform convergence* on Ω, even under the stronger assumption that the sequence is *uniformly bounded and equicontinuous* on Ω.

2. Discuss the converse of the Ascoli-Arzelà Theorem 9.1.2 and also that of the Montel Theorem 9.1.3.

3. Show that the real counterpart of the Montel Theorem 9.1.3 is not true. *Hint:* Consider $f_k(x) = \frac{kx}{1+(kx)^2}$ on $\mathbb{R}$.

4. (a) Let $\mathcal{F} = \{f_\alpha(z)\}_{\alpha \in I}$ (where I is an arbitrary nonempty index set) be a normal family of analytic functions in a region Ω, and g an entire function. Show that $\mathcal{F}_g = \{g \circ f_\alpha(z)\}_{\alpha \in I}$ is also a normal family of analytic functions in Ω.

   (b) Is the family $\mathcal{F}' = \{f'_\alpha(z)\}_{\alpha \in I}$ of derivatives a normal family?

5. (a) Let $\mathcal{F}$ be the family of all polynomials

$$f(z) = a_0 + a_1 z + a_2 z^2 + \cdots + a_{19} z^{19}$$

   of degree at most 19 such that

$$|a_n| \le 96^n \quad (n = 0, 1, 2, \ldots, 19).$$

   Show that $\mathcal{F}$ is a normal family.

   (b) Show that the family of *all* polynomials is not normal in *any* region.

6. (a) Let $\mathcal{F}$ be the family of all analytic functions $f(z) = \sum_{k=0}^{\infty} a_k z^k$ in the unit disc D satisfying $\sum_{k=0}^{\infty} |a_k|^2 \leq 1$. Is this a normal family?

   (b) What if the condition $\sum_{k=0}^{\infty} |a_k|^2 \leq 1$ is replaced by $\sum_{k=0}^{\infty} \frac{|a_k|}{k+1} \leq 1$?

7. Let $\mathcal{F}$ be the family of all analytic functions in the unit disc D satisfying

$$\int \int_D |f(z)|^2 dx \, dy \leq 1.$$

Is this a normal family?

8. Let z_0 be an arbitrary point in a region Ω, and

$$\mathcal{F} = \{f \in H(\Omega) \, ; \, f(z_0) = 1 \text{ and } \Re f(z) > 0 \text{ for all } z \in \Omega\}.$$

   (a) Show that $\mathcal{F}$ is a normal family.
   (b) What if the condition $f(z_0) = 1$ is deleted?
   (c) What if the condition $f(z_0) = 1$ is replaced by $|f(z_0)| \leq 1$?

9. Let Ω be a region and Γ a rectifiable arc. Suppose $f(z, \zeta)$ is continuous for $(z, \zeta) \in \Omega \times \Gamma$ and analytic for $z \in \Omega$ for each fixed $\zeta \in \Gamma$. Show that

$$g(z) = \int_\Gamma f(z, \zeta) \, d\zeta$$

is analytic in Ω
   (a) by using the Morera Theorem 4.5.4;
   (b) by using the Montel Theorem 9.1.3. (*Hint:* Consider Riemann sums.)
   (c) Show that

$$g^{(k)}(z) = \int_\Gamma \frac{\partial^k f(z, \zeta)}{\partial z^k} d\zeta.$$

10. Deduce the Montel Theorem 9.1.3. from the Vitali Theorem 9.1.5.

11. Let $f(z)$ be analytic and bounded in the half-strip $\{z \in \mathbb{C} \, ; \, a < \Re z < b, \, \Im z > 0\}$, and suppose, for some $x_0 \in (a, b)$, $\lim_{y \to \infty} f(x_0 + iy)$ exists; call it ℓ. Show that, if x_0 is replaced by any other number, x', in the interval (a, b), then $\lim_{y \to \infty} f(x' + iy)$ also exists, and it also equals ℓ. Furthermore, if x' is confined to any closed subinterval of (a, b), then the convergence of $f(x' + iy)$ to its limit is uniform.

12. A sequence of functions $\{f_k(z)\}_{k=0}^{\infty}$ is defined by

$$f_0(z) = z, \quad f_k(z) = z - \frac{1}{f_{k-1}(z)} \quad (k \in \mathbb{N}).$$

(a) Show that these functions are analytic in the upper half-plane $\{z \in \mathbb{C}; \Im z > 0\}$ and that the values of these functions are also in the upper half-plane.

(b) Verify that the sequence converges for $z = iy$ $(y > 1)$.

(c) Use the result in (b) to prove the local uniform convergence of the sequence in the upper half-plane.

(d) Determine the limit function.

13. Show that there exists a function analytic in the unit disc D that has radial limit nowhere. In fact, more is true: there exists a function $f(z) = \sum_{k=0}^{\infty} a_k z^k$ with $\sum_{k=0}^{\infty} |a_k| < \infty$, and yet its derivative $f'(z)$ has radial limit nowhere.

Chapter 10

Conformal Mapping

10.1 Classification of Regions

Let Ω and Ω' be regions in the z-plane and the w-plane, respectively. Suppose there exists an analytic function $w = f(z)$ that is a bijection of the region Ω onto the region Ω'; then, because $f(z)$ is univalent in Ω, $f'(z)$ never vanishes in Ω, and hence the mapping preserves angles (Section 3.3). We say the regions Ω and Ω' are *conformally equivalent*. (It is simple to verify that this is actually an equivalence relation.)

Clearly, a multiply connected region cannot be conformally equivalent to a simply connected region. (See Exercise 10.4.) Before we discuss conformal mappings of simply connected regions, we classify such regions on the Riemann sphere. There are three cases:

(a) The region has no boundary point; that is, it is the whole Riemann sphere.

(b) There exists a single boundary point; that is, one point is deleted from the Riemann sphere.

(c) There exist at least two boundary points; in this case, because of the simple connectedness, the boundary is a continuum (hence it consists of uncountably many points).

We have seen (Section 3.4) that a Möbius transformation maps the whole Riemann sphere conformally onto itself; similarly, any two punctured Riemann spheres, each with one point removed, can be mapped conformally by a Möbius transformation onto one another; by removing the point at infinity from one of the spheres, we see that any sphere of this type is conformally equivalent to $\mathbb{C}$.

In Section 10.2 we shall prove the important *Riemann mapping theorem*, which asserts that an arbitrary region in the class (c) is conformally equivalent to the unit disc. Because conformal equivalence is an equivalence relation, it is symmetric and transitive, and so the Riemann mapping theorem implies that any two regions in the class (c) are conformally equivalent, for we can use the unit disc as an intermediate stage.

On the other hand, two regions belonging to different classes cannot be conformally equivalent. This can be seen quickly as follows: Without loss of generality (since we are temporarily assuming the Riemann mapping theorem), we may confine attention to the sphere, $\mathbb{C}$, and the unit disc D. The sphere is compact, while neither $\mathbb{C}$ nor D is compact; since the continuous (*a fortiori* the conformal) image of a compact set is compact, we are left only with the question whether $\mathbb{C}$ and D are conformally equivalent; by the Liouville Theorem 4.4.9, they are *not* conformally equivalent, and so the above assertion is confirmed.

10.2   The Riemann Mapping Theorem

Riemann is justly credited with the theorem that bears his name, despite the fact that, as was pointed out by Weierstrass, the proposed proof contains a serious flaw. A long sequence of modified proofs were then developed during the following half-century, but all of them imposed assumptions concerning the smoothness of the boundary. Finally, H. Poincaré succeeded in mapping a region conformally onto a unit disc, without regard for boundary behavior (thus avoiding completely any assumption about the boundary). Since then numerous contributors have provided alternative proofs and have investigated delicate problems concerning the behavior of the mapping function near the boundary. Since 1851, when Riemann presented his flawed proof, many major analysts have participated in what has become a substantial part of complex analysis. A very partial listing of names with which the reader should become familiar is the following: C. Neumann, H.A. Schwarz, D. Hilbert, R. Courant, P. Koebe, E. Lindelöf, P. Painlevé, and C. Carathéodory (plus, of course, Riemann, Weierstrass, and Poincaré).

THEOREM 10.2.1 (Riemann Mapping Theorem) *Every simply connected region Ω in the Riemann sphere $\hat{\mathbb{C}}$ whose boundary consists of at least two points is conformally equivalent to the unit disc. More specifically, let z_0 be an arbitrary point in a region Ω. Then there exists a unique function $f(z) \in H(\Omega)$ which is a bijection of Ω onto the unit disc D satisfying*

$$f(z_0) = 0, \quad f'(z_0) > 0.$$

(The last condition states that directions at z_0 are not changed.)

Proof. Let us first establish the uniqueness of the mapping function. Suppose there exist two mapping functions $g_1(z)$ and $g_2(z)$ that satisfy all the requirements. Then $g_2^{-1}(w) \in \Omega$ if $w \in D$. Hence $h(w) = g_1\left(g_2^{-1}(w)\right)$ is an analytic function mapping the unit disc D onto itself; that is,

$$|h(w)| < 1 \quad \text{if} \quad w \in D$$

and

$$h(0) = g_1\left(g_2^{-1}(0)\right) = g_1(z_0) = 0.$$

Thus the conditions in the Schwarz Lemma 6.2.1 are satisfied, and we have

$$|h(w)| \le |w| \quad (w \in D).$$

Substituting $w = g_2(z)$, we obtain

$$|g_1(z)| \le |g_2(z)| \quad (z \in \Omega).$$

Interchanging the roles of g_1 and g_2, we obtain

$$|g_2(z)| \le |g_1(z)| \quad (z \in \Omega).$$

It follows that

$$|g_1(z)| = |g_2(z)|, \quad \text{and hence} \quad \left|\frac{g_1(z)}{g_2(z)}\right| = 1 \quad (z \in \Omega).$$

This implies that the analytic function $\frac{g_1(z)}{g_2(z)}$ (after removal of the singularity at $z = z_0$) has a constant modulus in Ω, hence is a constant (of modulus 1). Therefore, for some real constant α, we have

$$g_1(z) = e^{i\alpha} g_2(z) \quad (z \in \Omega).$$

But then $g_1'(z_0) = e^{i\alpha} g_2'(z_0)$, and because $g_1'(z_0) > 0$, $g_2'(z_0) > 0$, we have $\alpha \equiv 0 (\text{mod } 2\pi)$, which finally gives

$$g_1(z) \equiv g_2(z) \quad (z \in \Omega).$$

Thus, there can be at most one mapping function satisfying all the requirements.

Having established the uniqueness of the mapping function, we now turn to the problem of existence. Without loss of generality, we may assume that z_0 is not the point at infinity, for this can always be achieved by applying a Möbius transformation. Let $\mathcal{F}$ be the family of all functions $f(z) \in H(\Omega)$ with the following properties:

(a) $f(z)$ is univalent in Ω;

(b) $\|f\|_\Omega \leq 1$;

(c) $f(z_0) = 0$ and $f'(z_0) > 0$.

In other words, functions in $\mathcal{F}$ satisfy all the requirements of the mapping function except possibly the surjectivity. The rest of the proof is carried out in three steps.

1^0. $\mathcal{F}$ is not empty. The proof that $\mathcal{F} \neq \phi$ is immediate if Ω is bounded, $\Omega \subset D(z_0, R)$, say. For in this case, simply set $f(z) = \frac{z-z_0}{R}$; then it is trivial to verify that $f(z) \in \mathcal{F}$. Thus, when the region Ω is unbounded, we try to map it first to a bounded region. Because the square-root function maps the whole complex plane to a "half of it," it leaves a vast "empty space," allowing us to apply reciprocation $\left(w = \frac{1}{z}\right)$ to map our region to a bounded region. We carry out our plan as follows: By assumption, there exist at least two distinct points a and b on the boundary $\partial\Omega$ of Ω. Set $\varphi(z) = \sqrt{\frac{z-a}{z-b}}$. (If one of the points, say b, is the point at infinity, let $\varphi(z) = \sqrt{z - a}$.) Then the only possible singularities of this function are the points a and b. However, they are not in Ω, and so $\varphi(z)$ is analytic in Ω. Moreover, because Ω is simply connected, $\varphi(z)$ is single-valued there, by the Monodromy Theorem 8.6.1. Furthermore, $\varphi(z)$ is univalent in Ω. For if there were z_1 and z_2 in Ω such that $\varphi(z_1) = \varphi(z_2)$, then $\frac{z_1-a}{z_1-b} = \frac{z_2-a}{z_2-b}$, which implies that $z_1 = z_2$ (i.e., $\varphi(z)$ satisfies the condition (a)). The same argument shows that there are no two points z_1 and z_2 in Ω such that $\varphi(z_1) = -\varphi(z_2)$. By the Open Mapping Theorem 5.6.9, the image of Ω under $\varphi(z)$ contains a disc $D(\varphi(z_0), \rho)$ $(\rho > 0)$. By what we have just shown, the image of Ω under $\varphi(z)$ does not intersect the disc $D(-\varphi(z_0), \rho)$; that is, $|\varphi(z) + \varphi(z_0)| \geq \rho$ $(z \in \Omega)$. Therefore, the function

$$\psi(z) = \frac{\rho}{\varphi(z) + \varphi(z_0)}$$

satisfies $\|\psi\|_\Omega \leq 1$; in particular, the image of Ω under $\psi(z)$ is bounded. (Note that $\psi(z)$ satisfies conditions (a) and (b).) Now let $f(z)$ be the composite of $\psi(z)$ with a Möbius transformation that maps the unit disc onto itself and carries the point $\psi(z_0)$ to the origin, followed by a suitable rotation (multiplication by a complex number of modulus 1) to achieve that

$f'(z_0) > 0$. (The last step is possible because the derivative of a univalent analytic function never vanishes.) Thus $\mathcal{F}$ is not empty.

2^0. Now, temporarily assuming the existence of the (unique) mapping function, let $g(z)$ be this function. Then for an arbitrary function $f(z) \in \mathcal{F}$, as in the proof of uniqueness, the composition $(f \circ g^{-1})(w)$ satisfies the conditions in the Schwarz Lemma 6.2.1, and we have

$$|(f \circ g^{-1})'(0)| \leq 1, \quad \text{that is,} \quad f'(z_0) \leq g'(z_0).$$

This observation suggests that we look for a function $g(z) \in \mathcal{F}$ for which $g'(z_0)$ has the largest value. (Note that the roles of $f(z)$ and $g(z)$ may not be reversed here. (Why?))

We observe first that the set $\{f'(z_0) ; \ f(z) \in \mathcal{F}\}$ is bounded. For, if Ω contains the closed disc $\bar{D}(z_0, \ r) \ (r > 0)$, then by the Cauchy Estimate 4.4.7 and the condition (b), we have, for every $f(z) \in \mathcal{F}$,

$$|f'(z_0)| \leq \frac{1}{2\pi} \int_{C(z_0,r)} \left| \frac{f(z)}{(z - z_0)^2} \right| \cdot |dz| \leq \frac{1}{2\pi} \cdot \frac{2\pi r}{r^2} = \frac{1}{r}.$$

Let $M = \sup\{f'(z_0) ; \ f(z) \in \mathcal{F}\}$. Clearly, $M > 0$. We want to show that $\mathcal{F}$ has an element $g(z)$ such that $g'(z_0) = M$. (In the next step 3^0, we shall show that $g(z)$ is the desired mapping function.) Let $\{f_k(z)\}_{k=1}^{\infty}$ be a sequence of functions in $\mathcal{F}$ such that

$$\lim_{k \to \infty} f_k'(z_0) = M.$$

Because $\|f_k\|_\Omega \leq 1$, the Montel Selection Theorem 9.1.3 guarantees that $\{f_k(z)\}_{k=1}^{\infty}$ has a subsequence that converges locally uniformly on Ω (to $g(z)$, say). Being the local uniform limit of a sequence of univalent analytic functions, $g(z)$ is also analytic and univalent, for $g(z)$ is not a constant $(\because \ g'(z_0) = M > 0)$. (Cf. Corollary 5.6.6.) Clearly, $\|g\|_\Omega \leq 1$. Thus $g(z) \in \mathcal{F}$.

3^0. It remains to show that $g(z)$ maps Ω *onto*, not merely *into*, the unit disc. Suppose there were a point w_0 in the unit disc D that is not in the image of Ω under the mapping function $g(z)$. Then, because of the simple connectedness of Ω, we can determine a single-valued branch of

$$G(z) = \sqrt{\frac{g(z) - w_0}{1 - \bar{w}_0 g(z)}} \quad (z \in \Omega).$$

Clearly, $G(z)$ is univalent in Ω, and $\|G\|_\Omega \leq 1$. (That is, $G(z)$ satisfies the conditions (a) and (b).) We now perform the composition of $G(z)$ with an

appropriate Möbius transformation so that the resulting function satisfies
all the conditions (a), (b), (c), and thus belongs to $\mathcal{F}$. Set

$$F(z) = \frac{|G'(z_0)|}{G'(z_0)} \cdot \frac{G(z) - G(z_0)}{1 - \overline{G(z_0)}G(z)} \quad (z \in \Omega).$$

Then, clearly, $F(z)$ is analytic and univalent in Ω,

$$\|F\|_\Omega \leq 1, \quad F(z_0) = 0,$$

and

$$F'(z_0) = \frac{|G'(z_0)|}{1 - |G(z_0)|^2} = \frac{1 + |w_0|}{2|w_0|^{\frac{1}{2}}} M > M;$$

the inequality follows from $(1 - |w_0|^{\frac{1}{2}})^2 > 0$ (for $|w_0| < 1$). We have
shown that $F(z)$ is an element of $\mathcal{F}$ whose derivative at z_0 is greater than
M, which contradicts the definition of M. Thus $g(z)$ maps Ω *onto* the unit
disc. (The last part of this proof is attributed to Koebe.)   □

Now we obtain the following corollary (which is often also called the
Riemann Mapping Theorem). A formal proof is not needed.

COROLLARY 10.2.2 *Suppose Ω and Ω' are simply connected regions
in $\hat{\mathbb{C}}$ whose boundaries consist of at least two points, and let z_0, w_0 be
arbitrary points in Ω and Ω', respectively. Then there exists a unique
analytic function in $H(\Omega)$ that maps Ω conformally onto Ω', mapping z_0 to
w_0 and a preassigned direction at z_0 to a preassigned direction at w_0.*

In the Riemann Mapping Theorem 10.2.1, the correspondence between
the boundaries is not considered. In this connection, we mention the following

THEOREM 10.2.3 (Carathéodory) *Let Ω be a simply connected region
whose boundary is a curve (i.e., Ω is a Jordan region). If $f(z)$ is an
analytic function that maps Ω conformally onto the unit disc, then $f(z)$
can be extended so that it becomes a homeomorphism between the closure
$\bar{\Omega}$ of Ω and the closed unit disc $\bar{D}$. Moreover, if the boundary curve has
a tangent at z_1, then the mapping $w = f(z)$ is angle-preserving at the
point $w_1 = f(z_1)$. Furthermore, any 3 points on the boundary $\partial\Omega$ may
be mapped to any 3 preassigned points on the unit circle, provided they
determine the same orientation.*

Because the proof needs fairly lengthy preparation, we do not present it here.

10.3 Examples

Even though every simply connected region with at least two boundary points is conformally equivalent to the unit disc, it is usually rather difficult to work out the mapping function. In this section we illustrate some of the basic techniques.

EXAMPLE 1 $w = \frac{1}{2}\left(z + \frac{1}{z}\right)$.

Setting $z = re^{i\theta}$, $w = u + iv$, we obtain

$$u = \frac{1}{2}\left(r + \frac{1}{r}\right)\cos\theta, \quad v = \frac{1}{2}\left(r - \frac{1}{r}\right)\sin\theta.$$

Thus, the image of the circle $C(0, r)$ $(r \neq 1)$ is

$$\frac{u^2}{\left\{\frac{1}{2}\left(r + \frac{1}{r}\right)\right\}^2} + \frac{v^2}{\left\{\frac{1}{2}\left(r - \frac{1}{r}\right)\right\}^2} = 1,$$

which is an ellipse with foci at $w = \pm 1$, for $\left\{\frac{1}{2}\left(r + \frac{1}{r}\right)\right\}^2 - \left\{\frac{1}{2}\left(r - \frac{1}{r}\right)\right\}^2 = 1$.

If $r = 1$, then the ellipse degenerates to the line segment $[-1, 1]$. Note that the circles $C(0, r)$ and $C(0, \frac{1}{r})$ are mapped to the same ellipse (if $r \neq 1$) (Figure 10.1).

On the other hand, the image of the ray $\arg z = \theta$ is a branch of the hyperbola

$$\frac{u^2}{\cos^2\theta} - \frac{v^2}{\sin^2\theta} = 1,$$

whose foci are also at $w = \pm 1$, for $\cos^2\theta + \sin^2\theta = 1$. Being the conformal images of concentric circles and the rays from the common center, these two families of conics are orthogonal to each other.

Question: Find the images (in the w-plane) of the coordinate axes (in the z-plane).

Because $\frac{dw}{dz} = \frac{1}{2}\left(1 - \frac{1}{z^2}\right)$, the mapping ceases to be conformal at the points $z = \pm 1$, which are mapped to $w = \pm 1$.

It is simple to verify that the mapping is univalent inside and outside the unit circle, and each of these two regions is mapped conformally onto the extended w-plane with the real interval $[-1, 1]$ deleted.

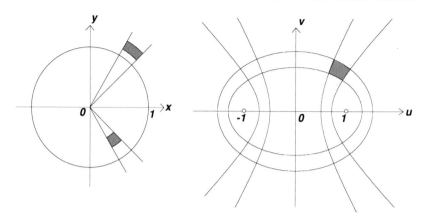

Figure 10.1

Figure 10.2

Now, consider the image of a circle C passing through the point $z = 1$ and containing the point $z = -1$ in its interior. Because $\frac{dw}{dz}$ has a simple root at $z = 1$, the image of the exterior of the circle has an *"angle"* 2π at $w(1) = 1$. If the circle C is symmetric with respect to the real axis, then C contains the unit circle $C(0, 1)$ in its interior (except the point $z = 1$). Because the image of the unit circle $C(0, 1)$ degenerates to the interval $[-1, 1]$ on the real axis, if the circle C is sufficiently close to the unit circle, then its image is like that in Figure 10.2 on the left.

If the circle C is not symmetric to the real axis, then its image is like that in Figure 10.2 on the right. Such a curve is known as a *Joukowski aerofoil*. As suggested by this name, the mapping which we have been discussing has been used very effectively in the investigation of airflow past a wing.

EXAMPLE 2 $w = \dfrac{z}{(1 - z)^2}.$

Because $\frac{z}{(1-z)^2} = \frac{1}{\left(z+\frac{1}{z}\right)-2}$, by setting $\zeta = z + \frac{1}{z}$, $w = \frac{1}{\zeta-2}$ we see that the unit disc in the z-plane is mapped to the ζ-plane with the real interval

Figure 10.3

$[-2, 2]$ deleted, and this region in turn is mapped to the w-plane with a cut along the negative real axis from ∞ to $-\frac{1}{4}$ (Figure 10.3).

We remark that

$$w = \frac{z}{(1 - z)^2} = z + 2z^2 + 3z^3 + \cdots + nz^n + \cdots \quad (z \in D)$$

is a Koebe function that has many extremal properties connected with the theory of functions univalent in the unit disc. (Cf. Theorem 9.2.6.)

EXAMPLE 3 $w = e^z$.

Set $z = x + iy$, $w = Re^{i\varphi}$; then $R = e^x$, $\varphi \equiv y \pmod{2\pi}$.

That is, a rectangle

$$\Omega = \{z \in \mathbb{C}; a < \Re z < b, \ c < \Im z < d\}$$

is mapped to

$$\{w \in \mathbb{C}; e^a < |w| < c^b, \ c < \arg w < d\}.$$

(Figure 10.4.) In particular, the strip

$$\{z \in \mathbb{C}; 0 < \Im z < \frac{\pi}{2}\}$$

is mapped to the first quadrant

$$\{w \in \mathbb{C}; \Re w > 0, \ \Im w > 0\}.$$

EXAMPLE 4 $w = \sin z = \frac{1}{2i}\left(e^{iz} - e^{-iz}\right)$.

Set $\zeta_1 = iz$, $\zeta_2 = e^{\zeta_1}$, $\zeta_3 = \frac{1}{i}\zeta_2$, $w = \frac{1}{2}\left(\zeta_3 + \frac{1}{\zeta_3}\right)$. Then we see that a horizontal line in the z-plane is mapped to a vertical line in the ζ_1-plane, which is then mapped to a circle with center at the origin in the ζ_2-plane, which is then mapped to a circle with the same radius in the ζ_3-plane, and

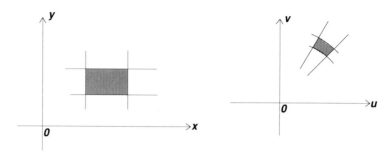

Figure 10.4

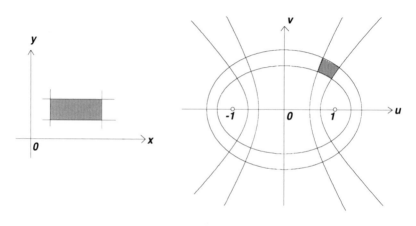

Figure 10.5

finally this circle is mapped to an ellipse in the w-plane with $w = \pm 1$ as its foci (Figure 10.5).

On the other hand, a vertical line in the z-plane is mapped to a horizontal line in the ζ_1-plane, then to a ray from the origin in the ζ_2-plane, then to another ray in the ζ_3-plane, and finally to a branch of a hyperbola in the w-plane with $w = \pm 1$ as its foci.

EXAMPLE 5   Find a conformal mapping of the crescent $\Omega = D(2, 2) \setminus D(1, 1)$ to the unit disc.

Because the two boundary circles are tangent to each other at the origin, the Möbius transformation $\zeta_1 = \frac{1}{z}$ maps these two boundary circles to a pair of parallel lines, which pass through the points $\frac{1}{4}$ and $\frac{1}{2}$, respectively (Figure 10.6). They are perpendicular to the real axis, for the Möbius transformation maps the positive real axis to itself, and the two boundary circles are orthogonal to the real axis. Next, the mapping $\zeta_2 = 4\pi i \zeta_1$ maps the

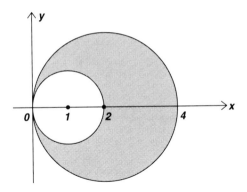

Figure 10.6

vertical strip to a horizontal strip with width π: $\{\zeta_2 \in \mathbb{C}; \; \pi < \Im\zeta_2 < 2\pi\}$. The mapping $\zeta_3 = e^{\zeta_2}$ maps the horizontal strip to the lower half-plane $\{\zeta_3 \in \mathbb{C}; \; \Im\zeta_3 < 0\}$. Finally, the Möbius transformation $w = \frac{\zeta_3+i}{\zeta_3-i}$ maps the lower half-plane to the unit disc.

$$\therefore \; w = \frac{\zeta_3 + i}{\zeta_3 - i} = \frac{e^{\zeta_2} + i}{e^{\zeta_2} - i} = \frac{e^{4\pi i \zeta_1} + i}{e^{4\pi i \zeta_1} - i} = \frac{e^{\frac{4\pi i}{z}} + i}{e^{\frac{4\pi i}{z}} - i}.$$

Question: What point in the region Ω is mapped to the origin by this mapping? Verify that at this point $\frac{dw}{dz} = \frac{9\pi}{32} > 0$.

EXAMPLE 6   Let us now find a conformal mapping of a lune region $D(1, 1) \setminus D(0, \sqrt{3})$ onto the unit disc (Figure 10.7).
    Because the two circular arcs intersect at $z = \sqrt{3}e^{\pm i \frac{\pi}{6}}$ with the angle of intersection $\frac{\pi}{6}$, the Möbius transformation

$$\zeta_1 = e^{i\theta} \frac{z - \sqrt{3}e^{i\frac{\pi}{6}}}{z - \sqrt{3}e^{-i\frac{\pi}{6}}}$$

(which maps one of the intersections, $\sqrt{3}e^{i\frac{\pi}{6}}$, to the origin and the other, $\sqrt{3}e^{-i\frac{\pi}{6}}$, to the point at infinity) maps the lune region to an infinite sector with angle $\frac{\pi}{6}$ at the origin (Figure 10.8). We choose θ in the above definition of ζ_1 such that this sector is in the first quadrant with one of the boundary rays coinciding with the positive real axis. To accomplish this, we find the image of the midpoint $z = \frac{3}{2}$ of the common chord (the line segment joining the two intersection points of the circular arcs) to be $-e^{i\theta}$. Because this common chord and the two circular arcs intersect (at the point $\sqrt{3}e^{i\frac{\pi}{6}}$)

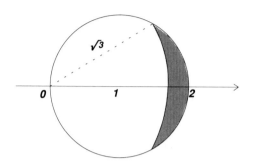

Figure 10.7

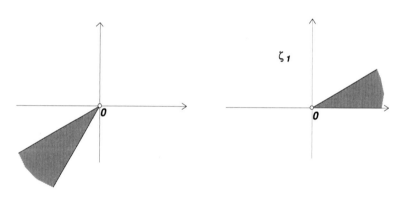

Figure 10.8

at angles $\frac{\pi}{6}$ and $\frac{\pi}{3}$, respectively, we see that the Möbius transformation *without* the factor $e^{i\theta}$ maps the lune region onto the sector

$$\{\zeta_1 \in \mathbb{C};\ \frac{7\pi}{6} < \arg \zeta_1 < \frac{4\pi}{3}\}.$$

Thus the choice $\theta = \frac{5\pi}{6}$ will serve our purpose; that is,

$$\zeta_1 = e^{i\frac{5\pi}{6}} \frac{z - \sqrt{3}e^{i\frac{\pi}{6}}}{z - \sqrt{3}e^{-i\frac{\pi}{6}}}.$$

Next, set $\zeta_2 = \zeta_1^6$. This transformation maps the sector onto the upper half-plane. It now remains to map the upper half-plane to the unit disc.

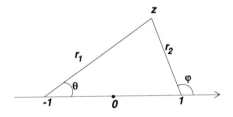

Figure 10.9

This step is accomplished by

$$w = \frac{\zeta_2 - i}{\zeta_2 + i}.$$

We obtain, finally,

$$w = \frac{\zeta_1^6 - i}{\zeta_1^6 + i} = \frac{\left(z - \sqrt{3}e^{i\frac{\pi}{6}}\right)^6 + i\left(z - \sqrt{3}e^{-i\frac{\pi}{6}}\right)^6}{\left(z - \sqrt{3}e^{i\frac{\pi}{6}}\right)^6 - i\left(z - \sqrt{3}e^{-i\frac{\pi}{6}}\right)^6}.$$

EXAMPLE 7 $\quad w = \displaystyle\int_0^z \frac{dt}{\sqrt{1 - t^2}}.$

Of course this is the arcsine function, and so we know its mapping behavior by reversing the mapping in Example 4, but here we find the image of the upper half-plane directly. We choose the principal branch for the square-root function (Figure 10.9). That is, for $\Im z > 0$,

$$\sqrt{1 + z} = r_1^{\frac{1}{2}} e^{i\frac{\theta}{2}} \quad (r_1 > 0,\ 0 < \theta < \pi),$$
$$\sqrt{1 - z} = r_2^{\frac{1}{2}} e^{i\frac{\varphi - \pi}{2}} \quad (r_2 > 0,\ 0 < \varphi < \pi).$$

It follows that, for $z < -1$, we have $\varphi = \theta = \pi$, and so

$$\frac{1}{\sqrt{1 - z^2}} = (r_1 r_2)^{-\frac{1}{2}} e^{-\frac{i\pi}{2}},$$

and the value of the integrand is on the negative imaginary axis. For $-1 < z < 1$, we have $\theta = 0$, $\varphi = \pi$, and so

$$\frac{1}{\sqrt{1 - z^2}} = (r_1 r_2)^{-\frac{1}{2}},$$

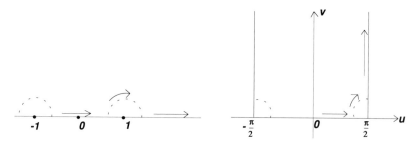

Figure 10.10

and the value of the integrand is on the positive real axis. For $z > 1$, we have $\theta = \varphi = 0$, and so

$$\frac{1}{\sqrt{1 - z^2}} = (r_1 r_2)^{-\frac{1}{2}} e^{\frac{i\pi}{2}},$$

and the value of the integrand is on the positive imaginary axis. Therefore, as z starts from the origin and moves along the positive real axis to 1 its image w moves from the origin along the positive real axis to

$$\int_0^1 \frac{dt}{\sqrt{1 - t^2}} = \frac{\pi}{2}.$$

Now we let the path of integration pass above the point $z = 1$ along a small semicircle. Then, as we have already discussed, the integrand takes values on the positive imaginary axis, and we obtain

$$w = \frac{\pi}{2} + i \int_1^z \frac{dt}{\sqrt{t^2 - 1}}.$$

As z tends to infinity along the positive real axis, its image w tends to infinity along the line $\Re w = \frac{\pi}{2}$ ($\Im w > 0$) (Figure 10.10). A similar discussion shows that the real interval $-1 < z < 0$ is mapped to the real interval $-\frac{\pi}{2} < w < 0$, and $\infty < z < -1$ is mapped to the line $\Re w = -\frac{\pi}{2}$ ($\Im w > 0$). We conclude that the upper half z-plane is mapped to the upper half-strip $\{w \in \mathbb{C}; |\Re w| < \frac{\pi}{2}, \Im w > 0\}$.

EXAMPLE 8   Finally, we consider the mapping of the upper half z-plane by

$$w = \int_0^z (t + 1)^{-\frac{5}{6}} \cdot t^{-\frac{1}{2}} \cdot (1 - t)^{-\frac{2}{3}} \, dt,$$

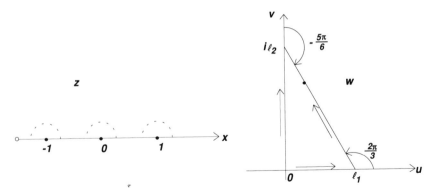

Figure 10.11

where all radicals are interpreted as the principal values. As z starts from the origin and moves toward 1 along the (positive) real axis, its image w also starts from the origin and moves along the positive real axis to

$$\ell_1 = \int_0^1 (t+1)^{-\frac{5}{6}} \cdot t^{-\frac{1}{2}} \cdot (1-t)^{-\frac{2}{3}} dt.$$

(We could actually express this value ℓ_1 in terms of the Euler gamma function, but this is inessential for our purpose here.) As z moves along a small semicircle above the point $z = 1$ the argument of $1 - z$ decreases by π, so that of $(1 - z)^{-\frac{2}{3}}$ increases by $\frac{2\pi}{3}$ (Figure 10.11). Because the arguments of the other factors are not affected in a neighborhood of $z = 1$, we see that as z moves beyond 1 toward ∞ along the positive real axis, its image w moves along a ray (from the point $w = \ell_1$) forming an angle $\frac{2\pi}{3}$ with the positive real axis. Note that it stops at a finite point because the improper integral converges:

$$\int_1^\infty (t+1)^{-\frac{5}{6}} \cdot t^{-\frac{1}{2}} \cdot (t-1)^{-\frac{2}{3}} dt < \infty.$$

On the other hand, as z moves along a small semicircle above the origin from the positive side to the negative side, the argument of z increases by π, and so that of $z^{-\frac{1}{2}}$ decreases by $\frac{\pi}{2}$. Therefore,

$$\int_0^{-1} (t+1)^{-\frac{5}{6}} \cdot t^{-\frac{1}{2}} \cdot (1-t)^{-\frac{2}{3}} dt = i\ell_2,$$

for some $\ell_2 > 0$; that is, the image of $-1 < z < 0$ is on the positive imaginary axis.

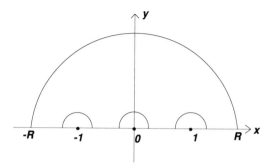

Figure 10.12

Finally, as z moves along a small semicircle above the point $z = -1$, the argument of $(z + 1)^{-\frac{5}{6}}$ decreases by $\frac{5\pi}{6}$. Therefore, as z moves from -1 to ∞ along the negative real axis, its image w moves on the ray (from the point $i\ell_2$) forming an angle $-\frac{5\pi}{6}$ with the positive imaginary axis (i.e., $-\frac{\pi}{3}$ with the positive real axis), stopping at some finite point.

Because the integrand is single-valued in the upper half-plane $\{z \in \mathbb{C} \, ; \, \Im z > 0\}$ (by the Monodromy Theorem 8.6.1), it follows from the Cauchy-Goursat Theorem 4.3.1 that

$$\int_{C_{r,R}} (t + 1)^{-\frac{5}{6}} \cdot t^{-\frac{1}{2}} \cdot (1 - t)^{-\frac{2}{3}} dt = 0,$$

where $C_{r,R}$ is the curve (consisting of four semicircles and four line segments) shown in Figure 10.12.

This implies that the image of the curve $C_{r,R}$ under the mapping

$$w = \int_0^z (t + 1)^{-\frac{5}{6}} \cdot t^{-\frac{1}{2}} \cdot (1 - t)^{-\frac{2}{3}} dt$$

is a closed curve in the w-plane. Because all four integrals over the semicircles tend to zero (when $r \to 0$, $R \to \infty$), we conclude that the image of the real axis in the z-plane must be a closed curve in the w-plane. We have shown that the image of the upper half z-plane is a triangle with angles $\frac{\pi}{2}$, $\frac{\pi}{3}$, $\frac{\pi}{6}$, in the w-plane (right half of Figure 10.11).

By magnification (or contraction) (i.e., by multiplying by a suitable positive real number) this triangle can be made congruent to any triangle with angles $\frac{\pi}{2}$, $\frac{\pi}{3}$, $\frac{\pi}{6}$ (with the same orientation). Now, a rotation followed by a translation makes the triangle coincide with any given triangle with angles $\frac{\pi}{2}$, $\frac{\pi}{3}$, $\frac{\pi}{6}$.

We have shown that, with suitable choice of constants k and c,

$$w = k \int_0^z (t+1)^{-\frac{5}{6}} \cdot t^{-\frac{1}{2}} \cdot (1-t)^{-\frac{2}{3}} dt + c$$

maps the upper half z-plane to (the interior of) a preassigned triangle with angles $\frac{\pi}{2}, \frac{\pi}{3}, \frac{\pi}{6}$ in the w-plane.

Obviously, the above reasoning can be generalized. Given any triangle in the w-plane whose *exterior* angles are $\alpha\pi, \beta\pi, \gamma\pi$ ($\alpha + \beta + \gamma = 2$, $\alpha > 0, \beta > 0, \gamma > 0$), there exist constants k and c such that

$$w = k \int_0^z (t+1)^{-\alpha} \cdot t^{-\beta} \cdot (1-t)^{-\gamma} dt + c,$$

maps the upper half z-plane conformally onto that triangular region in the w-plane. More generally, for any choice of 3 distinct real numbers a_1, a_2, a_3, and an arbitrary point z_0 in the upper half z-plane,

$$w = k \int_{z_0}^z (t - a_1)^{-\alpha} \cdot (t - a_2)^{-\beta} \cdot (t - a_3)^{-\gamma} dt + c$$

gives a desired mapping with appropriate choice of the constants k and c. Actually, this is a particular case of the *Schwarz-Christoffel formula*:

$$w = k \int_{z_0}^z (t - a_1)^{-\alpha_1} \cdot (t - a_2)^{-\alpha_2} \cdots (t - a_n)^{-\alpha_n} dt + c$$

(where $\Im z_0 > 0$, $a_j \in \mathbb{R}$ ($j = 1, 2, \ldots, n$) and $\alpha_1 + \alpha_2 + \cdots + \alpha_n = 2$), which maps the upper half z-plane onto an n-gon in the w-plane.

10.4   Conformal Mappings of Multiply Connected Regions

So far we have been discussing conformal mappings of simply connected regions. Let us now turn to the next simplest case. A natural candidate for the standard region for doubly connected regions is an annulus (a region bounded by two concentric circles). Contrary to what one might expect, there is no conformal mapping between two annuli unless they happen to have the same ratio of radii (Exercise 10.12). However, it *is* true that every doubly connected region is conformally equivalent to *some* annulus. Here we prove the following analogue of the Riemann Mapping Theorem 10.2.1 for multiply connected regions.

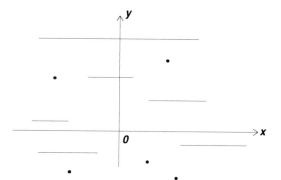

Figure 10.13

THEOREM 10.4.1 *Let Ω be a multiply connected region in the Riemann sphere $\hat{\mathbb{C}}$ containing the point at infinity in its interior. Then there exists a function $f(z)$, with Laurent expansion of the form*

$$w = f(z) = z + \frac{a_1}{z} + \frac{a_2}{z^2} + \cdots$$

near the point at infinity, which maps the region Ω conformally onto a parallel-slit region in the w-plane; that is, a region whose boundary components are either points or slits parallel to the real axis (Figure 10.13).

(The condition that the region Ω contains the point at infinity is no restriction at all, for it can always be achieved through a Möbius transformation.)

Proof. Let $\mathcal{F}$ be the family of all functions $f(z)$ univalent and analytic in Ω except at the point at infinity, where they have Laurent expansions of the form

$$w = f(z) = z + \frac{a_f}{z} + \cdots.$$

This family $\mathcal{F}$ is nonempty, because $f(z) = z$ is certainly in $\mathcal{F}$.

(1) There exists a function $g(z) \in \mathcal{F}$ such that

$$\Re a_g = \mu = \sup\{\Re a_f ;\ f \in \mathcal{F}\}.$$

By the assumption that the point at infinity belongs to Ω, it follows that we may take r so large that all the boundary points of Ω are in the disc $D(0, r)$. Then the function

$$\frac{f(rz)}{r} = z + \frac{a_f}{r^2 z} + \cdots$$

is univalent and analytic for $|z| > 1$. Thus, by the Bieberbach Area Theorem 9.2.4,

$$|a_f| \leq r^2, \quad f \in \mathcal{F}.$$

Hence $\mu = \sup\{\Re a_f \,;\, f \in \mathcal{F}\}$ is finite.

(2) Next we show that $\mathcal{F}$ is a normal family of analytic functions in Ω. Let r be as before. Any function $w = f(z)$ in $\mathcal{F}$ maps the region $\{z \in \mathbb{C}; |z| > r\}$ to a region outside some simple closed curve Γ in the w-plane. (Every function in $\mathcal{F}$ behaves like the identity transformation near the point at infinity.) Choose an arbitrary point w_0 inside the curve Γ. Then the function

$$F(\zeta) = \frac{r}{f(z) - w_0} \quad \left(\zeta = \frac{r}{z}\right),$$

is analytic and univalent for ζ in the unit disc D, and it has the Taylor expansion

$$F(\zeta) = \zeta + \frac{w_0}{r}\zeta^2 + \cdots, \quad \zeta \in D.$$

Therefore, by the Bieberbach Theorem 9.2.5,

$$\left|\frac{w_0}{r}\right| \leq 2; \quad \text{that is,} \quad |w_0| \leq 2r.$$

This result implies that if $z \in \Omega \cap D(0, r)$, then $|f(z)| \leq 2r$. Because r can be arbitrarily large, we conclude that $\mathcal{F}$ is uniformly bounded on compact subsets of Ω; hence, by the Montel Theorem 9.1.3, $\mathcal{F}$ is normal.

(3) Now let $\{f_k(z)\}_{k=1}^{\infty}$ be a sequence of functions in $\mathcal{F}$ such that $\Re a_{f_k} \longrightarrow \mu$ $(k \to \infty)$. Because $\mathcal{F}$ is a normal family, $\{f_k(z)\}_{k=1}^{\infty}$ has a subsequence that converges locally uniformly in the region Ω. Without loss of generality, we may assume that the original sequence converges locally uniformly, to $g(z)$, say. Clearly, $g(z)$ is analytic and univalent in $\Omega \setminus \{\infty\}$. Claim: the function $g(z)$ has a simple pole at the point at infinity. By assumption

$$f_k(z) = z + \frac{a_1^{(k)}}{z} + \frac{a_2^{(k)}}{z^2} + \cdots \quad (|z| > r).$$

Therefore, $f_k(z) - z$ is analytic for $r < |z| \leq \infty$ and converges uniformly to $g(z) - z$ on $C(0, 2r)$, hence uniformly for $|z| \geq 2r$. Thus $g(z) - z$ is analytic at the point at infinity (and vanishes there because all $f_k(z) - z$ do), and hence $g(z)$ has a Laurent expansion

$$g(z) = z + \frac{a_1}{z} + \frac{a_2}{z^2} + \cdots$$

for $|z|$ sufficiently large; we have therefore shown that $g(z) \in \mathcal{F}$. Because $\{f_k(z)\}_{k=1}^{\infty}$ converges uniformly to $g(z)$ on $C(0, r)$, we have

$$\frac{1}{2\pi i} \int_{C(0,r)} f_k(z)\, dz \longrightarrow \frac{1}{2\pi i} \int_{C(0,r)} g(z)\, dz \quad (k \to \infty);$$

that is,

$$\lim_{k \to \infty} a_{f_k} = a_g. \quad \therefore \Re a_g = \lim_{k \to \infty} \Re a_{f_k} = \mu.$$

(4) To show that this function $g(z)$ actually maps the region Ω conformally onto a parallel-slit region, we temporarily assume the following

LEMMA 10.4.2 *Let ω be a simply connected region in $\hat{\mathbb{C}}$ containing the point at infinity in its interior whose boundary is neither a point nor a slit parallel to the real axis. Then there exists a function $\varphi(\zeta)$ mapping the region ω conformally onto a parallel-slit region having an expansion*

$$w = \varphi(\zeta) = \zeta + \frac{a}{\zeta} + \cdots$$

near the point at infinity. Moreover, $\Re a > 0$.

Assume the lemma for a moment, and suppose that the function $g(z)$ obtained above does not map the region Ω conformally onto a parallel-slit region. Then the boundary of the image of $g(z)$ must have at least one component that is not a point and is not a slit parallel to the real axis. Let ω be a simply connected region containing the image of the region under the mapping $g(z)$ whose boundary consists of any boundary component of the image of Ω under $g(z)$ that is not a slit parallel to the real axis. By the lemma, let

$$\varphi(\zeta) = \zeta + \frac{a}{\zeta} + \cdots \quad (\text{for } |\zeta| \text{ sufficiently large})$$

be a function mapping ω conformally onto a parallel-slit region with $\Re a > 0$. Consider the composite mapping: $g(z)$ followed by $\varphi(\zeta)$. Because $\varphi \circ g$ is univalent and analytic on Ω, and

$$w = \varphi\left(g(z)\right) = g(z) + \frac{a}{g(z)} + \cdots$$

$$= \left(z + \frac{a_g}{z} + \cdots\right) + \frac{a}{\left(z + \dfrac{a_g}{z} + \cdots\right)} + \cdots$$

$$= z + \frac{a_g + a}{z} + \cdots.$$

$$\therefore \varphi \circ g \in \mathcal{F}.$$

However, $\Re \left(a_g + a\right) = \Re a_g + \Re a > \Re a_g$, but this result contradicts the choice of $g(z)$ as a solution of the extremum problem posed earlier. Therefore, the image of Ω under $g(z)$ must be a parallel-slit region.

(5) It remains to prove the lemma. By the Riemann Mapping Theorem 10.2.1 (and using a function like the one in Section 10.3, Example 1), there exists a univalent analytic function that maps ω conformally onto a parallel-slit region while keeping the point at infinity fixed and the directions there unchanged. Such a function must be of the form

$$\tilde{w} = c_1 \zeta + c_0 + \frac{c_{-1}}{\zeta} + \cdots \quad (c_1 \text{ is real and positive})$$

near the point at infinity. Then we set

$$\varphi(\zeta) = w = \frac{\tilde{w} - c_0}{c_1} = \zeta + \frac{a}{\zeta} + \cdots.$$

Because c_1 is real, the image of ω in the w-plane is still a parallel-slit region. This shows the existence of such a mapping.

We must now show that $\Re a > 0$. The proof involves a clever use of the Bieberbach Area Theorem 9.2.4. Let the slit boundary of the region that is the image of ω under $\varphi(\zeta)$ be $\overline{PG}$ (Figure 10.14). The exterior ($|z| > 1$) of the unit disc is mapped by

$$w_1 = z + \frac{1}{z}$$

conformally onto the parallel-slit region with the slit

$$\{w_1 \in \mathbb{C}; \; -2 \leq \Re w_1 \leq 2, \; \Im w_1 = 0\},$$

and this parallel-slit region is mapped by

$$w = A w_1 + B \quad (A \in \mathbb{R}, \; B \in \mathbb{C})$$

conformally onto the parallel-slit region that is the image of ω under $\varphi(z)$.

Thus,

$$\begin{aligned}
\zeta &= \varphi^{-1}(w) = w - \frac{a}{w} + \cdots \\
&= A w_1 + B - \frac{a}{A w_1 + B} + \cdots \\
&= A w_1 + B - \frac{a}{A} \cdot \frac{1}{w_1} + \cdots.
\end{aligned}$$

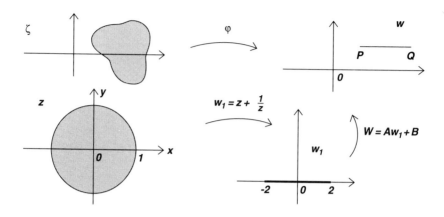

Figure 10.14

Setting $\zeta_1 = \frac{\zeta - B}{A}$ ($A \neq 0$, for the boundary of ω is not a point), we obtain

$$\zeta_1 = w_1 - \frac{a}{A^2}\frac{1}{w_1} + \cdots$$

$$= z + \frac{1}{z} - \frac{a}{A^2\left(z + \frac{1}{z}\right)} + \cdots$$

$$= z + \left(1 - \frac{a}{A^2}\right)\frac{1}{z} + \cdots.$$

Because $\zeta_1(z)$ is univalent for $|z| > 1$, the Bieberbach Area Theorem 9.2.4 now furnishes the inequality

$$\left|1 - \frac{a}{A^2}\right| \leq 1.$$

If equality holds here, we have, again by the Bieberbach Area Theorem 9.2.4,

$$\zeta_1 = z + \left(1 - \frac{a}{A^2}\right)\frac{1}{z}.$$

In this case $a \neq 0$; for otherwise the last equality becomes $\zeta_1 = z + \frac{1}{z}$; hence $\zeta_1 = w_1$, that is, $\zeta = w$, and therefore ω must be a slit region, contradicting the assumption. If the inequality holds in

$$\left|1 - \frac{a}{A^2}\right| \leq 1,$$

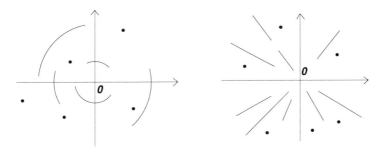

Figure 10.15

then certainly $a \neq 0$. Hence, $a \neq 0$ in any case, and this implies, in turn, $\Re a > 0$ (because $A \in \mathbb{R}$) and

$$\left| 1 - \frac{a}{A^2} \right| \leq 1. \qquad \qquad \square$$

Remarks. (a) If the region Ω has finite connectivity, then the mapping function is unique. However, Koebe proved (1918) that this assertion is no longer true if Ω is infinitely connected.

(b) Instead of taking a parallel-slit region as a standard region, we may use a circular-slit region (a region whose boundary consists of circular arcs with center at the origin) or radial-slit region (a region whose boundary consists of straight line segments whose extensions pass through the origin) (Figure 10.15).

Exercises

1. Show that conformal equivalence is an equivalence relation.

2. Show that the complex plane $\mathbb{C}$ is homeomorphic to the unit disc D, although these two regions are *not* conformally equivalent.

3. Show that the family of all analytic functions mapping a region conformally onto itself forms a group under composition.

4. Show that a multiply connected region is not conformally equivalent to a simply connected region.

5. (a) How many analytic functions are there that map the upper half-plane conformally onto itself and interchange two preassigned points (in the upper half-plane)? (Cf. Exercise 3.33.)

(b) What if the "upper half-plane" is replaced by a "simply connected region with at least two boundary points"?

6. (a) Show that the image of the unit disc under the mapping $w = (z + 1)^2$ is the interior of a cardioid, and find the equation of this cardioid.

   (b) Show that the image of the unit disc under the mapping $w = \frac{1}{(z+1)^2}$ is the exterior of a parabola with focus at the origin, and find the equation of this parabola.

7. Find the image under the mapping $w = \cosh z$ of a rectangle whose sides are parallel to the coordinate axes in the z-plane.

8. Find the image of the strip $\{z \in \mathbb{C} ; \ -\frac{\pi}{4} < \Re z < \frac{\pi}{4}\}$ under the mapping $w = \tan z$.

9. For each of the following regions find a conformal mapping of the region onto the unit disc:
   (a) $\{z \in \mathbb{C} ; \ |z| < 1, \ \Im z > 0\}$;
   (b) $\{z \in \mathbb{C} ; \ |z| < 1, \ \Re z > 0, \ \Im z > 0\}$;
   (c) $\{z \in \mathbb{C} ; \ |z| < 1, \ \Re z > \frac{1}{2}\}$;
   (d) $D(1, \ 1) \cap D(i, \ 1)$;
   (e) $D(1, \ \sqrt{3}) \setminus D(0, \ 2)$.

10. Find the conformal mapping $w = f(z)$ of a convex lens $D(\sqrt{3}, \ 2) \cap D(-\sqrt{3}, \ 2)$ onto the unit disc satisfying

$$f(0) = 0, \quad f'(0) > 0.$$

11. Find the image of the half-strip $\{z \in \mathbb{C} ; \ |\Re z| < \frac{\pi}{2}, \ \Im z > 0\}$ under the mapping $w = \sin z$, and verify that the result is consistent with Example 4 in Section 10.3.

12. Show that the annuli $1 < |z| < r$ and $1 < |w| < R$ are conformally equivalent if and only if $r = R$.
    Hint: Apply the Schwarz symmetry principle (Exercise 8.11) repeatedly to show that the mapping function is a Möbius transformation.

13. Find the image of the upper half z-plane under the mapping

$$w = \int_0^z \frac{dt}{\sqrt{(1 - t^2)(1 - k^2 t^2)}} \quad (0 < k < 1).$$

(This is one of the basic types which appear in the theory of elliptic integrals.)

14. (a) Show that the function

$$w = \int_0^z \frac{dt}{\sqrt{1 - t^4}}$$

maps the unit disc in the z-plane conformally onto a square in the w-plane.

Hint: Set $z = \frac{\zeta - i}{\zeta + i}$.

(b) How about $w = \int_0^z \frac{dt}{\sqrt{1 - t^3}}$?

15. Map the upper half-plane conformally onto an equilateral triangle.

Chapter 11

Harmonic Functions

11.1 Harmonic Conjugate

Let $u(x, y)$ be a real-valued function of class C^2 (i.e., the second partial derivatives exist and are continuous) that satisfies the *Laplace equation*

$$\nabla^2 u = \frac{\partial^2 u}{\partial x^2} + \frac{\partial^2 u}{\partial y^2} = 0$$

in a region Ω. We say that $u(x, y)$ is *harmonic* in Ω.

If a function $f(z) = u(z) + iv(z)$ is analytic in a region Ω, then the Cauchy-Riemann equations

$$\frac{\partial u}{\partial x} = \frac{\partial v}{\partial y}, \quad \frac{\partial u}{\partial y} = -\frac{\partial v}{\partial x},$$

hold, and hence

$$\nabla^2 u = \frac{\partial^2 u}{\partial x^2} + \frac{\partial^2 u}{\partial y^2} = \frac{\partial^2 v}{\partial x \partial y} - \frac{\partial^2 v}{\partial y \partial x} = 0.$$

We have used the fact that $\frac{\partial^2 v}{\partial x \partial y} = \frac{\partial^2 v}{\partial y \partial x}$ for functions $v \in C^2$. Similarly,

$$\nabla^2 v = \frac{\partial^2 v}{\partial x^2} + \frac{\partial^2 v}{\partial y^2} = 0.$$

Therefore, both the real and imaginary parts of an analytic function are harmonic; we say $v(x, y)$ is a *conjugate harmonic function* of $u(x, y)$; similarly $u(x, y)$ is a conjugate harmonic function of $-v(x, y)$ (since $-v(x, y) + iu(x, y)$ is analytic).

We now show that in a simply connected region, every harmonic function has a harmonic conjugate; that is, given a harmonic function $u(x, y)$

in a simply connected region Ω, we want to find a single-valued harmonic function $v(x,\, y)$ such that

$$\frac{\partial u}{\partial x} = \frac{\partial v}{\partial y}, \quad \frac{\partial u}{\partial y} = -\frac{\partial v}{\partial x}.$$

By the Green theorem, for every piecewise smooth curve Γ in Ω,

$$\int_\Gamma \left(-\frac{\partial u}{\partial y} dx + \frac{\partial u}{\partial x} dy \right) = \int\int_{[\Gamma]} \left(\frac{\partial^2 u}{\partial x^2} + \frac{\partial^2 u}{\partial y^2} \right) dx\, dy = 0.$$

Therefore, the line integral

$$\int_{(a,b)}^{(x,y)} \left\{ -\frac{\partial u(\xi,\, \eta)}{\partial \eta} d\xi + \frac{\partial u(\xi,\, \eta)}{\partial \xi} d\eta \right\}$$

does not depend on the arc joining the points $(a,\, b)$ to $(x,\, y)$ in Ω; with fixed initial point $(a,\, b)$, this integral becomes a function of the terminal point $(x,\, y)$ only. We write

$$v(x,\, y) = \int_{(a,b)}^{(x,y)} \left\{ -\frac{\partial u(\xi,\, \eta)}{\partial \eta} d\xi + \frac{\partial u(\xi,\, \eta)}{\partial \xi} d\eta \right\}.$$

Then $v(x,\, y)$ is single-valued. It remains to show that $v(x,\, y)$ satisfies the required conditions. Now

$$v(x + \triangle x,\, y) - v(x,\, y) = \int_{(x,y)}^{(x+\triangle x,\, y)} \left\{ -\frac{\partial u(\xi,\, \eta)}{\partial \eta} d\xi + \frac{\partial u(\xi,\, \eta)}{\partial \xi} d\eta \right\},$$

where the integral can be taken along the line segment joining $(x,\, y)$ to $(x + \triangle x,\, y)$ (Figure 11.1). Because Ω is open, we can ensure that this line segment lies entirely inside the region Ω by taking $\triangle x$ sufficiently small.

On this line segment $d\eta = 0$, and so, by the mean-value theorem,

$$v(x + \triangle x,\, y) - v(x,\, y) = \int_x^{x+\triangle x} -\frac{\partial u(\xi,\, y)}{\partial \eta} d\xi$$

$$= \triangle x \cdot \left\{ -\frac{\partial u(x + \theta\triangle x,\, y)}{\partial y} \right\} \quad (0 < \theta < 1).$$

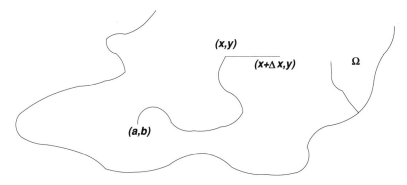

Figure 11.1

$$\therefore \frac{\partial v}{\partial x} = \lim_{\triangle x \to 0} \frac{v(x + \triangle x, y) - v(x, y)}{\triangle x}$$
$$= \lim_{\triangle x \to 0} \left\{ -\frac{\partial u(x + \theta \triangle x, y)}{\partial y} \right\} = -\frac{\partial u}{\partial y}.$$

Similarly,

$$\frac{\partial v}{\partial y} = \frac{\partial u}{\partial x}.$$

We have established the following

THEOREM 11.1.1 *Let $u(z)$ be a harmonic function defined in a simply connected region Ω. Then there exists a single-valued harmonic conjugate $v(z)$ of $u(z)$ in Ω.*

Remarks. (a) Two harmonic conjugates of the same harmonic function differ by a constant.

(b) If the region Ω is not simply connected, then there still exists a harmonic function $v(z)$ satisfying

$$\frac{\partial v}{\partial x} = -\frac{\partial u}{\partial y}, \quad \frac{\partial v}{\partial y} = \frac{\partial u}{\partial x}.$$

However, this statement must be understood in the sense that $v(z)$ may be multiple-valued, not a function in the strict sense. For example, if $u(z) = \log |z|$ in $\mathbb{C} \setminus \{0\}$, the $v(z)$ turns out to be arg z (plus an arbitrary real constant).

11.2   The Maximum and Minimum Principles for Harmonic Functions

THEOREM 11.2.1 *Let $u(z)$ be a nonconstant harmonic function in a region Ω. Then $u(z)$ has neither a (local) maximum nor a (local) minimum in Ω.*

Proof. Let $u(z)$ be a nonconstant harmonic function and a an arbitrary point in Ω. Choose r so small that the disc $D(a, r)$ is in Ω and let $v(z)$ be a harmonic conjugate of $u(z)$ in this disc. Then $u(z) + iv(z)$ is analytic there, and hence $f(z) = e^{u(z)+iv(z)}$ is also analytic. Because $f(z)$ is not constant, there exists, by the Maximum Modulus Principle 6.1.1, a point z_0 in the disc $D(a, r)$ such that

$$|f(z_0)| > |f(a)|.$$

However, because $|f(z)| = e^{u(z)}$, the above inequality implies that $u(z_0) > u(a)$, and so $u(z)$ does not have a local maximum at a. Similarly, replacing $u(z)$ by $-u(z)$, we conclude that $u(z)$ does not have a (local) minimum at a. The theorem is now proven.   □

Remark. Note that the minimum principle for harmonic functions holds without the assumption $u(z) \neq 0$, in contrast with the minimum modulus principle for analytic functions.

11.3   The Poisson Integral Formula

Let us restrict ourselves to the case that the region Ω is the unit disc. Suppose $f(z)$ is a function analytic in the unit disc D. Then, by the Cauchy Integral Formula 4.4.3,

$$f(0) = \frac{1}{2\pi i} \int_{C(0,r)} \frac{f(\zeta)}{\zeta} d\zeta = \frac{1}{2\pi} \int_0^{2\pi} f(re^{i\theta}) d\theta \quad (\zeta = re^{i\theta}, 0 < r < 1).$$

In addition, if $f(z)$ is continuous in the closed unit disc $\bar{D}$; then, because the closed disc is compact, $f(z)$ is uniformly continuous there, and so we may let $r \to 1-$, obtaining

$$f(0) = \frac{1}{2\pi} \int_0^{2\pi} f(e^{i\theta}) \, d\theta.$$

This formula says that under the assumptions that $f(z)$ is analytic in D and continuous in $\bar{D}$, the value of the function at the center of the circle

$C(0, 1)$ is the average of its values on the circle. But what about the points not at the center? Suppose a is an arbitrary point in D. Set

$$\zeta = Tz = \frac{z - a}{1 - \bar{a}z}.$$

Then the Möbius transformation T maps the unit disc conformally onto itself (and the boundary onto the boundary), mapping the point a to the origin. Therefore, $f(T^{-1}\zeta) = (f \circ T^{-1})(\zeta)$ is analytic in D and continuous on $\bar{D}$, and so we may apply the above result, obtaining

$$\left(f \circ T^{-1}\right)(0) = \frac{1}{2\pi} \int_0^{2\pi} \left(f \circ T^{-1}\right)\left(e^{i\varphi}\right) d\varphi;$$

that is,

$$f(a) = \frac{1}{2\pi} \int_0^{2\pi} f\left(T^{-1}e^{i\varphi}\right) d\varphi.$$

Observing that T is a bijection of the boundary circle to itself (and $f(T^{-1}e^{i\varphi})$ is 2π-periodic with respect to φ), we may change the variable of integration by setting $e^{i\theta} = T^{-1}e^{i\varphi}$; that is,

$$e^{i\varphi} = Te^{i\theta} = \frac{e^{i\theta} - a}{1 - \bar{a}e^{i\theta}}.$$

Differentiating both sides, we obtain

$$ie^{i\varphi}d\varphi = i\frac{e^{i\theta}\left(1 - \bar{a}e^{i\theta}\right) + \bar{a}e^{i\theta}\left(e^{i\theta} - a\right)}{\left(1 - \bar{a}e^{i\theta}\right)^2}d\theta$$

$$= i\frac{1 - a\bar{a}}{\left(1 - \bar{a}e^{i\theta}\right)^2}e^{i\theta}d\theta.$$

$$\therefore d\varphi = \frac{1 - a\bar{a}}{\left(1 - \bar{a}e^{i\theta}\right) \cdot \left(e^{i\theta} - a\right)}e^{i\theta}d\theta$$

$$= \frac{1 - |a|^2}{\left(1 - \bar{a}e^{i\theta}\right) \cdot \left(1 - ae^{-i\theta}\right)}d\theta = \frac{1 - |a|^2}{|1 - ae^{-i\theta}|^2}d\theta$$

$$= \frac{1 - r^2}{\left(1 - re^{-i(t-\theta)}\right) \cdot \left(1 - re^{i(t-\theta)}\right)}d\theta \quad (a = re^{it})$$

$$= \frac{1 - r^2}{1 - 2r\cos(t - \theta) + r^2}d\theta.$$

$$\therefore f(a) = \frac{1}{2\pi} \int_0^{2\pi} f\left(e^{i\theta}\right) \frac{1 - r^2}{1 - 2r\cos(t - \theta) + r^2} d\theta \quad (a = re^{it})$$

$$= \frac{1}{2\pi} \int_0^{2\pi} f\left(e^{i(t-\theta)}\right) \frac{1 - r^2}{1 - 2r\cos\theta + r^2} d\theta.$$

The function

$$P_r(\theta) = \frac{1 - r^2}{1 - 2r\cos\theta + r^2} \quad (0 \le r < 1)$$

is known as the *Poisson kernel*.

We have obtained the following

THEOREM 11.3.1 *If $f(z)$ is analytic in the unit disc D and continuous in the closed unit disc $\bar{D}$, then $f(z)$ satisfies the* **Poisson integral formula**

$$f(z) = \frac{1}{2\pi} \int_0^{2\pi} f\left(e^{i\theta}\right) \cdot P_r(t - \theta)\, d\theta \quad (z = re^{it}, \ 0 \le r < 1)$$

$$= \frac{1}{2\pi} \int_0^{2\pi} f\left(e^{i(t-\theta)}\right) \cdot P_r(\theta)\, d\theta.$$

Taking the real parts of both sides, we obtain

COROLLARY 11.3.2 *If $u(z)$ is harmonic in the unit disc D and continuous in the closed unit disc $\bar{D}$, then $u(z)$ satisfies the* **Poisson integral formula**

$$u(z) = \frac{1}{2\pi} \int_0^{2\pi} u\left(e^{i\theta}\right) \cdot P_r(t - \theta)\, d\theta \quad (z = re^{it}, \ 0 \le r < 1)$$

$$= \frac{1}{2\pi} \int_0^{2\pi} u\left(e^{i(t-\theta)}\right) \cdot P_r(\theta)\, d\theta.$$

11.4   The Dirichlet Problem

The Poisson integral formula shows that if $u(z)$ is harmonic in a disc and continuous in the closed disc, then its value at any interior point is completely determined by its values on the boundary circle. On the other hand, the Poisson integral is meaningful for every (bounded piecewise) continuous function $U\left(e^{i\theta}\right)$ on the circle (or even for Lebesgue integrable functions).

These facts suggest the following two questions:

1. Given a (real-valued) bounded piecewise continuous function $U\left(e^{i\theta}\right)$ on the unit circle, do we obtain a harmonic function $u(z)$ through the Poisson integral

$$u(z) = \frac{1}{2\pi} \int_0^{2\pi} U\left(e^{i\theta}\right) \frac{1 - r^2}{1 - 2r\cos(t - \theta) + r^2} d\theta \quad (z = re^{it})?$$

2. If so, do the boundary values of $u(z)$ agree with $U\left(e^{i\theta}\right)$?

We start with the first question. If $U\left(e^{i\theta}\right)$ is a (real-valued) bounded piecewise continuous function on the circle, then

$$
\begin{aligned}
u(z) &= \frac{1}{2\pi} \int_0^{2\pi} U\left(e^{i\theta}\right) \frac{1 - r^2}{1 - 2r\cos(t - \theta) + r^2} d\theta \quad (z = re^{it}) \\
&= \frac{1}{2\pi} \int_0^{2\pi} U\left(e^{i\theta}\right) \cdot \Re\left\{\frac{e^{i\theta} + z}{e^{i\theta} - z}\right\} d\theta \\
&= \Re\left\{\frac{1}{2\pi i} \int_{C(0,1)} U(\zeta) \cdot \frac{\zeta + z}{\zeta - z} \cdot \frac{d\zeta}{\zeta}\right\} \quad (\zeta = e^{i\theta}).
\end{aligned}
$$

Now let ·

$$
\begin{aligned}
f(z) &= \frac{1}{2\pi i} \int_{C(0,1)} U(\zeta) \cdot \frac{\zeta + z}{\zeta - z} \cdot \frac{d\zeta}{\zeta} \\
&= \frac{1}{2\pi i} \int_{C(0,1)} U(\zeta) \cdot \left\{\frac{-1}{\zeta} + \frac{2}{\zeta - z}\right\} d\zeta \\
&= -\frac{1}{2\pi i} \int_{C(0,1)} \frac{U(\zeta)}{\zeta} d\zeta + \frac{1}{2\pi i} \int_{C(0,1)} U(\zeta) \cdot 2 \sum_{k=0}^{\infty} \left(\frac{z}{\zeta}\right)^k \frac{d\zeta}{\zeta} \\
&= -\frac{1}{2\pi i} \int_{C(0,1)} \frac{U(\zeta)}{\zeta} d\zeta + 2 \sum_{k=0}^{\infty} \frac{z^k}{2\pi i} \int_{C(0,1)} \frac{U(\zeta)}{\zeta^{k+1}} d\zeta \\
&= a_0 + 2 \sum_{k=1}^{\infty} a_k z^k,
\end{aligned}
$$

where

$$a_k = \frac{1}{2\pi i} \int_{C(0,1)} \frac{U(\zeta)}{\zeta^{k+1}} d\zeta \quad (k = 0, 1, 2, \ldots).$$

The termwise integration is justified by the uniform convergence of the series for $\zeta \in C(0, 1)$ (with $z \in D$ fixed). Because

$$|a_k| \leq \frac{1}{2\pi} \int_{C(0,1)} \left|\frac{U(\zeta)}{\zeta^{k+1}}\right| \cdot |d\zeta| \leq \frac{M}{2\pi} \int_{C(0,1)} |d\zeta| = M,$$

where $M = \|U\|_{C(0, 1)}$, it follows that

$$\frac{1}{R} = \limsup_{k \to \infty} |2a_k|^{\frac{1}{k}} \leq \lim_{k \to \infty} (2M)^{\frac{1}{k}} = 1 \quad (\text{assuming } M \neq 0).$$

Thus $f(z)$ has a power-series expansion with radius of convergence not smaller than 1, and therefore $u(z)$, being the real part of an analytic function, is harmonic in the unit disc D.

We have established the first half of the following

THEOREM 11.4.1 *If $U\left(e^{i\theta}\right)$ is a bounded piecewise continuous function on the unit circle, then the function given by the Poisson integral*

$$u(z) = \frac{1}{2\pi} \int_0^{2\pi} U\left(e^{i\theta}\right) \frac{1 - r^2}{1 - 2r \cos(t - \theta) + r^2} d\theta$$

$$= \frac{1}{2\pi} \int_0^{2\pi} U\left(e^{i(t-\theta)}\right) \frac{1 - r^2}{1 - 2r \cos\theta + r^2} d\theta \quad (z = re^{it}, \; 0 \leq r < 1)$$

is harmonic in the unit disc. Furthermore, if $U\left(e^{i\theta}\right)$ is continuous at $\theta = t$, then

$$\lim_{r \to 1-} u\left(re^{it}\right) = U\left(e^{it}\right).$$

In particular, if $U\left(e^{it}\right)$ is continuous on the unit circle, then the convergence is uniform for $0 \leq t \leq 2\pi$.

Proof. It remains to prove the last half of the theorem. The proof is the standard *summability kernel argument* of Fourier analysis. In addition to the periodicity, we shall use only the following three properties of the Poisson kernel:

(a) $P_r(\theta) > 0$ for all θ, and $0 \leq r < 1$;

(b) $\frac{1}{2\pi} \int_{-\pi}^{\pi} P_r(\theta) \, d\theta = 1$ for all $0 \leq r < 1$;

(c) if $0 < \delta < \pi$, then

$$\lim_{r \to 1-} P_r(\theta) = 0 \quad \text{uniformly for } \delta \leq |\theta| \leq \pi.$$

The first property is immediate, for

$$1 - 2r \cos \theta + r^2 \geq 1 - 2r + r^2 = (1 - r)^2.$$

(It also follows from the expression

$$\frac{1 - r^2}{1 - 2r \cos \theta + r^2} = \frac{1 - r^2}{\left|1 - re^{i\theta}\right|^2}.)$$

Taking the Poisson integral of the constant function 1 at $z = r$ and using Corollary 11.3.2, we obtain the second property. The third property follows from the inequalities

$$0 \leq P_r(\theta) \leq \frac{1 - r^2}{1 - 2r \cos \delta + r^2} \quad \text{if } \delta \leq |\theta| \leq \pi.$$

Any family of functions that satisfies the above three properties is called a *summability kernel* (or an *approximate identity*, or a *Friedrichs mollifier*).

Now, by Property (b),

$$U\left(e^{it}\right) = \frac{1}{2\pi} \int_{-\pi}^{\pi} U\left(e^{it}\right) \cdot P_r(\theta) \, d\theta,$$

and using this equality with Property (a), we obtain

$$\left|u\left(re^{it}\right) - U\left(e^{it}\right)\right| \leq \frac{1}{2\pi} \int_{-\pi}^{\pi} \left|U\left(e^{i(t-\theta)}\right) - U\left(e^{it}\right)\right| \cdot P_r(\theta) \, d\theta.$$

Because we are assuming that $U\left(e^{i\theta}\right)$ is continuous at $\theta = t$, for any given $\epsilon > 0$ there exists $\delta > 0$ such that

$$\left|U\left(e^{i(t-\theta)}\right) - U\left(e^{it}\right)\right| \leq \frac{\epsilon}{2} \quad \text{if } |\theta| < \delta.$$

Set

$$I_1 = \frac{1}{2\pi} \int_{|\theta| < \delta} \left|U\left(e^{i(t-\theta)}\right) - U\left(e^{it}\right)\right| \cdot P_r(\theta) \, d\theta,$$

$$I_2 = \frac{1}{2\pi} \int_{|\theta| \geq \delta} \left|U\left(e^{i(t-\theta)}\right) - U\left(e^{it}\right)\right| \cdot P_r(\theta) \, d\theta.$$

Then

$$I_1 \leq \frac{1}{2\pi} \int_{|\theta|<\delta} \frac{\epsilon}{2} \cdot P_r(\theta)\, d\theta < \frac{\epsilon}{2} \cdot \frac{1}{2\pi} \int_{-\pi}^{\pi} P_r(\theta)\, d\theta = \frac{\epsilon}{2},$$

by Properties (a) and (b). On the other hand, because of Property (c),

$$0 < P_r(\theta) < \frac{\epsilon}{4M} \quad \left(M = \sup\left\{ \left| U\left(e^{i\theta}\right) \right|;\ -\pi \leq \theta \leq \pi \right\} \right)$$

for r sufficiently close to 1 if $\delta \leq |\theta| \leq \pi$. Therefore

$$I_2 \leq \frac{1}{2\pi} \int_{|\theta| \geq \delta} \left\{ \left| U\left(e^{i(t-\theta)}\right) \right| + \left| U\left(e^{it}\right) \right| \right\} \cdot \frac{\epsilon}{4M}\, d\theta$$

$$< \frac{1}{2\pi} \cdot 2M \cdot \frac{\epsilon}{4M} \cdot 2\pi = \frac{\epsilon}{2},$$

for r sufficiently close to 1. Thus, we have shown that

$$\left| u\left(re^{it}\right) - U\left(e^{it}\right) \right| \leq I_1 + I_2 < \frac{\epsilon}{2} + \frac{\epsilon}{2} = \epsilon,$$

for r sufficiently close to 1.

The last assertion of the theorem follows from the observation that if $U(e^{i\theta})$ is continuous on the unit circle, then, because the circle is compact, the function $U(e^{i\theta})$ is uniformly continuous there, and so the choice of δ, hence of r also, will not depend on the particular point $\theta = t$. The requirement of radial approach ($re^{it} \to e^{it}$) may be relaxed to the requirement of nontangential approach.   □

It is of interest to note that Abel's method of summing certain divergent series is closely related to Theorem 11.4.1. Actually, this theorem is a very particular case of the theorem of Fatou:

THEOREM 11.4.2 (Fatou) *Let μ be a finite (complex) Borel measure on the unit circle, and set*

$$u(z) = \frac{1}{2\pi} \int_{-\pi}^{\pi} P_r(t - \theta)\, d\mu\left(e^{i\theta}\right) \quad \left(z = re^{it} \right).$$

Then $u(z)$ is a (complex-valued) harmonic function in the unit disc D, and if e^{it_0} is any point where μ is differentiable with respect to Lebesgue measure (on the circle), then

$$\lim_{z} u(z) = \mu'\left(e^{it_0}\right)$$

as $z = re^{it}$ approaches e^{it_0} nontangentially. (Cf. Hoffman, *Banach Spaces of Analytic Functions*, Prentice-Hall, 1962.)

The problem of finding a function that is harmonic in a region and has preassigned values on the boundary is known as the *Dirichlet problem* (assuming, of course, that the prescribed boundary values satisfy suitable conditions). Our result shows that, for a disc, the Poisson integral formula provides the explicit solution of the Dirichlet problem. (The uniqueness of the solution is clear from the maximum and minimum principles.) In practice, however, it is often desirable to express the solution of a Dirichlet problem in series form. We now illustrate such a procedure in the case of a circular region.

If $U(e^{i\theta})$ is a (real-valued, bounded, piecewise) continuous function on the unit circle, and $u(z)$ is its harmonic extension to the unit disc D, then, because

$$P_r(\theta) = \sum_{k=-\infty}^{\infty} r^{|k|} e^{ik\theta} \quad (0 \leq r < 1),$$

$$u(z) = \frac{1}{2\pi} \int_{-\pi}^{\pi} U(e^{i\theta}) \cdot P_r(t - \theta) \, d\theta \quad \left(z = re^{it} \right)$$

$$= \frac{1}{2\pi} \int_{-\pi}^{\pi} U(e^{i\theta}) \cdot \left\{ \sum_{k=-\infty}^{\infty} r^{|k|} e^{ik(t-\theta)} \right\} d\theta$$

$$= \sum_{k=-\infty}^{\infty} \frac{r^{|k|} e^{ikt}}{2\pi} \int_{-\pi}^{\pi} U(e^{i\theta}) \cdot e^{-ik\theta} \, d\theta$$

$$= \sum_{k=-\infty}^{\infty} r^{|k|} c_k e^{ikt},$$

where

$$c_k = \frac{1}{2\pi} \int_{-\pi}^{\pi} U(e^{i\theta}) \cdot e^{-ik\theta} \, d\theta \quad (k \in \mathbb{Z})$$

is the kth *Fourier coefficient* of $U(e^{i\theta})$. Note that the termwise integration performed above is legitimate because the series in the integrand converges uniformly with respect to θ (for fixed r, $0 \leq r < 1$).

COROLLARY 11.4.3 *Given a (real-valued) continuous function $U(e^{i\theta})$ on the unit circle, the solution of the Dirichlet problem is given by*

$$u(z) = \sum_{k=-\infty}^{\infty} r^{|k|} c_k e^{ikt} \quad \left(z = re^{it}, \ 0 \leq r < 1 \right)$$

where

$$c_k = \frac{1}{2\pi} \int_{-\pi}^{\pi} U\left(e^{i\theta}\right) e^{-ik\theta} d\theta \quad (k \in \mathbb{Z}).$$

The series converges absolutely and locally uniformly in the unit disc D.

Remark. It is of interest that at no point in the discussion above was it necessary to assume that the Fourier series $\sum_{k=-\infty}^{\infty} c_k e^{ikt}$ of $U(e^{it})$ converges (to $U(e^{it})$ or any other function).

11.5   The Harnack Theorem

If the region under consideration is a disc $D(a, R)$, instead of $D(0, 1) = D$, then a simple change of variable furnishes the "generalized" Poisson integral formula:

$$\begin{aligned}
u\left(a + re^{it}\right) &= \frac{1}{2\pi} \int_{-\pi}^{\pi} u\left(a + Re^{i\theta}\right) \frac{R^2 - r^2}{R^2 - 2Rr\cos(t - \theta) + r^2} d\theta \\
&= \frac{1}{2\pi} \int_{-\pi}^{\pi} u\left(a + Re^{i(t-\theta)}\right) \frac{R^2 - r^2}{R^2 - 2Rr\cos\theta + r^2} d\theta \\
&\qquad\qquad\qquad\qquad\qquad\qquad\qquad\qquad (0 \le r < R).
\end{aligned}$$

Using this version, we now establish the following

THEOREM 11.5.1 (Harnack) *Let $\{u_k(z)\}_{k=1}^{\infty}$ be a sequence of harmonic functions defined in a region Ω.*

(a) If $\{u_k(z)\}_{k=1}^{\infty}$ converges locally uniformly to $u(z)$ on Ω, then $u(z)$ is harmonic on Ω;

(b) If $u_1(z) \le u_2(z) \le u_3(z) \le \cdots \le u_k(z) \le \cdots$, then either $\{u_k(z)\}_{k=1}^{\infty}$ converges locally uniformly on Ω or $u_k(z) \longrightarrow \infty$ ($k \to \infty$) everywhere in Ω.

Proof. (a) Choose any disc $\bar{D}(a, R)$ in Ω. Then, for $z = a + re^{it}$ $(0 \le r < R)$,

$$\begin{aligned}
\left| u(z) - \frac{1}{2\pi} \int_{-\pi}^{\pi} u\left(a + Re^{i\theta}\right) \frac{R^2 - r^2}{R^2 - 2Rr\cos(t - \theta) + r^2} d\theta \right| \\
\le |u(z) - u_k(z)| \\
+ \frac{1}{2\pi} \int_{-\pi}^{\pi} \left| u_k\left(a + Re^{i\theta}\right) - u\left(a + Re^{i\theta}\right) \right|
\end{aligned}$$

$$\times \; \frac{R^2 - r^2}{R^2 - 2Rr\cos(t - \theta) + r^2}\,d\theta$$

$$\leq |u(z) - u_k(z)|$$

$$+ \sup\left\{\left|u_k\left(a + Re^{i\theta}\right) - u\left(a + Re^{i\theta}\right)\right|; \; -\pi \leq \theta \leq \pi\right\}$$

$$\leq 2\|u_k - u\|_{\bar{D}(a, R)}.$$

The last term approaches zero as $k \to \infty$, for $\{u_k(z)\}_{k=1}^{\infty}$ converges to $u(z)$ uniformly on the closed disc $\bar{D}(a, R)$. We have, for $z = a + re^{it}$ $(0 \leq r < R)$,

$$u(z) = \frac{1}{2\pi}\int_{-\pi}^{\pi} u\left(a + Re^{i\theta}\right)\frac{R^2 - r^2}{R^2 - 2Rr\cos(t - \theta) + r^2}\,d\theta;$$

and $u(z)$ is harmonic by Theorem 11.4.1.

To prove (b), we may assume that $u_1(z) \geq 0$. (For otherwise, replace $u_k(z)$ by $u_k(z) - u_1(z)$.) Let

$$u(z) = \lim_{k\to\infty} u_k(z);$$

because the sequence is monotone, the pointwise limit always exists if we allow the limit to be ∞. Suppose the closed disc $\bar{D}(a, R)$ is in the region Ω. Then, noting the obvious but powerful *Harnack inequalities*

$$\frac{R - r}{R + r} \leq \frac{R^2 - r^2}{R^2 - 2Rr\cos(t - \theta) + r^2} \leq \frac{R + r}{R - r} \quad (0 \leq r < R),$$

multiplying by $\frac{1}{2\pi}u_k\left(a + Re^{i\theta}\right)$ (≥ 0) and integrating, we obtain

$$\frac{R - r}{R + r}u_k(a) \leq u_k\left(a + re^{it}\right) \leq \frac{R + r}{R - r}u_k(a).$$

Letting $k \to \infty$, we see that the corresponding inequalities hold for $u(z)$. It follows that either $u(z) = \infty$ for all z in the disc $D(a, R)$, or $u(z) < \infty$ for all z there. Therefore, the sets on which $u(z)$ is, respectively, finite or infinite, are both open; because Ω is connected, one of them must be empty.

If $u(z) = \infty$ at a single point, then the function is identically infinite. In the opposite case, the function $u(z)$ is finite everywhere. If $z \in \bar{D}\left(a, \frac{R}{2}\right)$, then the inequalities above imply that

$$0 \le u_m(z) - u_n(z) \le \frac{R + \frac{R}{2}}{R - \frac{R}{2}} \{u_m(a) - u_n(a)\}$$
$$= 3\{u_m(a) - u_n(a)\} \quad (m \ge n).$$

This inequality in turn implies that convergence at a point a implies uniform convergence in a (sufficiently small) neighborhood of a. This fact shows that the convergence is locally uniform in the region Ω, by Lemma 2.4.4, and hence the limit function $u(z)$ is harmonic in Ω by (a).   □

11.6   Green Functions

DEFINITION 11.6.1 *Let ζ be a fixed point in a region Ω. By a **Green function** of the region Ω with pole at ζ we mean a function $g(z, \zeta)$ with the properties:*

(a) *$g(z, \zeta)$ is a harmonic function of z in Ω except at $z = \zeta$, where it has a "logarithmic pole"; that is, $g(z, \zeta) + \mathrm{Log}\,|z - \zeta|$ is harmonic in Ω;*
(b) *$g(z, \zeta)$ is continuous in the closure $\bar{\Omega}$ of Ω except at $z = \zeta$ and equal to zero on the boundary $\partial\Omega$ of Ω; that is,*

$$\lim_{z \to z_0} g(z, \zeta) = 0 \quad \text{for all boundary points } z_0 \text{ of } \Omega.$$

It follows immediately from the definition that $g(z, \zeta)$ is a Green function of the region Ω if and only if $g(z, \zeta) + \mathrm{Log}\,|z - \zeta|$ is the solution of the Dirichlet problem in Ω with boundary values $\mathrm{Log}\,|z - \zeta|$ ($z \in \partial\Omega$). In particular, a Green function is unique whenever it exists, by the Maximum and Minimum Principles 11.2.1, and so we can refer to *the* Green function. Hence, the Green function exists in any region where the aforementioned Dirichlet problem has a solution for every ζ in Ω.

For simply connected regions, the relation between the Green function and conformal mapping is given by the following

THEOREM 11.6.2 *A simply connected region Ω has a Green function with a pole at ζ for all $\zeta \in \Omega$ if and only if it can be mapped conformally onto the unit disc D with ζ mapped to the origin.*

Proof. Suppose the region Ω is mapped conformally onto the unit disc by the function $f(z, \zeta)$, with the point ζ mapped to the origin. Then $f(z, \zeta)$

must be of the form

$$f(z, \zeta) = (z - \zeta) \cdot F(z),$$

where $F(z)$ never vanishes in Ω, because of the univalence of $f(z, \zeta)$. Therefore, the left-hand side of

$$-\mathrm{Log}\,|F(z)| = -\mathrm{Log}\,|f(z, \zeta)| + \mathrm{Log}\,|z - \zeta|$$

is the real part of an analytic function, $-\mathrm{Log}\,F(z)$, and hence is harmonic in Ω. To establish that $-\mathrm{Log}\,|f(z, \zeta)|$ is the Green function of the region Ω with a pole at ζ, it remains to verify the second condition in the definition. Because $f(z, \zeta)$ defines a homeomorphism of Ω onto the unit disc, $|f(z, \zeta)|$ tends to 1 as z approaches the boundary of Ω, and hence $-\mathrm{Log}\,|f(z, \zeta)|$ tends to zero as z approaches the boundary. By the uniqueness of the Green function, we obtain

$$g(z, \zeta) = -\mathrm{Log}\,|f(z, \zeta)|.$$

(If $f(z, \zeta)$ is a Riemann mapping function of the region Ω to the unit disc that maps ζ to the origin, then $|f(z, \zeta)|$ can be extended to the closure $\bar{\Omega}$ of Ω so that it is continuous there, even if $f(z, \zeta)$ itself may not be.)

Conversely, suppose that the region Ω has the Green function $g(z, \zeta)$ with a pole at $\zeta \in \Omega$. Then

$$u(z) = g(z, \zeta) + \mathrm{Log}\,|z - \zeta| \quad (z \in \Omega)$$

is harmonic in Ω. Because Ω is simply connected, by Theorem 11.1.1 (or by the Monodromy Theorem 8.6.1) there exists a single-valued analytic function $\varphi(z)$ in Ω with $\Re\varphi(z) = u(z)$. It is evident that

$$f(z, \zeta) = (z - \zeta) \cdot e^{-\varphi(z)} \quad (z \in \Omega)$$

is a single-valued analytic function in Ω. We want to show that this function maps the region Ω conformally onto the unit disc. Clearly, it maps the point ζ to the origin.

(a) The function $f(z, \zeta)$ maps the region Ω into the unit disc D. Indeed,

$$|f(z, \zeta)| = |z - \zeta| \cdot e^{-\Re\{\varphi(z)\}} = |z - \zeta| \cdot e^{-u(z)}$$
$$= |z - \zeta| \cdot e^{-g(z, \zeta) - \mathrm{Log}|z-\zeta|} = e^{-g(z, \zeta)} < 1 \quad \text{for } z \in \Omega,$$

for the Green function is positive in Ω. (Why?) Near the boundary $\partial\Omega$, $|f(z, \zeta)|$ tends to 1, for $g(z, \zeta)$ tends to zero.

(b) The function $f(z, \zeta)$ is univalent in Ω. It is sufficient to show that for $\tau \in \Omega$, the composition of $f(z, \zeta)$ with the Möbius transformation $\frac{w - f(\tau, \zeta)}{1 - \overline{f(\tau, \zeta)}w}$; that is,

$$T(z; \zeta, \tau) = \frac{f(z, \zeta) - f(\tau, \zeta)}{1 - \overline{f(\tau, \zeta)} \cdot f(z, \zeta)} \quad (z \in \Omega),$$

never vanishes in Ω except when $z = \tau$. If $\tau = \zeta$, then $T(z; \zeta, \zeta) = f(z, \zeta)$, which vanishes only when $z = \zeta$; that is, when $z = \tau$; hence we may exclude the case $\tau = \zeta$. Because of (a), we have

$$|T(z; \zeta, \tau)| < 1 \quad (z \in \Omega).$$

Note that if $f(z, \zeta)$ is indeed the mapping function we are seeking, then $T(z; \zeta, \tau)$ must be an analytic function that maps the region Ω conformally onto the unit disc D mapping the point τ to the origin. Thus, let $f(z, \tau)$ be the function obtained by replacing the pole ζ by τ in the definition of $f(z, \zeta)$. Because the point $z = \tau$ is the only root (which turns out to be simple) of $f(z, \tau)$ in Ω, the function $\frac{T(z; \zeta, \tau)}{f(z, \tau)}$ is analytic in Ω. Near the boundary, the modulus of this quotient tends to 1. Hence, by the maximum modulus principle,

$$\left\| \frac{T(z; \zeta, \tau)}{f(z, \tau)} \right\|_\Omega \leq 1.$$

For $z = \zeta$ this expression reduces to

$$\left| \frac{f(\tau, \zeta)}{f(\zeta, \tau)} \right| \leq 1.$$

Because ζ and τ are two arbitrary points in Ω, we may interchange the roles of ζ and τ, obtaining

$$\left| \frac{f(\tau, \zeta)}{f(\zeta, \tau)} \right| = 1.$$

But this equality means that the analytic function $\frac{T(z; \zeta, \tau)}{f(z, \tau)}$ attains its maximum modulus 1 at the (interior) point $z = \zeta$ of Ω, and hence must be a constant of modulus 1. Thus

$$|T(z; \zeta, \tau)| = |f(z, \tau)| \quad \text{for all } z \in \Omega.$$

Because $f(z, \tau)$ vanishes (in Ω) only at $z = \tau$, we obtain

$$T(z; \zeta, \tau) \neq 0 \quad \text{for } z \neq \tau,$$

which is what we wanted to show.

(c) The range of $f(z, \zeta)$ is the whole unit disc D. Suppose there were a point w_0 in the unit disc that is not the image of any point in Ω under the mapping $f(z, \zeta)$. Then

$$\frac{f(z, \zeta) - w_0}{1 - \bar{w}_0 \cdot f(z, \zeta)}$$

is a nonvanishing analytic function in Ω. Near the boundary, its modulus is close to 1; therefore, applying the Minimum Modulus Principle 6.1.1 to this function yields

$$|w_0| < \left| \frac{f(z, \zeta) - w_0}{1 - \bar{w}_0 \cdot f(z, \zeta)} \right| \quad \text{for all } z \in \Omega.$$

Then, for $z = \zeta$, we would obtain $|w_0| < |w_0|$. This contradiction proves the assertion.   □

The remainder of this section is devoted to the Dirichlet problem via Green functions. We recall the *Green identities*: If $u(x, y)$ and $v(x, y)$ are functions of class C^2 in the closure, $\bar{\Omega}$, of a region Ω whose boundary $\partial\Omega$ consists of a finite number of piecewise smooth curves, then

$$\int\!\!\int_\Omega \left(u \nabla^2 v + \frac{\partial u}{\partial x} \cdot \frac{\partial v}{\partial x} + \frac{\partial u}{\partial y} \cdot \frac{\partial v}{\partial y} \right) dx\, dy = \int_{\partial\Omega} u \frac{\partial v}{\partial n} ds,$$

and

$$\int\!\!\int_\Omega \left(u \nabla^2 v - v \nabla^2 u \right) dx\, dy = \int_{\partial\Omega} \left(u \frac{\partial v}{\partial n} - v \frac{\partial u}{\partial n} \right) ds.$$

Here the line integrals on the right are taken in the positive sense with respect to the region Ω,

$$\nabla^2 u = \frac{\partial^2 u}{\partial x^2} + \frac{\partial^2 u}{\partial y^2}, \quad \nabla^2 v = \frac{\partial^2 v}{\partial x^2} + \frac{\partial^2 v}{\partial y^2},$$

and $\frac{\partial u}{\partial n}$ and $\frac{\partial v}{\partial n}$ are the directional derivatives of u and v, respectively, in the direction of the exterior normal to the boundary $\partial\Omega$.

Now if, in addition, $u(x, y)$ and $v(x, y)$ are harmonic in Ω, then

$$\nabla^2 u = \nabla^2 v = 0 \quad \text{in } \Omega,$$

and the second identity yields

$$\int_{\partial\Omega} \left(u \frac{\partial v}{\partial n} - v \frac{\partial u}{\partial n} \right) ds = 0.$$

Suppose that the region Ω has the Green function $g(z, \zeta)$ with pole at $\zeta \in \Omega$ and that it belongs to the class C^2 in the closure $\bar{\Omega}$ of Ω except at ζ. Taking v to be the Green function $g(z, \zeta)$ and applying the last equality to the region obtained by deleting the disc $D(\zeta, \epsilon)$ ($\epsilon > 0$) from Ω, we obtain

$$\int_{\partial\Omega} \left(u \frac{\partial g}{\partial n} - g \frac{\partial u}{\partial n} \right) ds = \int_{C(\zeta,\epsilon)} \left(u \frac{\partial g}{\partial n} - g \frac{\partial u}{\partial n} \right) ds,$$

where the integral over the circle $C(\zeta, \epsilon)$ is taken counterclockwise. Because $g(z, \zeta) = 0$ on the boundary $\partial\Omega$, the left-hand side reduces to

$$\int_{\partial\Omega} u \frac{\partial g}{\partial n} ds.$$

Now, in the integration over the circle $C(\zeta, \epsilon)$, the directional derivative with respect to the exterior normal is just the partial derivative with respect to the radius; hence, introducing polar coordinates and writing

$$w(z) = g(z, \zeta) + \text{Log } |z - \zeta|,$$

we obtain

$$\int_{C(\zeta,\epsilon)} \left(u \frac{\partial g}{\partial n} - g \frac{\partial u}{\partial n} \right) ds$$

$$= \epsilon \int_0^{2\pi} \left\{ u\left(-\frac{1}{\epsilon} + \frac{\partial w}{\partial r} \right) + (\text{Log } \epsilon - w) \frac{\partial u}{\partial r} \right\} d\theta$$

$$= -\int_0^{2\pi} u \, d\theta + \epsilon \int_0^{2\pi} \left(u \frac{\partial w}{\partial r} - w \frac{\partial u}{\partial r} \right) d\theta + \epsilon \cdot \text{Log } \epsilon \cdot \int_0^{2\pi} \frac{\partial u}{\partial r} d\theta.$$

By the mean-value theorem for harmonic functions, the first term is $-2\pi u(\zeta)$; the remaining two integrals are bounded in a neighborhood of ζ; hence, letting $\epsilon \to 0$, we obtain

$$\lim_{\epsilon \to 0} \int_{C(\zeta,\epsilon)} \left(u \frac{\partial g}{\partial n} - g \frac{\partial u}{\partial n} \right) ds = -2\pi \cdot u(\zeta).$$

It follows that

$$u(\zeta) = -\frac{1}{2\pi} \int_{\partial\Omega} u(z) \cdot \frac{\partial g(z, \zeta)}{\partial n} ds.$$

This formula allows us to solve the Dirichlet problem for the region Ω, provided the Green function is known. Note the resemblance between the last representation (for *harmonic* functions) and the Cauchy Integral Formula 4.4.3 (for *analytic* functions).

Exercises

1. (a) Suppose $u(x, y)$ and $v(x, y)$ are harmonic in a region Ω. Show that, for arbitrary constants a and b, $au(x, y) + bv(x, y)$ is harmonic.

   (b) Suppose $v(x, y)$ is a harmonic conjugate of $u(x, y)$. Find a harmonic conjugate of $au(x, y) + bv(x, y)$, where a and b are constants.

2. (a) Show that $u(x, y) = e^x(x \sin y + y \cos y)$ is harmonic in the complex plane $\mathbb{C}$.

   (b) Find an entire function $f(z)$ such that

   $$\Re f(z) = u(x, y) = e^x(x \sin y + y \cos y).$$

3. Suppose a function $f(z)$ is analytic in a region Ω and never vanishes there; show that $\mathrm{Log}\,|f(z)|$ is harmonic in the region Ω.

4. Prove the maximum and minimum principles for harmonic functions by using the Gauss mean-value theorem for harmonic functions:

   $$u(a) = \frac{1}{2\pi} \int_0^{2\pi} u\left(a + re^{i\theta}\right) d\theta.$$

   Hint: Imitate the argument employed in the proof of Theorem 6.1.1.

5. Let $u(z)$ be harmonic in a region Ω.

   (a) Suppose $u(z)$ vanishes in a nonempty open subset of Ω; show that $u(z) \equiv 0$ in Ω.

   (b) Suppose $u(z)$ vanishes at a sequence of points having an accumulation point in Ω. Is it necessary that $u(z) \equiv 0$ in Ω?

6. Suppose $u(z)$ is a bounded harmonic function in the complex plane $\mathbb{C}$. Is $u(z)$ necessarily constant?

7. Suggest and justify analogues for harmonic functions of the Montel Selection Theorem 9.1.3 and the Vitali Theorem 9.1.5.

8. Suppose $\{u_k(z)\}_{k=1}^{\infty}$ is a sequence of harmonic functions converging locally uniformly to $u(z)$ in Ω. Show that $\left\{\frac{\partial u_k(z)}{\partial x}\right\}_{k=1}^{\infty}$ and $\left\{\frac{\partial u_k(z)}{\partial y}\right\}_{k=1}^{\infty}$ converge locally uniformly in Ω to $\frac{\partial u(z)}{\partial x}$ and $\frac{\partial u(z)}{\partial y}$, respectively.

9. Let Ω be a region in the upper half-plane having an interval (a, b) on the real axis as a part of its boundary. Suppose $u(z)$ is harmonic in Ω and continuous on $\Omega \cup (a, b)$ with $u(z) = 0$ on the interval (a, b). Show that $u(z)$ can be continued harmonically across the interval (a, b) by setting

$$U(z) = \begin{cases} u(z) & \text{for} \quad z \in \Omega \cup (a, b); \\ -u(\bar{z}) & \text{for} \quad \bar{z} \in \Omega. \end{cases}$$

10. Show that

$$\Re\left\{\frac{\zeta + z}{\zeta - z}\right\} = \frac{R^2 - r^2}{R^2 - 2Rr\cos(t - \theta) + r^2}$$

$$\left(\zeta = Re^{i\theta}, \; z = re^{it}, \; 0 \leq r < R\right).$$

11. Suppose $u(z)$ is harmonic in the unit disc D and $0 \leq u(re^{i\theta}) \leq P_r(\theta)$ for every point in the unit disc D. Show that $u(re^{i\theta}) = cP_r(\theta)$ for some constant c. Show also that this property characterizes the Poisson kernel $P_r(\theta)$.

12. (a) Let μ be a complex finite Borel measure on the unit circle. Show that the function defined by

$$f(z) = \int_{C(0,1)} \frac{d\mu}{\zeta - z} \quad (z \in D)$$

is analytic in the unit disc D.
   (b) More generally, if μ is a complex finite Borel measure with support $S \subset \mathbb{C}$, then

$$f(z) = \int_S \frac{d\mu}{\zeta - z}$$

is analytic in each component of the complement of S.

Hint: Imitate the argument in the proof of Theorem 11.4.1, or use the Morera Theorem 4.5.4.

13. Compute the Poisson integral for $U(e^{i\theta}) = \cos n\theta$.

14. (a) Find the harmonic function $u(z)$ in the unit disc with boundary values $\equiv 1$ on the upper semicircle and $\equiv 0$ on the lower semicircle.

    (b) Verify that $u(z) \longrightarrow 1$ as $z \to i$, and $u(z) \longrightarrow 0$ as $z \to -i$, both along the imaginary axis.

    (c) Find $\lim u(z)$ when z tends to 1 along the positive real axis.

15. Prove that a continuous function satisfying the mean-value property in a region must be harmonic there.

16. Let $f(z)$ be a meromorphic function in the disc $\bar{D}(0, R)$, with roots $a_1, a_2, \ldots, a_n$, and poles $b_1, b_2, \ldots, b_m$ in the disc (repeated as often as their multiplicities). If $z_0 = re^{it}$ $(0 \leq r < R)$ is neither a root nor a pole of the function $f(z)$, prove the *Poisson-Jensen formula*:

$$\frac{1}{2\pi} \int_{-\pi}^{\pi} \text{Log} \, |f(Re^{i\theta})| \cdot \frac{R^2 - r^2}{R^2 - 2Rr\cos(t - \theta) + r^2} d\theta$$

$$= \text{Log} \, |f(z_0)| - \sum_{k=1}^{n} \text{Log} \, \left| \frac{R(a_k - z_0)}{R^2 - \bar{a}_k z_0} \right| + \sum_{j=1}^{m} \text{Log} \, \left| \frac{R(b_j - z_0)}{R^2 - \bar{b}_j z_0} \right|.$$

Also discuss what modification is needed if z_0 is a root or a pole.

17. Suppose $U\left(e^{i\theta}\right)$ is Lebesgue integrable on the unit circle and

$$u\left(re^{it}\right) = \frac{1}{2\pi} \int_{-\pi}^{\pi} U\left(e^{i\theta}\right) \cdot P_r(t - \theta) \, d\theta.$$

Show that $u(re^{it}) \longrightarrow U(e^{it})$ in L^1-norm as $r \to 1-$.
Hint : Consider the summability kernel argument.

18. (a) Suppose that $w = f(z)$ maps a region Ω_z conformally onto a region Ω_w and that $u(w)$ is harmonic in Ω_w. Show that $U(z) = (u \circ f)(z) = u(f(z))$ is harmonic in Ω_z. (In brief, harmonicity is preserved under conformal mapping.)

    (b) Assuming the Carathéodory Theorem 10.2.3, show that the Dirichlet problem is solvable in a region bounded by a Jordan curve.

19. Let $C[-\pi, \pi]$ be the vector space of all (complex-valued) continuous functions in the interval $[-\pi, \pi]$, with $f(-\pi) = f(\pi)$. For $f(\theta) \in C[-\pi, \pi]$,

$$\hat{f}(n) = \frac{1}{2\pi} \int_{-\pi}^{\pi} f(\theta) e^{-in\theta} \, d\theta \quad (n \in \mathbb{Z})$$

is called the nth *Fourier coefficient* of the function $f(\theta)$. If $f(\theta) = \sum_{n=-N}^{N} a_n e^{in\theta}$, show that

$$\hat{f}(n) = \begin{cases} a_n, & |n| \leq N; \\ 0, & |n| > N, \end{cases}$$

and

$$\sum_{n=-N}^{N} \left| \hat{f}(n) \right|^2 = \frac{1}{2\pi} \int_{-\pi}^{\pi} |f(\theta)|^2 d\theta \quad (Parseval).$$

20. The integral

$$(f * g)(t) = \frac{1}{2\pi} \int_{-\pi}^{\pi} f(t - \theta) \cdot g(\theta) \, d\theta \quad (f(\theta), g(\theta) \in C[-\pi, \pi])$$

is called the *convolution* of the functions $f(\theta)$ and $g(\theta)$.
(a) Show that convolution is commutative and associative.
(b) Show also that

$$\widehat{f * g}(n) = \hat{f}(n) \cdot \hat{g}(n) \quad (n \in \mathbb{Z}).$$

21. Let

$$F_n(\theta) = \sum_{k=-n}^{n} \left(1 - \frac{|k|}{n+1} \right) e^{ik\theta}.$$

(a) Show that

$$F_n(\theta) = \frac{1}{n+1} \left\{ \frac{\sin \frac{n+1}{2} \theta}{\sin \frac{\theta}{2}} \right\}^2.$$

(b) Show that $\{F_n(\theta)\}_{n=1}^{\infty}$ is a summability kernel with discrete parameter $n \in \mathbb{N}$. (This is known as the *Fejér kernel*.)
(c) Let $f(z)$ be a function analytic in the unit disc D and continuous in the closed unit disc $\bar{D}$. Show that

$$(f * F_n)(z) = \frac{1}{2\pi} \int_{-\pi}^{\pi} f\left(r e^{i\theta} \right) \cdot F_n(t - \theta) \, d\theta \quad \left(z = r e^{it} \right)$$

is a polynomial of degree (at most) n.

(d) Show that the class of all polynomials is dense in the *Banach space*, equipped with the supremum norm, of all functions analytic in the unit disc D, and continuous in the closed unit disc $\bar{D}$. (This is a complex version of the *Stone-Weierstrass theorem*.)

(e) Can you extend the Parseval equality to the case $f(\theta) \in C[-\pi, \pi]$? What if $f(\theta)$ is a 2π-periodic, piecewise continuous function?

(f) Show that

$$\frac{\pi^2}{\sin^2 \pi t} = \sum_{k=-\infty}^{\infty} \frac{1}{(t-k)^2},$$

by applying the Parseval equality to the 2π-periodic extension of the function $f(t; x) = e^{itx}$ ($t, x \in \mathbb{R}, -\pi < t \le \pi$). What if $t \in \mathbb{C}$?

22. (a) Find the Green function of the unit disc D with pole $\zeta \in D$.

(b) Verify the relation

$$u(\zeta) = -\frac{1}{2\pi} \int_{\partial\Omega} u(z) \cdot \frac{\partial g(z, \zeta)}{\partial n} ds$$

for the unit disc D, and compare with the Poisson Integral Formula 11.4.1.

23. (a) Show that

$$g(z, \zeta) > 0 \quad \text{for all } z \in \Omega.$$

(b) Show that the Green function has the symmetry property:

$$g(\tau, \zeta) = g(\zeta, \tau) \quad (\zeta, \tau \in \Omega).$$

Hint: Apply the identity

$$\int_{\partial\Omega} \left(u \frac{\partial v}{\partial n} - v \frac{\partial u}{\partial n} \right) ds = 0$$

to the functions $g(z, \zeta)$ and $g(z, \tau)$ in the region obtained from Ω by deleting the discs $D(\zeta, \epsilon)$ and $D(\tau, \epsilon)$.

24. Show that the Green function of the right half-plane $\{z \in \mathbb{C}; \Re z > 0\}$ is given by

$$\begin{aligned}
g(z, \zeta) &= -\text{Log}\,|z - \zeta| + \text{Log}\,|z + \bar{\zeta}| \\
&= \frac{1}{2}\text{Log}\,\frac{(x+\xi)^2 + (y-\eta)^2}{(x-\xi)^2 + (y-\eta)^2} \quad (z = x + iy, \ \zeta = \xi + i\eta).
\end{aligned}$$

25. Let Ω_1 be a subregion of a (bounded) region Ω_2, and let $g_1(z, \zeta)$ and $g_2(z, \zeta)$ be the Green functions of the regions Ω_1 and Ω_2, respectively. Show that

$$g_1(z, \zeta) \leq g_2(z, \zeta),$$

where z and ζ are points in Ω_1. Also, if $\Omega_1 \neq \Omega_2$ (i.e., if Ω_1 is a *proper* subregion of Ω_2) then the strict inequality holds.

Chapter 12

The Picard Theorems

12.1 The Bloch Theorem

The purpose of this chapter is to prove the epoch-making Picard theorems that we have previously mentioned several times. The original proof of Picard is illuminating, but it depends on the existence of the *elliptic modular function*. This function, which is customarily denoted $\lambda(z)$, has the following properties: it is analytic in the upper half-plane with the real axis as its natural boundary and assumes every value except 0 and 1 in $\mathbb{C}$ (in fact, infinitely many times). Thus, its inverse function $\lambda^{-1}(w)$ is a (multiple-valued) function defined on $\mathbb{C} \setminus \{0, 1\}$ with values in the upper half-plane. Assuming the existence of such a function $\lambda^{-1}(w)$, we can easily prove the "little" Picard theorem.

THEOREM 12.1.1 (Picard) *An entire function that omits two values must be constant.*

Suppose $f(z)$ is an entire function that does not assume either of the values a or b, where $a \neq b$. Then

$$g(z) = \frac{f(z) - a}{b - a}$$

is an entire function that does not assume either of the values 0 or 1. Consider the function $\lambda^{-1}(g(z))$. With an appropriate choice of a branch of the inverse function $\lambda^{-1}(w)$, this is an entire function with values in the upper half-plane. (Note that the Monodromy Theorem 8.6.1 is applicable here.) But then $h(z) = \exp\{i\lambda^{-1}(g(z))\}$ is a bounded entire function, and hence, by the Liouville Theorem 4.4.9, is constant. Because $\lambda^{-1}(w)$ is not constant, $g(z)$ must be constant; hence $f(z)$ must also be constant, and the proof is complete.

Because we have not discussed the construction of the modular function, we shall not complete this proof. Instead, we present a more elementary proof based on the Bloch Theorem 12.1.3. (All we need in this chapter are the Cauchy Estimates 4.4.7, the Liouville Theorem 4.4.9, the Maximum Modulus Principle 6.1.1, the Schwarz Lemma 6.2.1, the Rouché Theorem 5.6.8, the Riemann Theorem 5.2.1, and the Casorati-Weierstrass Theorem 5.2.4.)

We begin with the following

LEMMA 12.1.2 *Let $f(z)$ be a function analytic in the disc $D(0, R)$ and satisfy the conditions*

$$f(0) = 0, \quad f'(0) \neq 0, \quad \text{and} \quad |f(z)| \leq M \quad \text{in} \ D(0, R).$$

Set

$$r = \frac{|f'(0)| \cdot R^2}{4M}, \quad \rho = \frac{|f'(0)|^2 \cdot R^2}{6M}.$$

Then, for an arbitrary $w_0 \in D(0, \rho)$, there exists a unique $z_0 \in D(0, r)$ such that $f(z_0) = w_0$. In other words, in the disc $D(0, r)$ $f(z)$ assumes each value whose modulus is less than ρ exactly once. (The equation $f(z) = w$, when $|w| \geq \rho$, may have no solution, one solution, or more than one solution in $D(0, r)$.)

Proof. First, the Cauchy Estimate 4.4.7 furnishes the inequality $|f'(0)| \leq \frac{M}{R}$ from which we obtain $r \leq \frac{R}{4}$, so that $f(z)$ is certainly defined in (and beyond) the closed disc $\bar{D}(0, r)$.

It may be helpful at this point to sketch the basic idea of the proof.

(a) It is shown that $f(z)$ has no root in $\bar{D}(0, r)$ other than the one at the origin.

(b) It is then shown that $|f(z)| \geq \rho$ everywhere on $C(0, r)$.

(c) An appeal to the Rouché Theorem 5.6.8 completes the proof.

(a) Let

$$f(z) = \sum_{k=1}^{\infty} c_k z^k \quad (c_1 = f'(0) \neq 0)$$

be the Taylor-series expansion of $f(z)$ in the disc $D(0, R)$. By the Schwarz Lemma 6.2.1,

$$|f(z) - c_1 z| \leq M(r) \cdot \frac{|z|}{r} \quad \text{in} \ \bar{D}(0, r),$$

where

$$M(r) = \max\{|f(z) - c_1 z|; \ |z| = r\}.$$

Then for $z \in \bar{D}(0, r)$, the triangle inequality yields

$$
\begin{aligned}
|f(z)| &= |c_1 z + (f(z) - c_1 z)| \\
&\geq |c_1 z| - |f(z) - c_1 z| \\
&\geq |c_1| \cdot |z| - M(r) \cdot \frac{|z|}{r} \\
&= |c_1| \cdot |z| \cdot \left(1 - \frac{M(r)}{|c_1| \cdot r}\right).
\end{aligned}
$$

Accepting momentarily the inequality $M(r) < |c_1| r$, we conclude that in $\bar{D}(0, r)$ the function $f(z)$ vanishes only at the origin (where it has a simple root).

(b) The inequality $M(r) < |c_1| r$ is demonstrated as follows: By the Cauchy Estimates 4.4.7 we obtain $|c_k| \leq \frac{M}{R^k}$ $(k \geq 2)$, and therefore

$$
\begin{aligned}
|c_1| \cdot r - M(r) &\geq |c_1| \cdot r - \sum_{k=2}^{\infty} |c_k| r^k \\
&\geq |c_1| \cdot r - \sum_{k=2}^{\infty} M \left(\frac{r}{R}\right)^k \\
&\geq |f'(0)| \cdot r - M \left(\frac{r}{R}\right)^2 \cdot \sum_{n=0}^{\infty} \left(\frac{1}{4}\right)^n \\
&= \frac{|f'(0)|^2 \cdot R^2}{4M} - \frac{|f'(0)|^2 \cdot R^2}{16M} \cdot \frac{4}{3} \\
&= \frac{|f'(0)|^2 \cdot R^2}{6M} = \rho.
\end{aligned}
$$

Thus $|c_1| \cdot r - M(r) > 0$ and, by the computation in (a), $|f(z)| \geq \rho$ on $C(0, r)$.

(c) The last inequality guarantees, by the Rouché Theorem 5.6.8, that for $w_0 \in D(0, \rho)$, $f(z)$ and $f(z) - w_0$ have the same number of roots — namely one — inside $C(0, r)$ (and none on $C(0, r)$), and so the proof is complete.   □

THEOREM 12.1.3 (Bloch) *Let $f(z)$ be analytic in the unit disc D and satisfy $f'(0) = 1$. Then there exists a subregion Ω of D that is mapped univalently by $f(z)$ onto a disc of radius B, where the constant B is independent of the function $f(z)$. (That is, B is an "absolute" constant.)*

Proof. We temporarily assume that $f(z)$ is analytic in the closed unit disc $\bar{D}$. Set

$$M_1 = \max\{(1 - |z|^2) \cdot |f'(z)| \,;\, z \in \bar{D}\}.$$

Then, because $f'(0) = 1$, we have $M_1 \geq 1$. Let α be a point such that

$$(1 - |\alpha|^2) \cdot |f'(\alpha)| = M_1.$$

Then, clearly, we have $|\alpha| < 1$. Define

$$F(\zeta) = f(z), \quad \text{where } z = \frac{\alpha - \zeta}{1 - \bar{\alpha}\zeta}.$$

Then

$$1 - |z|^2 = \frac{(1 - |\alpha|^2) \cdot (1 - |\zeta|^2)}{|1 - \bar{\alpha}\zeta|^2},$$

and so

$$\left|\frac{dz}{d\zeta}\right| = \frac{1 - |\alpha|^2}{|1 - \bar{\alpha}\zeta|^2} = \frac{1 - |z|^2}{1 - |\zeta|^2}.$$

$$\therefore (1 - |\zeta|^2) \cdot |F'(\zeta)| = (1 - |\zeta|^2) \cdot |f'(z)| \cdot \left|\frac{dz}{d\zeta}\right|$$
$$= (1 - |z|^2) \cdot |f'(z)|.$$

It follows that

$$(1 - |\zeta|^2) \cdot |F'(\zeta)| \leq M_1 \quad \text{for all } \zeta \in D_\zeta.$$

Because $\zeta = 0$ corresponds to $z = \alpha$, we obtain

$$|F'(0)| = (1 - |\alpha|^2)|f'(\alpha)| = M_1.$$

Therefore, the function

$$g(\zeta) = \frac{F(\zeta) - F(0)}{F'(0)}$$

satisfies

$$g(0) = 0, \quad g'(0) = 1, \quad |g'(\zeta)| \leq \frac{1}{1 - |\zeta|^2} \quad \text{for all } \zeta \in D_\zeta.$$

We now want to find an upper bound of $|g(\zeta)|$ for $\zeta \in \bar{D}(0, R)$ $(0 < R < 1)$. Let $\zeta_0 = Re^{i\theta}$; then

$$g(\zeta_0) = \int_0^{\zeta_0} g'(\zeta)\, d\zeta.$$

Integrating along a radius of the unit disc, we obtain

$$|g(\zeta_0)| \le \int_0^R \left| g'\left(re^{i\theta}\right) \right| dr$$

$$\le \int_0^R \frac{dr}{1 - r^2} = \frac{1}{2}\mathrm{Log}\, \frac{1 + R}{1 - R}.$$

We have shown that, for $\zeta \in \bar{D}(0, R)$,

$$|g(\zeta)| \le \frac{1}{2}\mathrm{Log}\, \frac{1 + R}{1 - R}.$$

Let M be the quantity on the right-hand side of this inequality (which obviously depends only on R). Then, by the previous lemma, the image of the disc $D\left(0, \frac{R^2}{4M}\right)$ under the mapping $g(\zeta)$ contains the disc of radius $\frac{R^2}{6M}$ centered at the origin. Because $|F'(0)| = M_1 \ge 1$ and[1] $\frac{R^2}{4M} < 1$ for $0 < R < 1$, we conclude that the image of the unit disc D under the mapping $F(\zeta)$ contains a disc of radius $\frac{R^2}{6M}$, and this result implies that the image of the unit disc D under the mapping $f(z)$ does also.

To eliminate the assumption made above that $f(z)$ is analytic in $\bar{D}$ (not only in D) we employ a routine approximation argument as follows. Let

$$f_\epsilon(z) = \frac{f((1 - \epsilon)z)}{1 - \epsilon} \qquad (0 < \epsilon < 1).$$

Then $f_\epsilon(z)$ is analytic in the closed unit disc and satisfies $f_\epsilon'(0) = 1$, so that the result above is applicable to $f_\epsilon(z)$. It follows that the image of the unit disc under the mapping $f((1 - \epsilon)z)$ contains a disc of radius $(1 - \epsilon)\frac{R^2}{6M}$. Now let $\epsilon \downarrow 0$. $\qquad \square$

1.
$$M = \frac{1}{2}\mathrm{Log}\frac{1 + R}{1 - R} = R + \frac{R^3}{3} + \frac{R^5}{5} + \cdots > R > \frac{R^2}{4} \qquad (0 < R < 1).$$

The supremum of the constant in the statement of the theorem, which is independent of the function $f(z)$, is called the *Bloch constant*. The precise value of the Bloch constant B is still unknown. The best known result is

$$0.433012\cdots = \frac{\sqrt{3}}{4} \le B \le \sqrt{\pi} 2^{\frac{1}{4}} \frac{\Gamma(\frac{1}{3})}{\Gamma(\frac{1}{4})} \cdot \left\{ \frac{\Gamma(\frac{11}{12})}{\Gamma(\frac{1}{12})} \right\}^{\frac{1}{2}} = 0.4719\cdots.$$

It is conjectured that the upper bound given here is the true value of the Bloch constant. The argument which we have presented shows that $B \ge \max_{0<R<1} \frac{R^2}{6M}$; the maximum is attained when $R \approx 0.8$, and then we obtain $B \ge \frac{0.64}{3\log 9} \approx 0.097$, which is far below the actual value of B.

12.2   The Schottky Theorem

We now use the Bloch Theorem 12.1.3 to prove the Schottky theorem, which is needed in the proof of the Picard theorems.

First recall that, if $f(z)$ is analytic and $f(z) \ne 0$ in a simply connected region Ω, then $f(z)$ can be expressed in the form

$$f(z) = e^{h(z)},$$

where $h(z) \in H(\Omega)$. The function $h(z)$ is not unique; however, if we impose the condition $h(z_0) = \text{Log } f(z_0)$ at a preassigned point $z_0 \in \Omega$, then $h(z)$ *is* uniquely determined. Furthermore, for $k \in \mathbb{N}$, if we set $\varphi(z) = e^{\frac{1}{k}h(z)}$, then we obtain $\varphi(z) \in H(\Omega)$, and it satisfies

$$f(z) = \{\varphi(z)\}^k.$$

Clearly, if $\varphi(z) \in H(\Omega)$ is one solution of the above equation, then so too is $\omega\varphi(z)$, where ω is any kth root of unity. Thus, $\varphi(z)$ is not uniquely determined. However, again by imposing a restriction on $\arg \varphi(z_0)$ at a preassigned point $z_0 \in \Omega$, we can determine $\varphi(z)$ uniquely.

THEOREM 12.2.1 (Schottky) *Suppose $f(z)$ is analytic in the disc $D(0, R)$ and does not assume either of the values 0 or 1; then for any θ $(0 \le \theta < 1)$ there exists a constant $S(f(0), \theta)$, depending only on θ and $f(0)$ (and not on the radius R or otherwise on $f(z)$), such that*

$$|f(z)| < S(f(0), \theta) \quad \text{for all } z \in \bar{D}(0, \theta R).$$

Proof. By the remark preceding the statement of the theorem, there exists a function $h(z)$ analytic in the disc $D(0, R)$ such that

$$f(z) = e^{2\pi i h(z)}.$$

Because $f(z)$ omits the value 1, $h(z)$ omits the values 0 and 1. Therefore, again by the above remark, there exist functions $\varphi_1(z)$ and $\varphi_2(z)$ analytic in the disc $D(0, R)$ satisfying

$$h(z) = \{\varphi_1(z)\}^2, \quad h(z) - 1 = \{\varphi_2(z)\}^2.$$

Because

$$1 = \{\varphi_1(z)\}^2 - \{\varphi_2(z)\}^2 = \{\varphi_1(z) + \varphi_2(z)\} \cdot \{\varphi_1(z) - \varphi_2(z)\},$$

we have $\varphi_1(z) - \varphi_2(z) \neq 0$, and so we may write

$$\varphi_1(z) - \varphi_2(z) = e^{\psi(z)}$$

for some function $\psi(z)$ analytic in the disc $D(0, R)$. Then

$$\varphi_1(z) + \varphi_2(z) = \frac{1}{\varphi_1(z) - \varphi_2(z)} = e^{-\psi(z)}.$$

It follows that

$$\varphi_1(z) = \frac{1}{2}\left\{e^{\psi(z)} + e^{-\psi(z)}\right\},$$

$$2\pi i h(z) = 2\pi i \{\varphi_1(z)\}^2$$

$$= \frac{\pi i}{2}\left\{e^{2\psi(z)} + e^{-2\psi(z)}\right\} + \pi i,$$

and

$$f(z) = e^{2\pi i h(z)} = -\exp\left[\frac{\pi i}{2}\left\{e^{2\psi(z)} + e^{-2\psi(z)}\right\}\right].$$

Note that the value of $h(0)$ is determined by that of $f(0)$ (up to an additive integer), those of $\varphi_1(0)$ and $\varphi_2(0)$ are determined by that of $h(0)$ (up to the $\pm$ sign), and finally, that of $\psi(0)$ is determined by those of $\varphi_1(0)$ and $\varphi_2(0)$ (up to an additive integer multiple of $2\pi i$). Therefore, for all functions $f(z)$ that satisfy the conditions of the theorem and have a *specified* value at the origin (say $f(0) = \alpha$ for some number α independent of $f(z)$), we may choose the corresponding functions $\psi(z)$

so that they have the same value at the origin, say $\psi(0) = \beta$. (Of course, $\psi(0) = \beta$ must be one of the possibilities consistent with $f(0) = \alpha$.)

We claim that the function $\psi(z)$ chosen above does not assume any of the values

$$\pm\text{Log}\left(\sqrt{m} + \sqrt{m-1}\right) + \frac{n\pi i}{2} \quad (m \in \mathbb{N},\ n \in \mathbb{Z}).$$

For, suppose

$$\psi(c) = \pm\text{Log}\left(\sqrt{m} + \sqrt{m-1}\right) + \frac{n\pi i}{2} \quad (m \in \mathbb{N},\ n \in \mathbb{Z}).$$

Then

$$e^{2\psi(c)} = (-1)^n \left(\sqrt{m} \pm \sqrt{m-1}\right)^2,$$
$$e^{-2\psi(c)} = (-1)^{-n} \left(\sqrt{m} \mp \sqrt{m-1}\right)^2,$$

and so

$$e^{2\psi(c)} + e^{-2\psi(c)} = (-1)^n\{2m + 2(m-1)\} = 2(-1)^n(2m-1).$$

Hence

$$f(c) = -e^{\pi i(-1)^n(2m-1)} = 1,$$

which contradicts our assumption that $f(z) \neq 1$.

Observe that, for $m \geq 2$,

$$\text{Log}\left(\sqrt{m+1} + \sqrt{m}\right) - \text{Log}\left(\sqrt{m} + \sqrt{m-1}\right)$$
$$< \text{Log}\left(2\sqrt{m+1}\right) - \text{Log}\left(2\sqrt{m-1}\right)$$
$$= \text{Log}\sqrt{\frac{m+1}{m-1}} \leq \text{Log}\sqrt{3} < 1;$$

while, when $m = 1$ we obtain directly

$$\text{Log}\left(\sqrt{m+1} + \sqrt{m}\right) - \text{Log}\left(\sqrt{m} + \sqrt{m-1}\right) = \text{Log}\left(\sqrt{2} + 1\right) < 1.$$

Moreover,

$$\left|\frac{(n+1)\pi i}{2} - \frac{n\pi i}{2}\right| = \frac{\pi}{2} < \sqrt{3}.$$

Thus, for any $z \in \mathbb{C}$, $\Re z$ is within $\frac{1}{2}$ of one of the numbers $\pm \mathrm{Log} \left(\sqrt{m+1} \right.$ $+\sqrt{m})$ and $\Im z$ is within $\frac{\sqrt{3}}{2}$ of one of the numbers $\frac{n\pi}{2}$. (Remember that $m \in \mathbb{N}$ and $n \in \mathbb{Z}$.) This shows that any disc in the complex plane $\mathbb{C}$ having the radius $\sqrt{(\frac{1}{2})^2 + (\frac{\sqrt{3}}{2})^2} = 1$ must contain a point that is not in the range of the function $\psi(z)$.

Suppose $\zeta \in D(0, \theta R)$ $(0 < \theta < 1)$, where $\psi'(\zeta) \neq 0$, and consider the function

$$F(z) = \frac{\psi(\zeta + (1-\theta)Rz)}{(1-\theta)R\psi'(\zeta)}.$$

The function $F(z)$ is well defined in the unit disc, and it satisfies the conditions of the Bloch theorem. Now, because the image of $D(0, R)$ under the mapping $\psi(z)$ cannot contain a disc of radius 1, and hence the image of D under $F(z)$ cannot contain a disc of radius $\frac{1}{(1-\theta)R|\psi'(\zeta)|}$, it follows that we have

$$B < \frac{1}{(1-\theta)R|\psi'(\zeta)|}; \quad \text{that is, } |\psi'(\zeta)| < \frac{1}{(1-\theta)RB}.$$

Trivially, the last inequality is valid even if $\psi'(\zeta) = 0$, and therefore the condition $\psi'(\zeta) \neq 0$ may now be disregarded. Therefore, using

$$\psi(z) = \psi(0) + \int_0^z \psi'(\zeta)\, d\zeta \quad z \in \bar{D}(0, \theta R),$$

we obtain

$$|\psi(z)| \leq |\psi(0)| + \int_0^{\theta R} |\psi'(\zeta)|\, |d\zeta|$$

$$\leq |\psi(0)| + \int_0^{\theta R} \frac{dr}{(1-\theta)RB}$$

$$\leq |\psi(0)| + \frac{\theta}{(1-\theta)B} \quad z \in \bar{D}(0, \theta R).$$

Denoting the quantity in the rightmost member by ρ, we obtain the estimate

$$|f(z)| < e^{\frac{\pi}{2}(e^{2\rho} + e^{2\rho})} = e^{\pi e^{2\rho}}.$$

Noting that the right-hand side depends on $|\psi(0)|$ (hence on $f(0)$) and θ only, we obtain the desired conclusion. $\square$

Remark. For given k and θ $(k > 2, 0 \le \theta < 1)$, it is evident now that there exists a constant $S^*(k, \theta)$, depending only on k and θ, satisfying

$$S(c_0, \theta) \le S^*(k, \theta) \quad \text{whenever } |c_0| \le k.$$

This remark will be used in the proof of the "great" Picard Theorem 12.3.2. (See Exercise 12.1.)

As a corollary, we obtain the *theorem of two constants.*[2]

THEOREM 12.2.2 (Landau) *Suppose*

$$f(z) = \sum_{k=0}^{\infty} c_k z^k$$

has radius of convergence R, does not assume either of the values 0 or 1, and $f'(0) = c_1 \ne 0$; then

$$R \le L(c_0, c_1),$$

where $L(c_0, c_1)$ is a constant depending on c_0 and c_1 only (and is otherwise independent of $f(z)$).

Proof. By the Schottky Theorem 12.2.1, we have

$$|f(z)| < S\left(c_0, \frac{1}{2}\right) \quad \text{for } z \in \bar{D}\left(0, \frac{R}{2}\right).$$

Therefore, by the Cauchy Estimate 4.4.7,

$$|c_1| < \frac{S\left(c_0, \dfrac{1}{2}\right)}{\dfrac{R}{2}}; \quad \text{that is, } R < \frac{2S\left(c_0, \dfrac{1}{2}\right)}{|c_1|}. \qquad \square$$

Remark. In particular, a polynomial

$$f(z) = \sum_{k=0}^{n} c_k z^k \quad (c_1 \ne 0)$$

---

[2] The second author was told by I.J. Schoenberg, who was Landau's son-in-law, that Landau could not believe this theorem and was convinced that there must be an error. He kept the paper locked in his desk for several years!

must assume at least one of the values 0 and 1 in the disc $D(0, L(c_0, c_1))$. An algebraic proof of this innocent-looking fact does not seem to be known.

12.3   The Picard Theorems

We now have all the tools needed for the proofs of the Picard theorems. We start with the "little" Picard theorem. This was stated as Theorem 12.1.1, but for convenience we state it again.

THEOREM 12.3.1 (Picard) *A nonconstant entire function assumes every value in the complex plane* $\mathbb{C}$ *with at most one exception.*

   Proof. As explained previously, we may form from the entire function $f(z)$ that omits two values the auxilliary entire function $g(z)$ that omits the values 0 and 1. Therefore, by the Schottky Theorem 12.2.1,

$$|g(z)| < S\left(g(0), \frac{1}{2}\right) \quad \text{for all } z \in D\left(0, \frac{R}{2}\right).$$

Because R can be arbitrarily large, the function $g(z)$, hence $f(z)$ also, must, by the Liouville Theorem 4.4.9, be constant.   □

We now come to our ultimate goal, the "great" Picard theorem.

THEOREM 12.3.2 (Picard) *Suppose $f(z)$ has an isolated essential singularity at z_0. Then in each punctured neighborhood of z_0 the function $f(z)$ assumes every value c in the complex plane $\mathbb{C}$, with perhaps one exception, infinitely many times.*

   Proof. Without loss of generality, we may assume $z_0 = 0$. Suppose that, in a punctured neighborhood $D'(0, \rho)$ of the origin, there are two distinct values a and b which the function $f(z)$ assumes at most finitely many times; then, for a sufficiently small ρ_0 $(0 < \rho_0 < \rho)$, the function $f(z)$ does not assume either of the values a or b in $D'(0, \rho_0)$. Therefore, the function

$$F(z) = \frac{f(z) - a}{b - a}$$

has an isolated essential singularity at the origin and does not assume either of the values 0 or 1 in the punctured neighborhood $D'(0, \rho_0)$. By the Casorati-Weierstrass Theorem 5.2.4 there exists a positive sequence $\{z_n\}_{n=1}^{\infty}$ converging to the origin such that

$$\lim_{n \to \infty} F(z_n) = 0.$$

In particular, there exists a positive constant k satisfying

$$|F(z_n)| \leq k \quad (n \in \mathbb{N}).$$

We may also assume, without loss of generality, that

$$\frac{\rho_0}{2} > r_1 > r_2 > \cdots > r_n \longrightarrow 0 \quad (n \to \infty),$$

where $r_n = |z_n|$ ($n \in \mathbb{N}$). Because $D(z_n, r_n) \subset D'(0, \rho_0)$, $F(z)$ is analytic and omits the values 0 and 1 in the disc $D(z_n, r_n)$. We may apply the Schottky Theorem 12.2.1 (and the remark after its proof), and so we obtain

$$|F(z)| < S^*\left(k, \frac{4}{5}\right) \quad \text{for } z \in \bar{D}\left(z_n, \frac{4}{5}r_n\right).$$

Note that

$$2\sin\frac{\pi}{8} = \sqrt{2 - \sqrt{2}} = 0.765\cdots < \frac{4}{5},$$

and so, if we choose a point z_n' on the circle $C(0, r_n)$ such that the angle between the radii $0z_n$ and $0z_n'$ is $\frac{\pi}{4}$, then $z_n' \in D(z_n, \frac{4}{5}r_n)$ (Figure 12.1). Hence

$$|F(z_n')| < S^*\left(k, \frac{4}{5}\right) \quad (= k', \text{ say}).$$

Now, as before, $D(z_n', r_n) \subset D'(0, \rho_0)$, $F(z)$ is analytic and omits the values 0 and 1 in $D(z_n', r_n)$. By the Schottky Theorem 12.2.1, we have

$$|F(z)| < S^*\left(k', \frac{4}{5}\right) \quad \text{for } z \in \bar{D}\left(z_n', \frac{4}{5}r_n\right).$$

Choose the point z_n'' (other than the point z_n) on the circle $C(0, r_n)$ such that the angle between the radii $0z_n'$ and $0z_n''$ is $\frac{\pi}{4}$; then $z_n'' \in D(z_n', \frac{4}{5}r_n)$, and so

$$|F(z_n'')| < S^*\left(k', \frac{4}{5}\right) \quad (= k'', \text{ say}),$$

and

$$|F(z)| < S^*\left(k'', \frac{4}{5}\right) \quad \text{for } z \in \bar{D}\left(z_n'', \frac{4}{5}r_n\right).$$

Repeating this procedure 7 times, we obtain 7 discs, with centers on the circle $C(0, r_n)$ and with radii all equal to $\frac{4}{5}r_n$, whose union covers the circle $C(0, r_n)$. It follows that there exists a constant M, independent of n, such that

$$|F(z)| < M \quad \text{for all } z \in C(0, r_n), \ n \in \mathbb{N}.$$

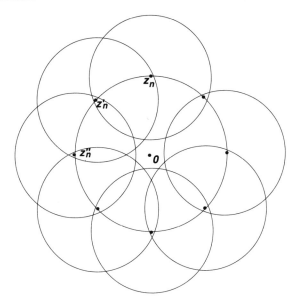

Figure 12.1

In other words, $F(z)$ is bounded on a sequence of circles converging to the origin. By the Maximum Modulus Principle 6.1.1 (or its Corollary 6.1.2), we have

$$|F(z)| < M, \quad r_{n+1} \le |z| \le r_n, \quad n \in \mathbb{N}.$$

Because $r_n \downarrow 0$ $(n \to \infty)$, this result means that $F(z)$ is bounded in the punctured disc $D'(0, r_1)$:

$$|F(z)| < M \quad \text{for } z \in D'(0, r_1).$$

It follows from the Riemann Theorem 5.2.1 that the singularity at the origin is removable. This result contradicts the assumption that the origin is an essential singularity, and so the proof is complete. □

In the theorem above, $f(z)$ is analytic in a punctured neighborhood of an isolated essential singularity, and so $f(z)$ does not assume the value ∞. If we now impose on $f(z)$ the condition that it is meromorphic in a punctured neighborhood of z_0, then we arrive at the following strengthened version.

THEOREM 12.3.3 *Suppose $f(z)$ is meromorphic in a punctured neighborhood of an essential singularity z_0. Then in each punctured neighborhood of the point z_0, $f(z)$ assumes every value (finite or infinite) in the* *extended complex plane* $\hat{\mathbb{C}}$ *infinitely many times with at most 2 exceptions* (*finite or infinite*).

   Proof. Suppose there are 3 exceptional values a, b, and c; then at least one of them (say, a) must be finite. But then the function

$$F(z) = \frac{1}{f(z) - a}$$

is analytic in a punctured neighborhood of the point z_0, and has an isolated essential singularity at z_0, yet it assumes the values $\frac{1}{b-a}$ and $\frac{1}{c-a}$ at most finitely many times. This result contradicts the "great" Picard Theorem 12.3.2, and so the number of exceptional values is at most 2.   □

   If $f(z)$ is a transcendental entire function (i.e., not a polynomial), then the function $F(\zeta) = f(z)$, where $z = \frac{1}{\zeta}$, has the origin $\zeta = 0$ as its isolated essential singularity, and so, by the "great" Picard Theorem 12.3.2, in each punctured neighborhood of $\zeta = 0$, $F(\zeta)$ assumes every value in the complex plane $\mathbb{C}$, with perhaps one exception, infinitely many times. In terms of the original function $f(z)$, this demonstrates

THEOREM 12.3.4 *A transcendental entire function assumes every value in the complex plane* $\mathbb{C}$, *with perhaps one exception, infinitely many times in each neighborhood of the point at infinity.*

   The "great" Picard Theorem 12.3.2 asserts that if $f(z)$ is a non-rational meromorphic function, then for any $\gamma \in \hat{\mathbb{C}}$, unless γ is an exceptional value, there are infinitely many γ-points of $f(z)$. Thus it is natural to inquire about the distribution of these γ-points. This question, which arises from the Picard theorems, spurred investigations that ultimately led to the beautiful Nevanlinna theory, which states, roughly speaking, that $f(z)$ has the same "affinity" for every γ (including ∞).

Exercises

1. Justify the remark immediately after the proof of the Schottky Theorem 12.2.1.

2. Deduce the "little" Picard Theorem 12.3.1 from the Landau Theorem 12.2.2.

   Hint: If $f(z)$ is not constant, then $f'(z_0) \neq 0$ for some $z_0 \in \mathbb{C}$. Consider $f(z + z_0)$ for $|z| < L(f(z_0), f'(z_0))$.

3. Suppose $f(z)$ has an isolated essential singularity at z_0. Prove, without appealing to the Picard theorems, that there is a point $c \in \mathbb{C}$ and a sequence $\{z_k\}_{k=1}^{\infty}$ of points converging to z_0 such that

$$f(z_k) = c \quad \text{for all } k \in \mathbb{N}.$$

4. The ray $\arg z = \theta$ is called a *Julia direction* of the entire function $f(z)$ if, for every $\epsilon > 0$, $f(z)$ assumes every complex value in $\mathbb{C}$, perhaps with one exception, infinitely many times in the (infinite) sector

$$\{z \in \mathbb{C}; \, |\arg z - \theta| < \epsilon\}.$$

Find a Julia direction of the exponential function e^z. Is it unique?

5. Determine the two exceptional values of $\dfrac{1}{e^{\frac{1}{z}} - 1}$ at the isolated essential singularity $z = 0$. (Here, of course, the value ∞ is allowed.)

Appendix:

Unsolved Problems

Here are some problems that we have not succeeded in solving. The authors would appreciate it very much if readers would let us know their solutions to any of the problems.

(1) It is well known[1] that there is a sequence $\{z_k\}_{k=1}^{\infty}$ with the property that all the series

$$\sum_{k=1}^{\infty} z_k^n \quad (n \in \mathbb{N})$$

converge, but all the series

$$\sum_{k=1}^{\infty} |z_k|^n \quad (n \in \mathbb{N})$$

diverge. That is, each of the original series converges conditionally, not absolutely. Is it true that for some $\epsilon > 0$, if

$$|\arg z_k| < \pi - \epsilon \quad (\text{for all } k \in \mathbb{N})$$

and all the series

$$\sum_{k=1}^{\infty} z_k^n \quad (n \in \mathbb{N})$$

converge, then

$$\sum_{k=1}^{\infty} |z_k|^m$$

converges for some $m \in \mathbb{N}$?

Conjecture: Suppose that there exists $m \in \mathbb{N}$ such that

$$|\arg z_k| \leq \left(1 - \frac{1}{m}\right) \pi \quad (\text{for all } k \in \mathbb{N}),$$

---

[1] Pólya-Szegö: *Problems and Theorems in Analysis*, Springer-Verlag, 1972, Vol. 1, Pt. III, Chapter 1, no. 38. See also Exercises 2.19, 2.20, and 2.45.

and all the series

$$\sum_{k=1}^{\infty} z_k^n \quad (1 \le n \le m)$$

converge; then

$$\sum_{k=1}^{\infty} |z_k|^m$$

converges.

(2) It follows from the Taylor Theorem 4.21.4 that an entire function is completely determined by the values of its derivatives at one point; that is, given $\{w_n\}_{n=0}^{\infty}$ and $a \in \mathbb{C}$, there exists a unique entire function $f(z)$ satisfying

$$f^{(n)}(a) = w_n \quad (n = 0, 1, 2, \ldots),$$

provided that $\{w_n\}_{n=0}^{\infty}$ does not grow too fast.

On the other hand, a simple application of the Mittag-Leffler Theorem 7.1.2 implies that there exist (infinitely many) entire functions that assume the preassigned values for finitely many derivatives at infinitely many points having no finite accumulation point;[2] that is, given a sequence $\{a_k\}_{k=1}^{\infty}$ of distinct points having no accumulation point in the finite complex plane, and arbitrary sets

$$\{w_{k,m} \in \mathbb{C}; \ m = 0, \ldots, n_k\} \quad (k \in \mathbb{N}),$$

there exist (infinitely many) entire functions $f(z)$ satisfying

$$f^{(m)}(a_k) = w_{k,m} \quad (\text{for all } k \in \mathbb{N}, \text{ and } m = 0, \ldots, n_k).$$

Thus, it is natural to ask: To what extent can we specify the values of infinitely many (but not all) derivatives of an entire function at finitely many points? In other words, given finitely many distinct points $\{a_k\}_{k=1}^{n}$, and sets

$$S_k \subset \mathbb{N} \cup \{0\} \quad \text{and} \quad \{w_{k,m} \in \mathbb{C}; \ m \in S_k\} \quad (1 \le k \le n),$$

when does there exist an entire function $f(z)$ satisfying

$$f^{(m)}(a_k) = w_{k,m} \quad (1 \le k \le n, \ m \in S_k)?$$

---

[2] Cf. Exercises 7.16 and 7.17.

For example, suppose S_1 is the set of all prime numbers, S_2 is the set of all Fibonacci numbers, and S_3 is the set of all natural numbers with $m \equiv 1$ (mod 5). Is there an entire function $f(z)$ satisfying

$$f^{(m)}(0) = e^{im} \quad (m \in S_1),$$
$$f^{(m)}(i) = e^m \quad (m \in S_2),$$
$$f^{(m)}(17) = \frac{i^m}{m} \quad (m \in S_3)?$$

(3) A theorem of Fatou states: *Given any power series with a finite radius of convergence, it is possible, by changing signs of the coefficients, to obtain a power series for which the circle of convergence is the natural boundary.* Pólya extended this result as follows: *Given any power series with a finite radius of convergence, form the family of power series obtained by making all possible sign changes. Then, with probability 1, an arbitrary power series selected from this family has the circle of convergence as its natural boundary.*[3]

The "little" Picard Theorem 12.3.1 says that a nonconstant entire function assumes every complex value with at most one exception. Suppose

$$f(z) = \sum_{k=0}^{\infty} a_k z^k$$

is an entire function with an exceptional value. Is there always a sequence $\{\epsilon_n\}_{k=0}^{\infty}$, with $\epsilon_k = \pm 1$ for all k, such that

$$g(z) = \sum_{k=0}^{\infty} \epsilon_k a_k z^k$$

has no exceptional value? (In the case of $e^z = \sum_{k=0}^{\infty} \frac{z^k}{n!}$, any finite number of changes of signs (but not just for the constant term) will do.) If so, does the conclusion hold for '*almost all*' choices of the signs $\{\epsilon_k\}_{k=0}^{\infty}$?

How about the converse? Given a transcendental entire function

$$f(z) = \sum_{k=0}^{\infty} a_k z^k$$

without exceptional value, can we always obtain an entire function with an exceptional value by suitably changing the signs of the Taylor coefficients?

---

[3] Cf. p. 288.

(4) Let $\mathcal{L}^2(D)$ be the Bergman space; that is, the space of all functions $f(z)$ analytic in the unit disc D such that

$$\|f\|_2 = \left(\frac{1}{\pi} \int \int_D |f(z)|^2 \, dx \, dy \right)^{1/2} < \infty.$$

Define the translation operator by $(Tf)(z) = z \cdot f(z)$. Find all the translation-invariant subspaces.

Index